Human Genetics and Society

Second Edition

Human Genetics and Society

Second Edition

RONNEE K. YASHON
Northeastern University

MICHAEL R. CUMMINGS
Illinois Institute of Technology

BROOKS/COLE
CENGAGE Learning

Australia • Brazil • Japan • Korea • Mexico • Singapore • Spain • United Kingdom • United States

Human Genetics and Society, **Second Edition**

Ronnee K. Yashon and Michael R. Cummings

Senior Acquisitions Editor, Life Sciences: Peggy Williams

Developmental Editor: Shelley Parlante

Assistant Editor: Shannon Holt

Editorial Assistant: Sean Cronin

Media Editor: Lauren Oliviera

Marketing Manager: Tom Ziolkowski

Marketing Coordinator: Jing Hu

Marketing Communications Manager: Linda Yip

Content Project Manager: Hal Humphrey

Design Director: Rob Hugel

Art Director: John Walker

Print Buyer: Karen Hunt

Rights Acquisitions Specialist: Dean Dauphinais

Production Service: Lachina Publishing Services

Text Designer: Yvo Riezebos

Photo Researcher: Chris Althof, Bill Smith Group

Text Researcher: Karyn Morrison

Copy Editor: Lachina Publishing Services

Illustrator: Lachina Publishing Services

Cover Designer: Joe Devine/Redhanger Design

Cover Images: Front: Michael Hall/Getty Images; Rear right:

© Everett Collection Inc /Alamy; Rear left: © Ocean/Corbis

Compositor: Lachina Publishing Services

Credit information:

Figure 16.15: Reich, et al., 2010. Genetic history of an archaic hominin group from Denisova cave in Siberia. *Nature* 468: 1053-1060; Fig.1, p. 1055.

Figure BB3.1: Source: From Phenylthiocarbamide: A 75-Year Adventure in Genetics and Natural Selection by Stephen Wooding from Genetics, Vol. 172, 2015–2023, April 2006, Copyright © 2006. Reprinted by permission of Genetics Society of America.

For product information and technology assistance, contact us at
Cengage Learning Customer & Sales Support, 1-800-354-9706.

For permission to use material from this text or product, submit all requests online at **www.cengage.com/permissions.**
Further permissions questions can be e-mailed to
permissionrequest@cengage.com.

Library of Congress Control Number: 2011931547
ISBN-13: 978-0-538-73321-2
ISBN-10: 0-538-73321-7

Brooks/Cole
20 Davis Drive
Belmont, CA 94002-3098
USA

Cengage Learning is a leading provider of customized learning solutions with office locations around the globe, including Singapore, the United Kingdom, Australia, Mexico, Brazil, and Japan. Locate your local office at **www.cengage.com/global**.

Cengage Learning products are represented in Canada by Nelson Education, Ltd.

To learn more about Brooks/Cole visit **www.cengage.com/brookscole**.
Purchase any of our products at your local college store or at our preferred online store **www.CengageBrain.com.**

Printed in the United States of America
1 2 3 4 5 6 7 15 14 13 12 11

About the Authors

Ronnee K. Yashon
Northeastern University

Ronnee K. Yashon is a nationally known expert in teaching genetics, ethics, and the law on all levels—including high school, undergraduate, graduate, and law school. A recipient of the Presidential Award for Excellence in Science Teaching and the Outstanding Biology Teacher Award in Illinois, she has directed numerous workshops for science teachers and disseminated interdisciplinary lessons at local and national conventions, including NSTA and NABT. A genetics seminar at Ball State University in 1985 sparked her interest in law and inspired her to study the field. Her case study methodology for introducing bioethics and law in the curriculum uses simple, personalized, and current scenarios that involve the student in decision-making. Yashon has presented this case study method all over the country. She has six case study books, including two mini-books that focus on genetics and environmental issues. The implementation of science-oriented law courses in current law school curriculum sparks her interest, and she has run workshops for jurists and attorneys on the subject of genetics. She now teaches at Northeastern University and Boston University School of Medicine.

Michael R. Cummings
Illinois Institute of Technology

Michael R. Cummings is the author or coauthor of several leading college textbooks, including *Human Heredity: Principles and Issues, Concepts of Genetics,* and *Essentials of Genetics.* He also authored an introductory biology text for non-majors, *Biology: Science and Life,* as well as articles on genetics for the *McGraw Encyclopedia of Science and Technology.* For several years, he wrote and published a newsletter on advances in human genetics for instructors and students. He was a faculty member at the University of Illinois at Chicago for over 25 years. While there, he was recognized as an outstanding teacher by his peers on the University faculty and mentored junior faculty in undergraduate teaching. He was twice named by graduating seniors as the best teacher in their in years at the University, and in several years, was selected as an outstanding instructor by the Biology Student Organization. His research interests center on the physical organization of repetitive DNA sequences in the heterochromatic short arm/centromere region of human chromosome 21 and the role of these sequences in generating chromosomal aberrations. He now teaches general biology, cell biology, and genetics at the Illinois Institute of Technology.

We dedicate this book to our teachers and students—
past, present, and future.

Brief Contents

Contents

Preface

Since the publication of the last edition of this book, research has led the way to many new successes in genetics, including using gene therapy to treat genetic disorders, gaining insight to the evolution of our species by decoding the genome of Neanderthals, and spurring legal decisions that have transformed the way genes are patented.

The rapid pace of advances in our knowledge of human genetics, along with our ability to translate this information into applications, means that the distance between research, new products, medicines, and foods in the marketplace gets shorter and shorter every day. An understanding of genetics will help today's students to benefit in their personal lives from these advances, and enable them to help shape the ways in which our society can use these advances for the benefit of everyone. As part of this discussion, an understanding of the new choices we face and the social consequences of using these technologies must be placed on an equal footing with the relevant genetic and scientific concepts. This book provides a foundation for making informed decisions about the use of genetics in our personal and professional lives and in our society.

New to this Edition

Here we highlight the many changes made to this edition to enhance the pedagogical effectiveness of the text, expand the coverage of topics, and incorporate the latest findings in genetics.

New Chapter on Genetic Change: Mutation and Epigenetics

The new Chapter 6 introduces the roles of mutation and epigenetics in genetic change, and includes recent findings on how the environment plays an important role in influencing the outcomes of both of these processes.

New Chapter on Human Evolution

The new Chapter 16 on human evolution introduces students to the use of the fossil record, genomics, and research on human migration in reconstructing the evolution of human species and their ancestors. Students gain a new perspective on the recent origin of our species and the close genetic relationship among different present-day human populations, as well as the genetic heritage we share with extinct human species and our primate ancestors.

New Topic Review Opens Every Chapter

A new short list of review topics at the start of each chapter helps students refresh their understanding of concepts which carry over into the new chapter.

New Expanded Coverage of Core Genetic Concepts

We have also expanded the number of topics in almost every chapter in order to deepen the student's understanding of genetic concepts and how they influence the social and legal interpretations of genetics. As a part of this expansion, we have provided more detail on many core biological processes such as mitosis and meiosis, control of gene expression, and transcription and translation, including annotated figures, which walk students through these crucial processes step by step.

New Terms Defined on the Text Page

Each bolded term and a short definition is presented at the foot of the page where the term first appears, helping students learn terminology in context. A more comprehensive glossary at the end of the book contains more detailed definitions.

New Enhanced Art Program

An enhanced art program provides more effective visual learning, with over 30 new photos and over 20 new figures. All figures are numbered with explanatory captions. Many complex figures have step-by-step annotations so that students can understand the process even without reference to the related text. We feel strongly that having detailed descriptions in both the figure and the related text helps students understand and interpret the concepts in the figures and reinforces their understanding of complex biological processes.

New Case Studies Engage Students with Real-Life problems

Every chapter engages students with case studies, which include questions asking them to use their knowledge of genetics to apply their own judgment to challenging dilemmas. New case studies include a family facing sickle-cell anemia in a newborn child (Ch. 6) and a teacher planning how best to present the theory of evolution (Ch. 16).

Enriched End-of-Chapter Material

We have added significant numbers of questions to both the Review Questions and Application Questions, and revised the content of those carried over from the first edition. We have done this to sharpen the focus of the review questions on the basic concepts and to expand the scope of the application questions, addressing social issues related to genetics from the past as well as the present. In both cases, we used Bloom's taxonomy as a guide to establishing learning objectives and helping students achieve them.

The In the Media sections have been updated to include recent items from the media that relate to genetics, both as a way of showing how pervasive genetics is in our society, and to provide more material for discussions, blogs, and research projects.

This edition is larger and more comprehensive than the first edition. However, as with the first edition, this new edition relies on context to hold the flow of ideas together, allowing the student to see that successes in areas like reproductive technology and prenatal genetic diagnosis often lead to unexpected problems in social, personal, and legal arenas.

Chapter-Specific Changes

The second edition contains a great deal of reworking. The most significant change was the addition of two new chapters, Chapters 6 and 16. Chapter 6 covers mutation and epigenetics, two topics of importance in genetics. Chapter 16, on human evolution, builds on the information in Chapter 15, on populations, and explores the origins of our species and that of other human species.

Every chapter has been updated in both text and art. Almost every figure has been re-worked, most have been enlarged, and dozens of new photographs have been added. In addition, figures that show important processes such as meiosis and translation have been spread across two pages.

Following is a summary of the significant changes.

Biology Basics: Cells and Cell Structure: New material added on the classes of macromolecules found in cells, and updated information on different cell types is included. A new table on genetic disorders related to organelles, and new figures that combine drawings and photographs have been added, along with more detailed information about the nuclear envelope.

Chapter 1 Sex and Development: Two new sections have been added—one on stem cells, and another on problems in embryonic development. Background coverage on sex ratios has been added, and the art program was thoroughly reworked.

Chapter 2 Assisted Reproductive Technology: A new section on age and infertility has been added; additional coverage of *in vitro* fertilization (IVF) and sperm and egg donation has also been added. The coverage of methods of assisted reproductive technologies has been expanded.

Chapter 3 Changes in Chromosome Number: New material on telomeres and centromeres was added, and artwork on chromosomes was reworked. A discussion of polar body production accompanies a new discussion of techniques and uses of prenatal genetic diagnosis. The coverage of Down syndrome was expanded to include translocations.

Chapter 4 How Genes Are Transmitted from Generation to Generation: Several new sections have been added, including material on multiple allele inheritance, incomplete dominance, codominance, and mitochondrial inheritance. New figures and photographs were added for these sections.

Biology Basics DNA and RNA: Coverage of the structure and functions of RNA was expanded. Figures on DNA replication were reworked and clarified.

Chapter 5 Gene Action: A section was added on gene regulation and protein folding with new figures, in addition to a major reworking of the art program in this chapter, with new illustrations of transcription and translation.

Chapter 6 Gene Changes: Mutation and Epigenetics: This is a new chapter covering two types of changes in DNA: mutation and epigenetic alterations. All new figures and photographs accompany this chapter. Two new cases and a spotlight on the legacy of diethylstilbesterol (DES) are included.

Chapter 7 Biotechnology: New section on the use of biotechnology in treating genetic diseases has been added; expanded coverage of gene therapy, its failures (Jesse Gelsinger) and successes (treating Leber's disease). New material on stem cell uses; discussion of landmark decision in Myriad Genetics case that may change how or if genes can be patented.

Chapter 8 Genetic Testing: Reworking of the art. Expansion of details on chorionic villi, and non-invasive methods of prenatal testing; coverage on impact of changes in the CLIAA law, recent developments in deCode genetics bankruptcy.

Chapter 9 DNA Forensics: Expanded coverage of PCR, microarrays, CODIS and DNA databases. New material has been added on the use of DNA to free wrongly-convicted individuals, Y chromosome mapping, mtDNA mapping, and familial DNA searches. Other new topics include use of forensics to identify victims of disasters, and a new section on paternity testing. New figures and photographs were also added.

Chapter 10 Genomics: Chapter was renamed and coverage broadened beyond human genome project. New section on classical methods of gene linkage, mapping, details on new methods of DNA sequencing and exome sequencing. New figures added; others revamped.

Biology Basics: Genes, Populations, and the Environment: Some material moved to Chapter 15. Two new sections, and new figures added on population genetics and evolution, and the use of DNA databases in population genetics.

Chapter 11 Complex Traits: This chapter was retitled. There are new sections on how complex traits are studied, how the genetic component of complex traits is identified through pedigrees and testing. A new section on the use of genome-wide association studies (GWAS) has been added.

Chapter 12 Cancer: This chapter has been significantly expanded with new coverage of clinical research trials and FDA supervision. New sections on the properties of cancer cells, and metastasis were added, along with new figures and photographs. Additional coverage of cancer treatment, including stem cells.

Chapter 13 Behavior Genetics: Coverage of behaviors was expanded, including addictive behavior and control of complex behaviors. A new section on the use of GWAS to study behavior was added as well as a new section on the treatment of behavioral problems.

Chapter 14 Immunogenetics: The chapter was retitled and revised to focus on the function of the immune system and the genetics of the system components. Art program was updated and expanded. New coverage of the histocompatibility complex, cell surface in the immune response, with new art.

Chapter 15 Population Genetics: New material was added about allele frequency, clarifying its definition. Two new sections were added: one on migration and drift and selection, another on how these factors are involved in evolution. New figures were also added.

Chapter 16 Human Evolution: A new chapter discusses human evolution and decoding of the Neanderthal genome and how comparative genomics provides insight into the evolution of our species. New art and new cases are presented. Discussion on the eugenics movement begins here, as well how past human migrations are studied, and how people dispersed across the globe.

Chapter 17 Past, Present and Future: This chapter was extensively reorganized. Some sections were moved, others deleted because of added coverage in earlier chapters. A section on the use of fetal DNA in the mother's blood as a form of prenatal testing was added to the future section and all sections and figures were updated.

Ancillary Materials

For Instructors

PowerLecture DVD with JoinIn and Examview

This one-stop digital library and presentation tool includes preassembled Microsoft® PowerPoint® lecture slides. In addition to a full Instructor's Manual and Test Bank, Power-Lecture also includes all of the media resources organized by chapter: an image library with art and photos from the book, animations and video clips, and more.

The JoinIn Student Response System allows you to pose book-specific questions and display students' answers within the Microsoft® PowerPoint® lecture slides, in conjunction with the "clicker" hardware of your choice.

CourseMate

Interested in a simple way to complement your text and course content with study and practice materials? Cengage Learning's Biology CourseMate brings course concepts to life with interactive learning, study, and exam preparation tools that support the printed textbook.

Biology CourseMate includes:

- an interactive eBook
- Quizzes
- Flashcards
- Videos
- Animations
- Engagement Tracker, a first-of-its-kind tool that monitors student engagement in the course
- and more

Go to CengageBrain.com to access these resources, and look for the lock icon, which denotes a resource available within CourseMate. Watch student comprehension soar as your class works with the printed textbook and the textbook-specific website. Biology CourseMate goes beyond the book to deliver what you need!

WebTutor

With WebTutors we can have the animations, quizzing, flash-cards, and more that accompany the book delivered to a range of Learning Management Systems! Also included is access to the interactive eBook.

Interactive eBook

This interactive eBook features highlighting, note taking, and an interactive glossary.

Online Instructor's Manual

Each chapter has suggestions for videos and news clips that can be incorporated into your class, in addition to suggested topics for discussions, papers, and activities. Also available on PowerLecture.

- Summary and Lecture outlines organize the material and have the key terms integrated and highlighted.
- Teaching tips give ideas on additional ways to engage the students in the material.
- Suggested answers are given for all the questions in the chapter.

Test Bank

The test bank contains a variety of questions, such as Multiple Choice, True/False, Matching, and Short Answer to help you create quizzes and exams easily.

- More than 750 new questions written specifically for this text.

ExamView

This full-featured program helps you create and deliver customized tests (both print and online) in minutes. Includes all the questions from the Test Bank.

Transparencies

This set features all the important figures and photos from the text making it easy to incorporate the art into your lecture.

ABC News: Genetics in the Headlines

The informative and short video clips cover current news stories in genetics. Covering topics such as the use of PGD to pick the sex of your child or living with the cancer gene, these clips can spark discussions in class and provide opportunities for students to analyze the news.

For Students

CourseMate

The more you study, the better the results. Make the most of your study time by accessing everything you need to succeed in one place. Read your textbook, take notes, review flashcards, watch videos, and take practice quizzes—online with CourseMate.

WebTutor

With WebTutors students can access the animations, quizzing, flashcards, and more in your classroom LMS! Also included is access to the interactive eBook.

Interactive eBook

This interactive eBook features highlighting, note taking, and an interactive glossary.

Study Guide

This study guide has been designed to complement the textbook by including additional case studies for every chapter to allow students more opportunities to think about how the science of genetics can apply to their own lives. A fill-in-the-blank summary of the chapter and various types of questions further challenge the students' grasp of the material, and new case studies that are not included in the textbook enhance learning. A fill-in-the-blank summary review helps students review the chapter, while allowing them to check their understanding of the content.

Acknowledgments

This book began with a conversation between one of us (MRC) and Peter Adams, who at the time was the executive editor at Brooks/Cole. Peter became an early and enthusiastic supporter of the idea, and was the original editor on this project. He was also responsible for getting us together as authors. Peter's input, gleaned from many years of experience and conversations with instructors and students, helped shape the book. We are very grateful for his efforts and in particular, his role as author matchmaker. For the first edition, Yolanda Cossio provided attention to quality and pedagogy and helped guide the project through its crucial stages of development. Also for the first edition, it was our good fortune to have Christopher Delgado as our development editor. His upbeat, can-do attitude and skills at nudging two often picky authors back on track were remarkable.

For the second edition, we are fortunate to be under the guidance of Peggy Williams, with her extensive background in editorial work and marketing. Her knowledge of what instructors want in a non-majors genetics text along with her drive to make this the best text available have helped shape a broader view of genetics and a sharper focus on the social issues related to the use of genetics. Our developmental editor Shelley Parlante brings experience with texts big and small, and helped keep us focused on carefully evaluating the reorganization and rewriting that is such a key part of this edition. Our assistant editor, Shannon Holt, the media editor, Lauren Olivera, and Sara Black, the copy editor, rounded out the editorial team. They adapted to the often rapidly evolving format and schedule deadlines and smoothly moved the book toward production. Finally, at Cengage Learning, we wish to thank Hal Humphrey, who was kind enough to listen to our ideas about the layout and format for this edition.

At Lachina Publishing Services, Bonnie Briggle and her colleagues took the diverse elements that make up a science text and brought them all together to produce a book that will serve both students and instructors as an effective learning tool.

The many tasks involved in photo research were handled by Chris Althof and others at the Bill Smith Group. Their attention to detail has provided us with a suite of photos that are not only pedagogically effective, but also often striking in their own right.

Once again, we extend thanks to all others who played a role in creating this edition, and as we move into future editions, we hope that instructors and students will help us improve the book as a resource for those who want to teach and learn about the interactions between human genetics and society.

Reviewers

In the first edition, the following reviewers were instrumental in helping us develop and refine our approach through their feedback and suggestions: William P. Baker, Midwestern University; Bruce Bowerman, University of Oregon; Cherif Boudaba, Tulane University; Jay L. Brewster, Pepperdine University; Mary Bryk, Texas A&M University; Barry Chess,

Pasadena City College; T.B. Cole, Kent State University; Patricia L. Conklin, SUNY-Cortland; Drew Cressman, Sarah Lawrence College; William Cushwa, Clark College; Thomas R. Danford, West Virginia Northern Community College; Kathleen Duncan, Foothill College; Cheryld L. Emmons, Alfred University; Michael L. Foster, Eastern Kentucky University; Gail E. Gasparich, Towson University; Urbi Ghosh, Triton College; Nabarun Ghosh, West Texas A&M University; Meredith Hamilton, Oklahoma State University; Deborah Han, Palomar College; Jennifer Herzog, Utica College; Barbara Hetrick, University of Northern Iowa; Robert Hinrichsen, Indiana University of Pennsylvania; Mary King Kananen, Penn State University-Altoona; Gwendolyn M. Kinebrew, John Carroll University; Jennifer Knight, University of Colorado; Karen Kurvink, Moravian College; Clint Magill, Texas A&M University-College Station; Gerard P. McNeil, York College CUNY; Tyre J. Proffer, Kent State University-Salem; Michael D. Quillen, Maysville Community and Technical College; Laura Rhoads, SUNY-Potsdam; Michael P. Robinson, University of Miami; Jeanine Sequin Santelli, Keuka College; Monica M. Skinner, Oregon State University; Sue Trammell, John A. Logan College; Jose Vazquez, New York University; Mary Ann Walkinshaw, Pima Community College; Dan Wells, University of Houston; Robert Wiggers, Stephen F. Austin State University; Denise Woodward, Pennsylvania State University-University Park; and Calvin Young, Fullterton College.

Second Edition Reviewers

For the second edition, the following reviewers took on the task of working through the first edition and letting us know what worked, what needed work, and where we should expand the coverage:

Edward Berger, Dartmouth College

Kimberly A. Carlson, University of Nebraska at Kearney

Sandra Gilchrist, New College of Florida

Virginia McDonough, Hope College

Sheila Gibbs Miracle, Southeast Kentucky Community and Technical College

Sharon K. Rittman, North Carolina Central University

Wendy A. Shuttleworth, Lewis Clark State College

Joel Adams-Stryker, Evergreen Valley College

Contacting the Authors

We want to hear from instructors and students about the book, about new findings they would like to see in future editions, and things they don't like, or they think are not working, with suggestions on how to fix the problem. Please contact Ronnee Yashon at **yashon.boston@gmail.com** and/or Michael R. Cummings at **cummings.chicago@gmail.com** and let us know how we can improve the text or its pedagogy.

Preview

- **Have you ever wondered why you look like your parents?**
- **Have you gone into a coffee shop and seen families talking and noticed how family members look similar to each other?**
- **Have you seen ads on television for the breast cancer gene test?**
- **Do you know any identical twins?**

If the answer to any of these questions is YES, **this is the book for you**.

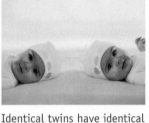

Identical twins have identical genes.

Some people say they don't like science. They have no interest in it. It doesn't have anything to do with their lives. But human genetics is more than a science. It is something that will touch you and the lives of your family and friends. Some of the ways that this will happen will be discussed throughout this book and once you have finished this course, you will be able to take what you learn and apply it to questions and issues in your personal and professional life.

DNA analysis is now part of medicine.

All through the book, we will examine genetics through the lens of case studies based on the lives of real people. Possibly even people you know. Hopefully you will talk to your friends and families about the cases you will be reading. They are interesting and thought provoking, as well as conversation starters.

These cases and the questions they generate will allow you to analyze your own knowledge, thoughts, feelings and opinions about many issues related to genetics and share them with others. This will include information about your own heredity and that of your family. Some of your questions will be answered, others will not. Science is like that, an ongoing process in which answers are the starting point for more questions.

Can you put yourself in the place of the following people?

- a woman about to give birth to her first child?
- a man having a test for infertility?
- a scientist with a great discovery he wants to share with the world?
- a physician giving sad news to a patient?

Physicians often wrestle with difficult issues in helping patients deal with genetic disorders.

If you can, you'll be able to gain new insight by using our cases. If you can't, the cases may bring questions to mind that will help you understand the problems faced by these people, and start you wondering what you would do in their situation.

In addition, as the course proceeds you will be analyzing your own opinions, the opinions of others and talking to others about issues related to genetics that cut across medicine, politics, laws and social norms. Hopefully, as you acquire more information and examine and discuss these issues, your opinions will be refined and you might even change your mind about some things. You'll find that many people have questions about genetics and the uses of these technologies.

After using this book, you'll be able to answer many of your own questions and those of others and perhaps even help members of your family understand their backgrounds. You will see references in social networking sites, on blogs, on Websites and in magazines that will intrigue you and remind you of many things from this book.

That is the goal of this book: to put human genetics into your everyday life.

It will equip you with the knowledge needed to make sound decisions about the use of genetics in your personal life and in our society That's why this book is called *Human Genetics and Society*.

To share your opinions and your knowledge and to discuss questions and issues with others, think about setting up a blog and invite your family, your friends, and others in your class to see what you are thinking and work through the issues together. Watch for some blogging ideas in chapter 1 and other chapters throughout the book.

Sorting out the reliable information on human genetics from the unreliable on the Internet is often difficult.

As you can see from Table P-1, the topics that we will cover throughout the book will include questions to get you thinking about human genetics and what

Table P-1 Some General Questions You May Encounter . . .

	One of Our Topics	A Personal Question	Chapter
Cheryl Casey/iStockphoto.com	Your own genetics	Do I carry a disease gene?	Chapter 4
Keith Brofsky/Stockbyte/Getty Images	Genetic testing now and in the future	Will I be asked to take a test to find out information about my genetics?	Chapter 8
Courtesy of Ifti Ahmed	Analyzing all the human genes and finding out what they do	When will we know all about human genes and how they work?	Chapter 10
Martin Shields/Photo Researchers, Inc.	Applying genetics to legal actions such as criminal and civil proceedings	Should a man in prison be released if DNA evidence proves he is innocent?	Chapter 9
Elyse Lewin/Getty Images	Applying genetics to laws enacted by a state	Should my state pass a law that all couples being married be tested for genetic conditions?	Chapter 8
Alice Edward/Getty Images	Applying genetics to infertility and the treatments for it	When a couple finds they are infertile, what can be done to help them?	Chapter 2
Billy Hustace/Getty Images	Applying genetics to society as a whole	Are certain genes more common in groups?	Chapter 15
WHAT SHOULD WE DO?	Looking at the ethical questions that surround human genetics.	What would you do if you had to answer the ethical questions raised by human genetics?	All chapters

you might do when you come face to face with personal or professional decisions related to genetics.

Genetics is everywhere! You see it on television, on advertising, on talk shows, and even in comedy routines. The problem is that a lot of the information you have been hearing is an oversimplification of the concepts and uses of human genetics. In this book, we have made every effort to provide essential information in a way that is understandable and accessible. To do this, we start each chapter with a case study that presents real problems faced by real people. The case is followed by a discussion of the genetic concepts related to these problems. After the concepts are presented, there are questions, discussions of legal, ethical and social issues that arise out of the use and application of genetics that apply to the case.

Families often explore their family history to understand their genetic makeup.

In the very near future, analysis of our genomes may affect our insurance rates, our medical treatment and our lifestyles. This is a book that you will find yourself using long after the course is over. Welcome to the world of *Human Genetics and Society*.

Genetic testing kits are available at many drug stores, but counseling with a health professional should be a part of understanding the results.

Will you want to be a science major when you are done with this book? Maybe not, but you will acquire useful skills and information about human genetics. You may be able to interest your family and friends in some of the ideas and related issues and provide them with information to use in their own decision-making.

We are sure that what you learn from this book will stay with you throughout your life. So read on . . .

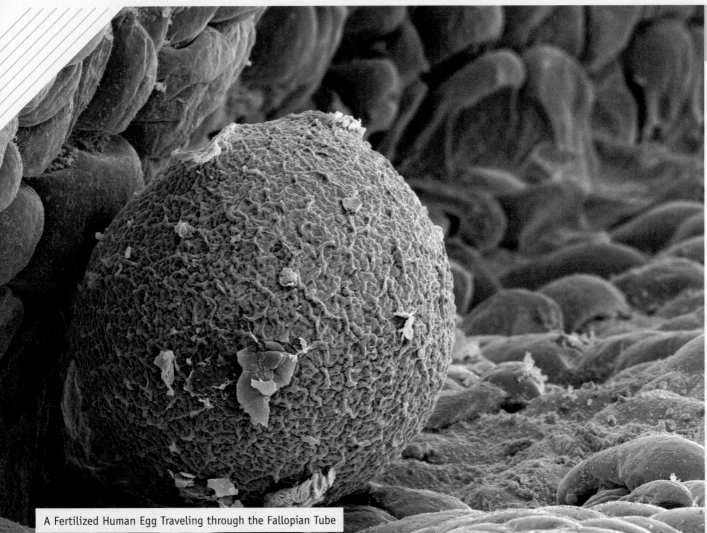

A Fertilized Human Egg Traveling through the Fallopian Tube

BIOLOGY BASICS 1

Cells and Cell Structure

BB1.1 Why Are Cells Important in Human Genetics?

Humans begin life as a single cell: a fertilized egg, or **zygote,** formed by the fusion of a sperm and an egg. The instructions for making all the cells, tissues, and organs of an adult are encoded in the DNA of that single cell. About 36 to 39 weeks after the zygote is formed, the newborn entering the world contains approximately 40 billion cells, made up of about 200 different cell types. All these cells originated from the zygote and were created by cell division. The infant's cells are organized into many organ systems, each associated with highly specialized functions, and all controlled by the genetic information inherited from each parent.

zygote fertilized egg that develops into a new individual

Figure BB1.1 **Some of the More Than 200 Types of Somatic Cells in the Human Body**

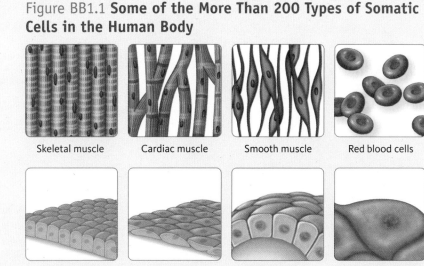

Skeletal muscle   Cardiac muscle   Smooth muscle   Red blood cells

Kidney cells   Lung cells   Thyroid cells   Pancreatic cells

The study of cells is a key part of human genetics because genetic information is carried in our cells, and errors in this information (called mutations) have an impact on the structure and function of cells. For example, sickle-cell anemia is a genetic disease that changes the shape of red blood cells. Most of the resulting physical symptoms, such as heart failure, stroke, pneumonia, brain damage, and pain, are a direct effect of the changes in cell shape. This disorder afflicts many of those who live in or have ancestors from West Africa, regions around the Mediterranean Sea, or parts of the Middle East and India.

Most of the cells in the adult are called **somatic** (body) **cells.** A small number of cells are **germ cells;** these form sperm and eggs. Other cells called **stem cells** remain unspecialized and, during repair processes, can divide to form many different cell types in the body. Figure BB1.1 shows some of the several hundred types of somatic cells in the body.

New cells are constantly needed in our body for growth and/or replacement of dead or worn-out cells. These new cells are formed by cell division, a process in which one cell divides to form two cells. As they prepare to divide, the genetic information they carry must be copied exactly so it can be distributed to the daughter cells formed by division. This is accomplished by structures in the nucleus known as **chromosomes.** Chromosomes carry the cell's genetic information in the form of DNA. The process of cell division that produces these exact copies of the chromosomes is known as **mitosis** (discussed in Chapter 12). Although most cells divide by mitosis, the sperm and egg that fuse to form the zygote are produced by a special form of cell division known as **meiosis**, which is discussed in detail in Chapter 3. Mistakes in either

form of cell division can have serious genetic consequences, many of which can cause life-threatening or fatal disorders.

BB1.2 How Is the Cell Organized?

Cells are the building blocks of the body. A description of the basic chemical and structural features of a human cell will provide a background for later discussions about what happens to cells and their contents in many genetic disorders. A diagram of a human cell is shown in Figure BB1.2, page 4.

somatic cells cells that form the body of an organism but do not form gametes

germ cells cells that divide to form sperm and eggs

stem cells cells in the embryo and adult that can divide to form many different cell types

chromosomes DNA-containing threadlike structures in the nucleus that carry genetic information

mitosis form of cell division that produces two daughter cells that are genetically and chromosomally identical to the parent cell

meiosis form of cell division in which haploid cells are produced

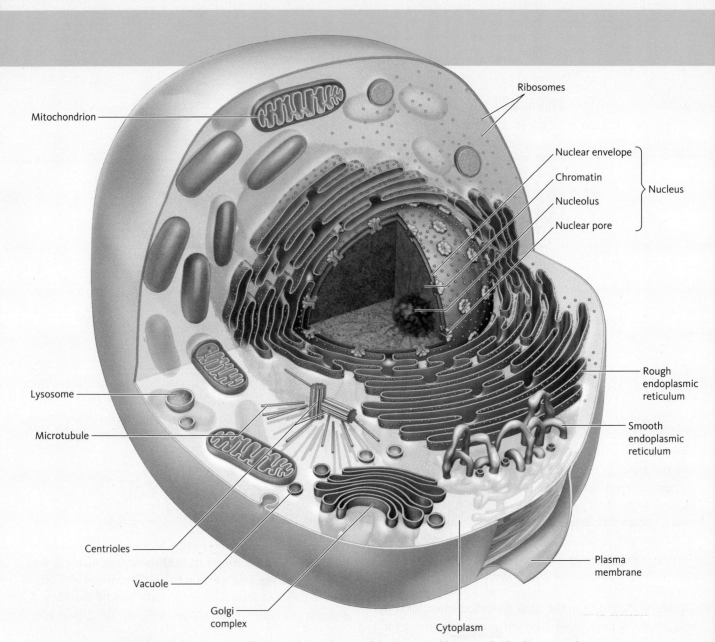

Figure BB1.2 **A Typical Human Cell Showing the Major Organelles and Their Locations**

Cells of all organisms contain four classes of large molecules called **macromolecules**: carbohydrates, lipids, proteins, and nucleic acids. These macromolecules are assembled by linking subunits together one at a time. Table BB1.1 lists these macromolecules, their subunits, and some of their functions.

Carbohydrates are made by linking small sugar molecules together to form long chains. These long chains serve as stored energy reserves for the cell. Subunits can be removed and broken down to release energy that the cell can use for many different functions. Shorter chains of sugar subunits are attached to proteins and lipids on the outer surface of cells.

These macromolecular complexes serve a variety of functions: they hold cells together and act as chemical markers to identify cells.

As a group, lipids are a diverse collection of molecules that do not mix with water but can be mixed with oils. There are three major types of lipids: fats, steroids, and waxes. Some fats are structural components of cells and form membranes; other fats are energy storage molecules. Steroids include cholesterol, an important component of cell membranes. Other steroids act as hormones; both estrogen and testosterone are steroids.

Proteins are composed of long chains of subunits called amino acids and are the most multifunctional and versatile macromolecule in the cell. Proteins act as enzymes, increasing the rate of chemical reactions necessary for important cellular functions. In addition, proteins are structural components found inside cells, surrounding cells, and in bone, skin, and hair.

macromolecules large molecules composed of subunits found in cells

Table BB1.1 **Classes of Macromolecules in the Cell**

Class	Subunit	Functions	Example
Carbohydrates	Sugars	Energy storage	Glycogen, starch
		Structure	Cellulose
Lipids	Fatty acids	Energy storage	Body fat
	Isoprenes	Membrane structure	Plasma membranes
		Hormones	Estrogens, testosterone
Proteins	Amino acids	Catalyze chemical reactions (enzymes)	Lysozyme
		Structure	Cartilage, hair
		Hormones	Insulin
Nucleic Acids	Nucleotides	Stores genetic information	DNA
		Directs protein synthesis	RNA

Nucleic acids are macromolecules assembled from subunits called nucleotides. There are two types of nucleic acids: DNA and RNA. DNA stores genetic information in the form of genes. One type of RNA carries working copies of genes that are used to assemble amino acids into proteins.

Although human cells differ widely in their size, shape, functions, and life cycles, almost all have three components in common: (1) the plasma membrane, which separates a cell from its external environment; (2) the cytoplasm; and (3) membrane-enclosed structures in the cytoplasm called organelles.

Figure BB1.3 **The Plasma Membrane** Proteins are embedded in a double layer of lipid molecules. These proteins have specific functions, including transport and reception of chemical signals. Short carbohydrate chains are attached to some proteins on the cell surface, giving the cell a molecular identity.

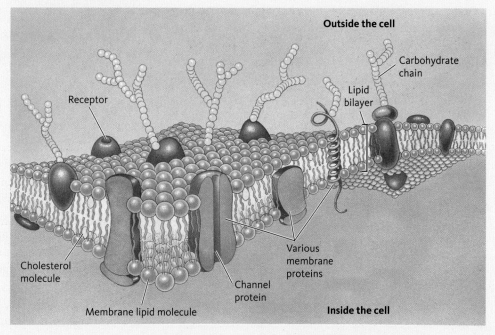

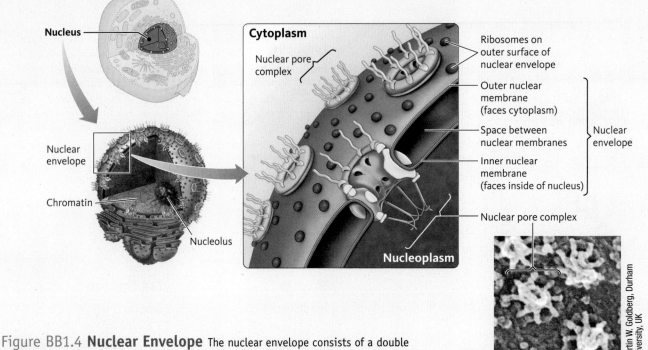

Nucleus

Nuclear envelope

Chromatin

Nucleolus

Cytoplasm

Nuclear pore complex

Nucleoplasm

Ribosomes on outer surface of nuclear envelope

Outer nuclear membrane (faces cytoplasm)

Space between nuclear membranes

Inner nuclear membrane (faces inside of nucleus)

Nuclear envelope

Nuclear pore complex

0.1 µm

Martin W. Goldberg, Durham University, UK

Figure BB1.4 **Nuclear Envelope** The nuclear envelope consists of a double membrane studded with pores. The pores control the exchange of molecules between the nucleus and the cytoplasm.

BB1.3 What Does the Plasma Membrane Do?

Follow along with the cell drawing in Figure BB1.2 as we discuss the major parts of a cell.

The double-layered **plasma membrane**, composed of lipids and embedded proteins (Figure BB1.3, page 5), has many important functions. It controls the passage of materials such as water, oxygen, and nutrients into the cell and the transfer of secreted molecules and waste products out of the cell. Several genetic disorders are associated with defects in plasma membrane structure or function. Chemical markers in the form of carbohydrates attached to proteins or lipids on the plasma membrane surface give the cell a molecular identity. These markers are determined by our genes and are responsible for many important properties of cells. In organ transplants, for

plasma membrane outer border of cells that serves as an interface between the cell and its environment

cytoplasm viscous material that is located between the inner surface of the plasma membrane and the outer nuclear envelope that contains organelles, each with specialized functions

organelles membrane-enclosed structures with specialized functions found in the cytoplasm of cells

nucleus membrane-enclosed organelle in cells that contains the chromosomes

example, the HLA (human leukocyte antigen) markers identify organs that are foreign to the body. Transplanted organs with mismatched HLA markers can be rejected by the recipient's body.

The plasma membrane also contains molecular sensors or receptors that receive and process chemical signals from the cell's environment. These signals are important in regulating many critical cell functions, including cell division. Genetic changes that affect these receptors play an important role in some diseases, including cancer. The plasma membrane encloses and protects the **cytoplasm**, a watery mixture that contains many types of molecules and structural components.

BB1.4 What Is in the Cytoplasm?

In addition to its molecular components, the cytoplasm contains several types of membrane-bound structures called **organelles** (small organs). Cell function is directly related to the number and type of organelles it contains. Some organelles are involved in making or chemically modifying proteins, while others transport them to their final destinations. The largest and most important organelle in a cell is the **nucleus**.

What is in the nucleus? The nucleus is enclosed by a double-layered membrane called the nuclear envelope. The

Figure BB1.5 **Chromatin** The chromosomes in the nucleus are uncoiled and dispersed throughout the nucleus and appear as clumps of darkly staining chromatin.

Figure BB1.6 **Chromosomes** Chromosomes are DNA/protein complexes found in the nucleus and exist in pairs. Each member of a chromosome pair is called a homologous chromosome. Most human cells carry 46 chromosomes (23 pairs). This is called the diploid number of chromosomes. Certain cells, such as sperm and eggs, carry only one copy of each chromosome (23 chromosomes); this is called the haploid number of chromosomes. At fertilization, fusion of a sperm and an egg form a diploid cell, the zygote, which gives rise to a new individual.

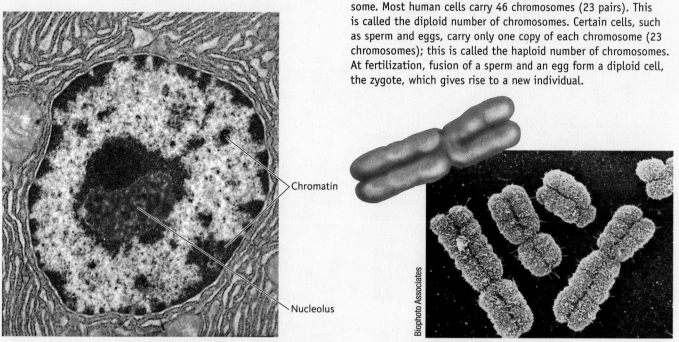

Chromatin

Nucleolus

© Kenneth Bart

Biophoto Associates

envelope is studded with a series of pores (Figure BB1.4) that allow materials to be transported in and out of the nucleus. Within the nucleus, dense regions called **nucleoli** synthesize ribosomes, organelles that move to the cytoplasm where they are involved in making proteins. The nucleus also contains chromatin, a combination of DNA and protein that is visible as dark spots (Figure BB1.5). Chromatin represents the collection of uncoiled structures known as chromosomes carried in the nucleus. As cells prepare to divide, the chromatin condenses and coils, making chromosomes visible (Figure BB1.6).

Chromosomes carry the genetic information stored in DNA that ultimately determines the structure and shape of the cell and the functions it has in the body. The genetic information carried by DNA is organized into units called **genes**. Chromosomes exist in pairs: each member of a chromosome pair is called a **homologous chromosome**. Most human cells carry 46 chromosomes (23 pairs). This is called the **diploid number ($2n$)** of chromosomes. Certain cells, such as sperm and eggs, carry only one copy of each chromosome (23 chromosomes). This is called the **haploid number (n)** of chromosomes. At fertilization, the fusion of a sperm with an egg (each with the haploid number of chromosomes) forms a diploid zygote. We will discuss the role of chromosomes in development in Chapter 1.

What do other organelles do?

The **endoplasmic reticulum (ER)** is a network of membranes that form channels in the cytoplasm (Figure BB1.7a, page 8). Parts of the outer surface of the ER are covered with **ribosomes**, another organelle, forming the rough endoplasmic reticulum (RER). It is known as rough endoplasmic reticulum because the ribosomes appear as small bumps on the surface of the RER. Proteins made by the ribosomes on the RER enter the ER (protein

nucleoli (singular, nucleolus) region in the nucleus that synthesizes ribosomes

genes carriers of genetic information in the form of DNA

homologous chromosomes members of a chromosome pair

diploid number ($2n$) condition in which chromosomes are present as pairs; in humans, the diploid number is 46

haploid number (n) condition in which each chromosome is present once, unpaired; in humans, the haploid number is 23

endoplasmic reticulum (ER) series of cytoplasmic membranes arranged as sheets and channels that function in the synthesis and transport of gene products

ribosomes cytoplasmic organelles that are the site of protein synthesis

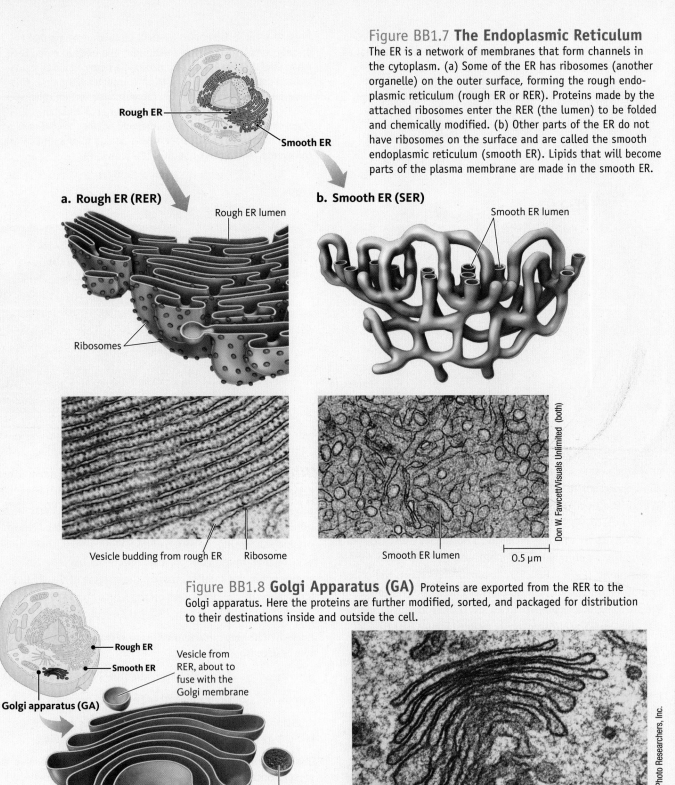

Figure BB1.7 **The Endoplasmic Reticulum**

The ER is a network of membranes that form channels in the cytoplasm. (a) Some of the ER has ribosomes (another organelle) on the outer surface, forming the rough endoplasmic reticulum (rough ER or RER). Proteins made by the attached ribosomes enter the RER (the lumen) to be folded and chemically modified. (b) Other parts of the ER do not have ribosomes on the surface and are called the smooth endoplasmic reticulum (smooth ER). Lipids that will become parts of the plasma membrane are made in the smooth ER.

Rough ER

Smooth ER

a. Rough ER (RER)

Rough ER lumen

Ribosomes

b. Smooth ER (SER)

Smooth ER lumen

Vesicle budding from rough ER Ribosome

Smooth ER lumen

0.5 μm

Don W. Fawcett/Visuals Unlimited (both)

Figure BB1.8 **Golgi Apparatus (GA)** Proteins are exported from the RER to the Golgi apparatus. Here the proteins are further modified, sorted, and packaged for distribution to their destinations inside and outside the cell.

Rough ER

Smooth ER

Vesicle from RER, about to fuse with the Golgi membrane

Golgi apparatus (GA)

Vesicles budded from Golgi containing finished product

Internal space

0.25 μm

Biophoto Associates/Photo Researchers, Inc.

Figure BB1.10 **A**

Mitochondrion Cells such as liver cells that require a lot of energy can contain more than 1000 mitochondria. Each mitochondrion carries its own genetic information that is used to make molecules involved in energy production.

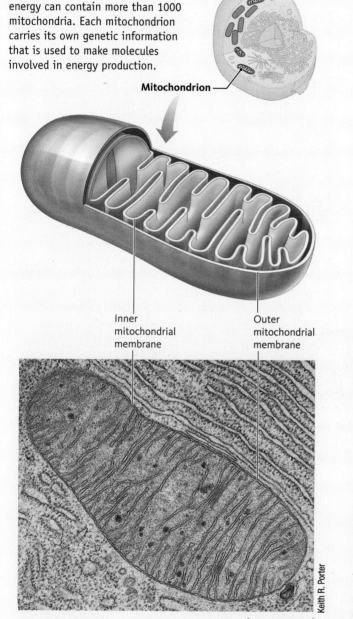

Mitochondrion

Inner mitochondrial membrane

Outer mitochondrial membrane

Keith R. Porter

0.5 μm

synthesis is discussed in Chapter 5). The space inside the ER is called the lumen. Here, proteins are folded and modified and packaged into vesicles to move to other locations inside the cell. Another type of ER (Figure BB1.7b), the smooth ER (SER), has no ribosomes on its surface. The SER functions in making lipids that become part of the plasma membrane. In the liver, SER converts drugs, toxins, and other substances into forms that can be more easily removed from the body.

After leaving the ER, proteins move to the **Golgi apparatus (GA)** (Figure BB1.8). In the Golgi, proteins are sorted, modified further, and distributed to their destinations inside and outside the cell.

Lysosomes are the processing centers of the cells. They are membrane-enclosed vesicles that contain digestive enzymes (Figure BB1.9). Worn-out organelles and other materials that have been brought into cells to be destroyed end up in the lysosomes, where they are broken down, and their molecular components are reused in making new molecules and organelles. More than 40 genetic disorders are associated with defects in lysosomes.

Mitochondria (Figure BB1.10) are energy-producing centers in the cell. Each mitochondrion contains an outer and an inner membrane. The inner membrane is folded to increase surface area for energy production. Cells such as those in the liver require a lot of energy and can contain more than 1000 mitochondria. Mitochondria carry their own genetic information in the form of circular DNA molecules, as well as their own ribosomes, used in the synthesis of mitochondrial proteins involved in making energy-containing molecules that the cell uses to carry out chemical reactions.

Golgi apparatus (GA) membranous organelle composed of a series of flattened sacs; it sorts, modifies, and packages proteins produced in the ER

lysosomes membrane-enclosed organelles that contain digestive enzymes

mitochondria (singular, mitochondrion) membrane-enclosed organelles that are the site of energy production

Figure BB1.9 **Lysosomes Are the Processing Centers of the Cell**

Lysosomes are membrane-enclosed vesicles that contain digestive enzymes. Materials marked for destruction and worn-out organelles end up in the lysosomes, where they are broken down.

Lysosome

Lysosome containing ingested material

Don W. Fawcett/Photo Researchers, Inc.

Table BB1.2 Organelles and Genetic Disorders

Organelle	Normal Function	Disorders Associated with Organelle Malfunction
Rough endoplasmic reticulum (RER)	Synthesis, folding of proteins	Inflammatory diseases related to arthritis; endocrine disorders, liver disease
Smooth endoplasmic reticulum (SER)	Lipid synthesis	SER accumulates in the brains of those with Alzheimer disease; may also play a role in diabetes
Lysosomes	Breakdown of molecules, worn-out organelles	Incomplete processing of molecules and cell components is associated with more than 40 disorders
Golgi apparatus	Sorting and modification of proteins	Bone growth disorders, degenerative bone diseases
Mitochondria	Energy production	Malfunctions in the nervous system and muscles

Organelles are involved in genetic disorders.
Heritable changes in the structural components or function of organelles can result in genetic disorders (Table BB1.2), once again emphasizing the importance of studying cells as part of human genetics.

Preview

Chapter 1 examines human development from fertilized egg to newborn and considers how new technology allows parents to select the sex of their children. Chapter 2 explores how reproductive technology is used to help couples with fertility problems who wish to have children. Chapter 3 follows chromosomes through meiosis and reviews the problems that arise when errors in meiosis result in children with an abnormal number of chromosomes. Chapter 4 explores how genes are transmitted from parents to offspring and how this information is used in human genetics.

1

Sex and Development

CENTRAL POINTS

- The chromosomal sex of a fetus is determined at fertilization.

- Fetal development has many stages.

- During development, a fetus becomes visibly male or female.

CASE A: Choice of Baby's Sex Now Available

Susanna Carter had been thinking for the last few months about having another baby. She was happy with her family of three boys but had always wanted a girl. Already in her late 30s, she would have to decide soon, before the beginning of menopause. She wanted to discuss the possibility with her husband at dinner because she had seen an article in the paper about a new medical procedure, **sex selection**. Using this method, Susanna and her husband, Bob, could pick the sex of their baby before it was even growing in her uterus.

After talking about it, the Carters made an appointment with their physician, Dr. George Leon, to discuss sex selection. Dr. Leon knew that some methods of sex selection are more successful than others. In the past, Dr. Leon had a number of patients who wanted this procedure for a variety of reasons. After meeting with the Carters, he thought about some of them:

The first patient was a color-blind male. His wife's family had no history of color blindness. Because of the way color blindness is inherited, none of their sons will be color blind, nor will any sons carry the color blindness gene. All their daughters will carry the color blindness gene they receive from their father, but because they will also get a normal copy of the gene from their mother, they will have normal vision. However, these daughters will pass the color blindness gene on to their children. The grandsons of this couple will have a 50% chance of being color blind, and their granddaughters will have a 50% chance of inheriting the color blindness gene, which they can pass on to future generations. The man and his wife do not want to pass this gene along to future generations of the family and so

A newborn male and female.

would prefer to have only boys.

The second patient was a woman who carries the gene for Hunter syndrome. Children with this disorder are normal at birth, but around six months of age, they begin to show progressive physical and mental deterioration and, eventually, affected infants become blind, deaf, and mentally retarded. Few live beyond the age of 5 years. Sons of this woman will have a 50% chance of getting this deadly disorder. All daughters will be unaffected, but each will have a 50% chance of inheriting the gene, which may be passed on to future generations. To ensure that this disorder does not affect her child, this patient wants to know that an embryo is healthy before implanting and carrying it to birth.

The husband of the third patient believes that men are physically and mentally superior to women. He wants only sons. His wife does not agree with him, but she worries about how he would treat a girl.

The fourth patient inherited several million dollars from an eccentric uncle. He will receive the money only if he and his wife have a son and name him after the uncle. The man already has a daughter and wants only two children. Obviously, he wants his next child to be a boy.

Some questions related to biology, law, and ethics come to mind when reading about these cases. Before we can address those questions, let's look at the biology behind sex selection.

A Developing Human Fetus

1.1 How Is the Sex of a Child Determined?

In humans, as in many other animals, we can see obvious physical differences between males and females. These differences begin to appear early in a fetus's development. Whether male or female sex organs form during development depends on a complex interaction between the genes of the fetus and its environment. Some differences— such as body size, muscle mass, patterns of fat distribution, and body hair—develop later in life and are not directly related to reproduction. These are called **secondary sex characteristics**.

sex selection a method of predetermining the sex of a child by using prefertilization or preimplantation methods

secondary sex characteristics changes in the body that occur in early teens that relate to reproduction

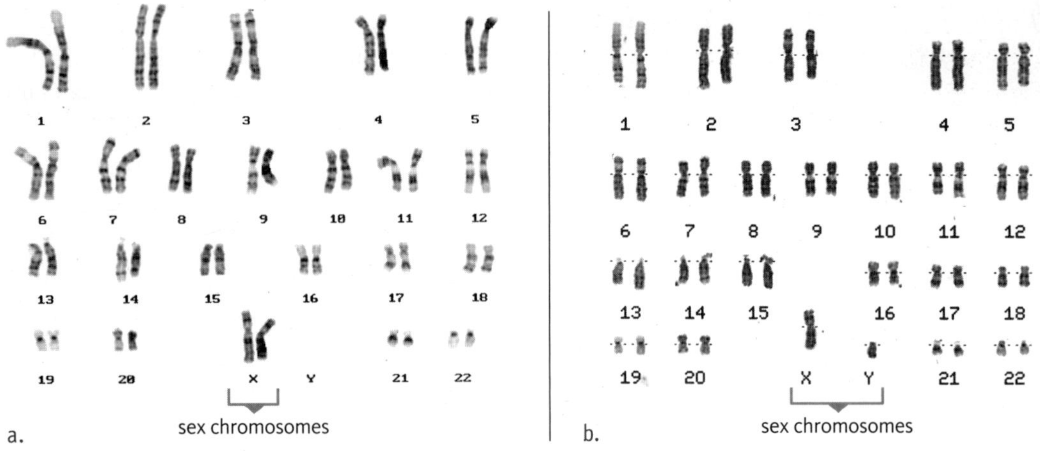

Figure 1.1
The Human Chromosome Set

(a) Female chromosome set with 46 chromosomes, including two X chromosomes. (b) Male chromosome set with 46 chromosomes, including an X and a Y chromosome.

a.

sex chromosomes

b.

sex chromosomes

How do chromosomes help determine sex?

In "Biology Basics: Cells and Cell Structure," we learned that human cells carry a set of 46 chromosomes (Figure 1.1a, b). Females have two **X chromosomes** (XX) and males have an X chromosome and a **Y chromosome** (XY). These chromosomes (X and Y) are the sex chromosomes because they help determine the sex of an individual. The other 22 pairs of chromosomes are called **autosomes**. In humans, sex determination begins at **fertilization** when a sperm, carrying an X or a Y chromosome, fuses with an egg (which carries an X chromosome). Fertilization produces a cell called a **zygote** that carries either an XX or XY set of sex chromosomes in addition to the 22 pairs of autosomes.

Although saying that females are XX and that males are XY seems straightforward, there is more to it than that. Is a male a male because he has a Y chromosome or because he does *not* have two X chromosomes? Can someone be XY and develop as a female? Can someone be XX and develop as a male? These questions are still not completely resolved, but about 40 years ago individuals with only one X chromosome (45,X)* were discovered. Those with only one X chromosome are female. At about the same time, males who carry two X chromosomes and a Y chromosome were discovered (47,XXY).*

---

*The number indicates the total number of chromosomes present; the letter indicates which of the sex chromosomes are present.

X chromosome the sex chromosome present in organisms (including humans) where female gametes carry one type of sex chromosome

Y chromosome the sex chromosome present in organisms (including humans) where male gametes carry one of two types of sex chromosomes

autosomes all chromosomes except the X and Y

fertilization when the sperm and egg join

zygote the diploid cell produced by the fusion of haploid gametes

sex ratio the proportion of males to females in a population

Anyone who carries a Y chromosome is almost always male, no matter how many X chromosomes he may have. However, the X and Y chromosomes are not the only determinants of one's sex. As we will see later in this chapter, the outcome depends on interactions between genes and the environment in the uterus.

How does the human sex ratio change over time?

The **sex ratio** is the proportion of males to females in a population. All eggs produced by human females carry a single X chromosome. Roughly half the sperm produced by males carry an X chromosome and half carry a Y chromosome (how eggs and sperm are formed will be explored in Chapter 3).

An egg (carrying an X chromosome) fertilized by an X-bearing sperm results in an XX zygote, which will develop as a female. Fertilization by a Y-bearing sperm produces an XY zygote, which will be male (Figure 1.2). Obviously, Susanna Carter wants to guarantee fertilization of one of her eggs by an X-bearing sperm to have the daughter she has always wanted. The couple who wishes to satisfy the terms of inheritance from the eccentric uncle needs to achieve fertilization by a Y-bearing sperm, or find a way to select XY embryos in order to have a son and inherit several million dollars.

Because males produce approximately equal numbers of X- and Y-bearing sperm (Figure 1.3), it seems obvious that about equal numbers of males and females should be conceived. If equal numbers of male and female zygotes are produced, the sex ratio will be 1:1. Although we cannot be completely certain about the numbers or the reasons, estimates indicate that more males are conceived than females. This does not mean that the sex ratio at birth reflects this imbalance. At birth the ratio is about 1:1.05 (100 females for every 105 males). As a generation ages, these numbers change. More males than females die in childhood, and when this generation reaches the age of 20, the ratio of males to females moves closer to 1:1. Beyond this age, females begin to outnumber males because men have a shorter life span than women.

Figure 1.2 **X and Y Chromosomes in Sex Determination**

The random combination of an X- or a Y-bearing sperm with an X-bearing egg produces, on average, a 1:1 ratio of males to females.

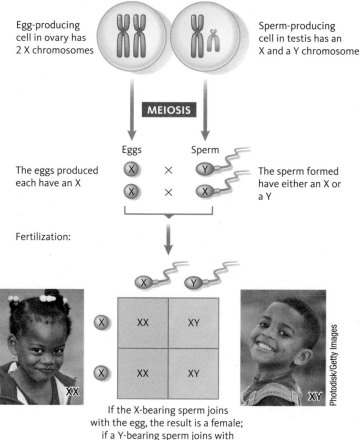

Egg-producing cell in ovary has 2 X chromosomes

Sperm-producing cell in testis has an X and a Y chromosome

MEIOSIS

Eggs   Sperm

The eggs produced each have an X

The sperm formed have either an X or a Y

Fertilization:

	X	Y
X	XX	XY
X	XX	XY

If the X-bearing sperm joins with the egg, the result is a female; if a Y-bearing sperm joins with the egg, the result is a male.

Eye Wire

Photodisk/Getty Images

Figure 1.3 **A Scanning Electron Micrograph of a Human Sperm** The sperm head carries one copy of each chromosome, which is either an X or Y chromosome.

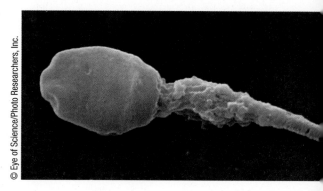

© Eye of Science/Photo Researchers, Inc.

Figure 1.4 **Comparing Chromosome Size** The human X chromosome (left) is much larger than the Y chromosome (right).

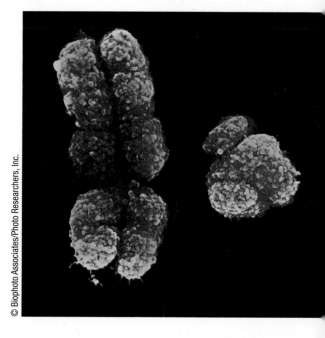

© Biophoto Associates/Photo Researchers, Inc.

What procedures are used to select the sex of a baby?

For centuries, people have wanted to control the sex of their children. Many rituals and traditions have been used to try to ensure that a child will be a boy or girl. Some of these cultural practices include having sex during certain phases of the moon, putting a knife or an egg under the bed before sex, or eating certain foods during pregnancy. Although many of these folklore-inspired methods are interesting and even ingenious, only two reliable, scientific procedures for sex selection exist: **sperm sorting** and **preimplantation genetic diagnosis (PGD)**.

Sperm sorting is possible because sperm carrying an X chromosome have about 2% more DNA than sperm carrying a Y chromosome. These sperm are heavier because the X chromosome is larger than the Y (Figure 1.4).

After collection (Figure 1.5, page 16), the sperm are concentrated using a centrifuge (1). They are treated with a nontoxic fluorescent dye that binds to the DNA in the chromosomes (2). The sperm are then sorted using a laser beam (3). As sperm flow past the laser, the dye glows. Sperm with X chromosomes contain slightly more DNA and glow a little brighter than those with Y chromosomes. Based on how brightly they fluoresce, the sperm are sorted into two test tubes (4). One tube collects sperm with an X chromosome; the other collects sperm with a Y chromosome. Sperm from either tube can be placed into the uterus to fertilize the egg. This procedure costs between $4,000 and $6,000. If eggs are removed from the body and fertilized in a dish by the addition of sperm, the cost increases greatly.

sperm sorting a method of separating sperm that carry Y chromosomes from sperm that carry X chromosomes; separated sperm are used in fertilization to determine the sex of the offspring

preimplantation genetic diagnosis (PGD) a procedure that removes one cell from an early embryo for testing

Figure 1.5 **The Process of Sperm Sorting for Sex Selection**

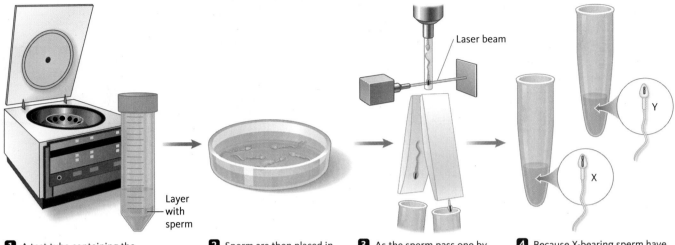

Laser beam

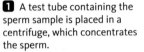
Layer with sperm

1 A test tube containing the sperm sample is placed in a centrifuge, which concentrates the sperm.

2 Sperm are then placed in a salt solution containing a fluorescent dye, which binds to the DNA in the sperm.

3 As the sperm pass one by one through a detector, a laser bounces light off the dyed DNA.

4 Because X-bearing sperm have more DNA and reflect more light than Y-bearing sperm, each can be separated into different test tubes.

Because some sperm can be misidentified during sorting, this method of sex selection is not 100% accurate. Sperm sorting has a success rate of about 90% for female births and about 73% for male births. This might be an acceptable method for Susanna Carter, but it probably will not be acceptable for the couple seeking to have a son so that they can inherit an uncle's money.

The second method of sex selection, preimplantation genetic diagnosis (PGD), is shown in Figure 1.6. PGD begins with hormone treatments to stimulate a woman's ovaries to produce many eggs at the same time. These eggs are surgically collected and placed in a dish with sperm for fertilization (a process called *in vitro* **fertilization (IVF)**, discussed in Chapter 2).

in vitro **fertilization (IVF)** fertilization of an egg in a laboratory dish

Figure 1.6 **Steps in the Process of Preimplantation Genetic Diagnosis**

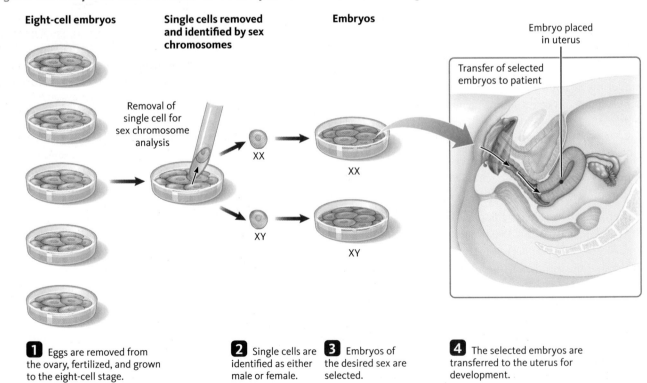

Eight-cell embryos

Single cells removed and identified by sex chromosomes

Embryos

Embryo placed in uterus

Removal of single cell for sex chromosome analysis

Transfer of selected embryos to patient

XX

XX

XY

XY

1 Eggs are removed from the ovary, fertilized, and grown to the eight-cell stage.

2 Single cells are identified as either male or female.

3 Embryos of the desired sex are selected.

4 The selected embryos are transferred to the uterus for development.

When the newly formed embryos have developed to the eight-cell stage (1), a single cell is removed, and its chromosomes are analyzed to determine the sex of the embryo (2). Once male and female embryos have been identified and separated (3), only embryos of the desired sex are placed into the mother's uterus (4). Unused embryos are discarded, frozen for later use, or donated to others. PGD is a more invasive procedure than sperm sorting and is riskier for the mother. PGD is also much more expensive, as it costs between $12,000 and $15,000. But, because the success rate for sex selection using PGD approaches 100%, many parents are willing to use it. PGD would likely be the method of choice for Susanna Carter and the other patients mentioned earlier.

As you can see from thinking about the Carters' situation and those of Dr. Leon's other patients, the use of sex selection and the fate of unused embryos created in PGD raise many social and ethical issues. To avoid these issues, some physicians refuse to use PGD for sex selection except when one spouse carries a genetic disease that affects one sex, whereas other physicians embrace the technology.

1.1 Essentials

- **Females carry two X chromosomes; males carry an X chromosome and a Y chromosome.**
- **Chromosomal sex is determined at fertilization.**
- **Sperm sorting and preimplantation genetic diagnosis (PGD) are two methods of sex selection.**

1.2 How Does a Baby Develop from Fertilization to Birth?

To understand other factors that determine maleness and femaleness, let's look at events that occur during development of the fetus. Figures 1.7 and 1.8 show the sequence of events that lead to these changes. Follow along by looking at the numbers in each figure.

Fertilization, the fusion of sperm and egg, usually occurs in the upper third of the fallopian tube (1). Sperm deposited in the vagina move into the cervix, up the uterus, and into the fallopian tube; this trip takes about 30 minutes. Sperm travel this distance of about 7 inches (approximately 17 cm) using whip-like contractions of their tails, assisted by muscular contractions of the uterus.

Usually only one sperm enters the egg, but other sperm (2) help by triggering chemical changes in the egg that prevent the entry of more than one sperm. Before the sperm enters the egg, the tail falls off and only the head enters the egg. The head is the sperm's nucleus and fuses with the egg's nucleus (3 and 4), forming a zygote with 46 chromosomes (22 pairs of autosomes and 1 pair of sex chromosomes) (4).

The zygote begins a series of cell divisions (Figure 1.8, page 18) that form an embryo; the divisions occur as the embryo travels to the uterus over a period of three

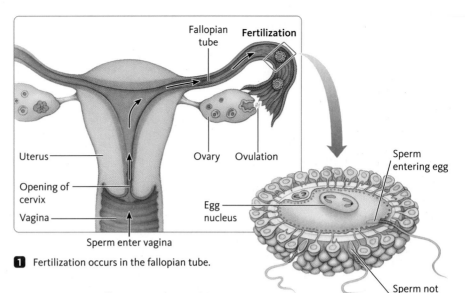

1 Fertilization occurs in the fallopian tube.

2 Several sperm surround the egg (above), but only one enters it.

3 The sperm nucleus fuses with the egg nucleus.

4 A diploid zygote forms.

Figure 1.7 **Stages in Human Fertilization** (1), (2) In fertilization, many sperm surround the egg and secrete enzymes that dissolve the outer barriers surrounding it. Only one sperm enters the egg. (3) The sperm tail degenerates, and its nucleus enlarges and fuses with the nucleus of the egg. (4) After fertilization, a zygote has formed.

Figure 1.8 Human Development from Fertilization through Implantation As a blastocyst forms, its inner cell mass gives rise to a disk-shaped early embryo. As the blastocyst implants into the uterus, cords of chorionic cells start to develop. When implantation is complete, the blastocyst is buried in the endometrium.

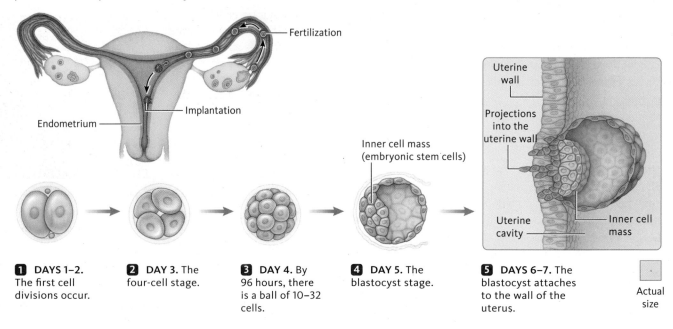

1 DAYS 1–2. The first cell divisions occur.

2 DAY 3. The four-cell stage.

3 DAY 4. By 96 hours, there is a ball of 10–32 cells.

4 DAY 5. The blastocyst stage.

5 DAYS 6–7. The blastocyst attaches to the wall of the uterus.

Actual size

to four days. First, the zygote divides to form two cells (1) and continues dividing as it moves down the fallopian tube. When the embryo reaches the uterus, it remains unattached for several days (2 and 3). Cell division continues during this time, and the embryo enters a new stage of development; it is now a large hollow ball of cells, a **blastocyst** (4). A blastocyst has several distinct parts: an outer layer of cells, an internal cavity, and a cluster of cells called the **inner cell mass**. The outer layer of cells will form the membranes that surround the embryo. The inner cell mass contains the human embryonic stem cells (hESC). Cells of the inner cell mass at this point are unspecialized, but will eventually differentiate to form all the cells of the fetus.

There has been a great deal of controversy about the isolation and use of human embryonic stem cells. Because they will develop into many different cell types, researchers have been working to specify which tissues they can form. It is hoped that this work will lead to treatments for conditions otherwise considered incurable (such as Parkinson's disease) and possibly to the growth of new organs to replace diseased ones. Removal of these cells from the embryo is one of the ways to obtain them for experimentation. In later chapters we will discuss the uses of stem cells and some of the ethical and legal issues surrounding their use.

While the embryonic cells are dividing, cells lining the uterus enlarge, preparing for the attachment or **implantation** of the embryo. About 12 days after fertilization, the embryo becomes firmly embedded in the wall of the uterus (5) and has formed a protective membrane, the **chorion**. The chorion makes and releases a hormone, **human chorionic gonadotropin (hCG)**, which prevents the lining of the uterus from breaking down and expelling the embryo. Excess hCG is eliminated in the urine. Home pregnancy tests work by detecting elevated hCG levels in urine.

A series of chorionic finger-like projections, the **villi**, extend into the spaces in the uterine wall that are filled with maternal blood. The villi and maternal tissues form the **placenta**, a disc-shaped structure that nourishes the embryo throughout pregnancy (Figure 1.9). Membranes connecting the embryo to the placenta form the umbilical cord.

The second trimester is a period of organ maturation. The skeleton forms, and the heartbeat can be heard with a stethoscope. By the fourth month, the mother can feel movement of the fetus. At the end of this trimester, the fetus weighs about 700 g (27 oz.) and is 30–40 cm (about 13 in.) long.

blastocyst a stage of development in the early embryo

inner cell mass group of cells in the early embryo

implantation the process by which the fertilized egg attaches to the uterine wall

chorion precursor for the placenta

human chorionic gonadotropin (hCG) a hormone created by the chorion that holds the lining of the uterus in place for implantation

villi (sing. villus) finger-like projections composed of cells

placenta a disk-shaped organ that nourishes the developing embryo; formed in the uterus by the interaction of maternal cells and fetal cells

Figure 1.9 **Healthy Human Placenta after Birth**

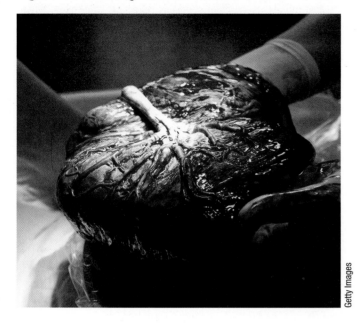

Getty Images

The fetus grows rapidly in the third trimester (weeks 25–36), and the circulatory system and respiratory system mature to prepare for breathing air.

The fetus doubles in size during the last 8 weeks (as shown in Figure 1.10), and chances for survival outside the uterus increase rapidly during this time. At the end of the third trimester, the fetus is about 19 inches (48 cm) long, weighs between 5.5 and 10.5 lb (2.5–4.7 kg), and is ready to be born.

The main stages of fetal development are summarized in Figure 1.11.

Figure 1.10 **In later stages of development, the fetus seems crowded into the uterus.**

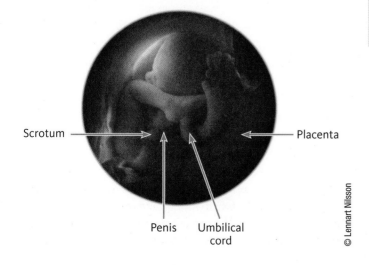

Scrotum ———————→   ←——— Placenta

Penis   Umbilical cord

© Lennart Nilsson

CASE A: QUESTIONS

Now that we understand how the XX and XY makeup of a fetus is normally determined and how sex selection works, let's look at some of the issues raised in the case of the Carters and Dr. Leon's other patients and apply the science we have discussed. A discussion of how courts of law interpret difficult scientific questions will follow.

A newborn male and female

Sandra Warick/Photonica/Getty Images

1. Knowing how sex selection is accomplished, what do you think the Carters should decide to do? Why?

2. Which one of the Carters should make this decision— Susanna, Bob, or both? Why?

3. If you were Dr. Leon, would you perform sex selection for the Carters? Why or why not?

4. If you were Dr. Leon, would you perform sex selection for the other patients? Give an answer and a reason for each patient.

5. Should sex selection be available to anyone, no matter his or her reason? Why or why not?

6. Should ability to pay for the procedure limit access to sex selection? Should government programs like Medicaid pay for this?

7. Should insurance companies pay for sex selection? In which of Dr. Leon's cases would this seem like a good idea and why?

8. In India and China, there are problems with the use of sex selection. Historically, families have favored boys over girls, and with the use of sex selection, the sex ratio is changing. In China's 2000 census, it was 100 girls to 120 boys. That census also showed 19 million more males than females in the 0–15 age group in China. What social problems might this create? If everyone could have access to sex selection, would this trend continue on a global basis and affect the world population? How?

Figure 1.11 **The Stages of Fetal Development**

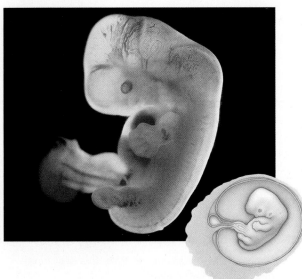

In weeks 5 and 6, the embryo grows dramatically to a length of about 11 inches (28 cm). Most of the major organ systems, including the heart, are formed by this time. Limb buds develop into arms and legs, complete with fingers and toes. The head is very large compared to the rest of the body because of the rapid development of the brain.

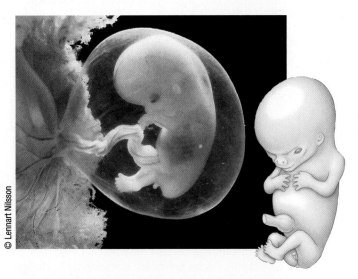

By about 8 weeks, the embryo is large enough to be called a fetus. All the major organs have formed at this point and are functional. Although chromosomal sex (XX in females and XY in males) is determined at the time of fertilization, the fetus appears to be neither male nor female at the beginning of the third month. The sex organs cannot be seen in ultrasound scans until the 12th to 15th week.

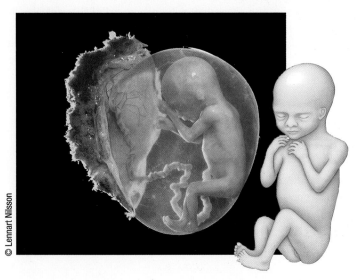

By 16 weeks, major changes include an increase in size and further development of organ systems. Bony parts of the skeleton begin to form, and the heartbeat can be heard with a stethoscope. Fetal movements begin in the third month, and by the fourth month the mother can feel movements of the fetus's arms and legs. It has a well-formed face, its eyes can open, and it has fingernails and toenails.

- **Sex is only one of the thousands of traits determined during the gestation period in humans.**
- **Many developmental changes occur in the fetus from fertilization to birth.**

1.3 What Are the Stages of Sex Development?

As discussed previously, having an XX or XY chromosome pair is only part of becoming a male or a female. The formation of male or female reproductive organs depends on several factors, including:

- which genes are activated, when they are activated, and how long they are turned on.

- the hormones present in the uterine environment.

- the maternal environment.

Interactions among these factors are important to the development of an individual's sex (Figure 1.12, page 22). Between fertilization and birth, sex can be defined at several levels: chromosomal sex, which internal sex organs develop, and which genitals form. In most cases, these events are consistent with the chromosomal sex, but in other situations, they are not. Understanding what can go wrong requires a look at what normally happens in development.

The sex of an individual is formed in stages. The first stage (1), chromosomal sex, begins at fertilization when the zygote has either an XX or an XY chromosome pair.

Although chromosomal sex is established at fertilization, the internal and external sex organs do not begin to develop until the seventh or eighth week of pregnancy. Until this time, the external genitals of the embryo are neither male nor female. Inside the body of the fetus, nonspecific **gonads** and two sets of ducts are present (2).

In the second stage (3), the presence and expression of one or more genes on the Y chromosome cause the nonspecific gonads to become testes. Cells in the newly formed testes secrete **testosterone**, a male hormone that controls the development of the male duct system, the degeneration of the female ducts, and the external genitals (4).

If two X chromosomes are present (and no Y chromosome is present), the gonads develop into ovaries (3). In other words, if a Y chromosome is not present, the default developmental pathway for the nonspecific gonads is the formation of ovaries. Once ovaries form, the female duct systems develop to form the fallopian tubes, uterus, and parts of the vagina, while the male duct system degenerates.

Then, in the third stage, the genitals form. In males, male hormones direct formation of the penis and the scrotum. In females, the clitoris, the labia minora, and the labia majora are formed from the same tissues (5 and 6). Near the end of this stage, the fetus can be identified as either male or female by ultrasound.

In summary, the development of internal and external sex organs results from different genetic pathways. In males, one pathway begins with the action of a gene (*SRY*) on the Y chromosome, the presence of at least one X chromosome, and expression of genes carried on the other 22 chromosomes. In females, another pathway begins with the presence of two X chromosomes, the absence of Y chromosome genes, and expression of a female-specific set of genes on the X chromosome and the other 22 chromosomes.

BY THE NUMBERS

3
Number of trimesters in a human pregnancy

© Brand X Pictures/Jupiter Images

12
Number of weeks in one trimester

44
Number of rounds of mitosis in a human pregnancy

Trillions
Number of cells in a human newborn

gonad an embryonic organ that gives rise to the testes or ovaries

testosterone a steroid hormone produced by the testes; the male sex hormone

Figure 1.12 **Stages of Sex Determination and Differentiation**

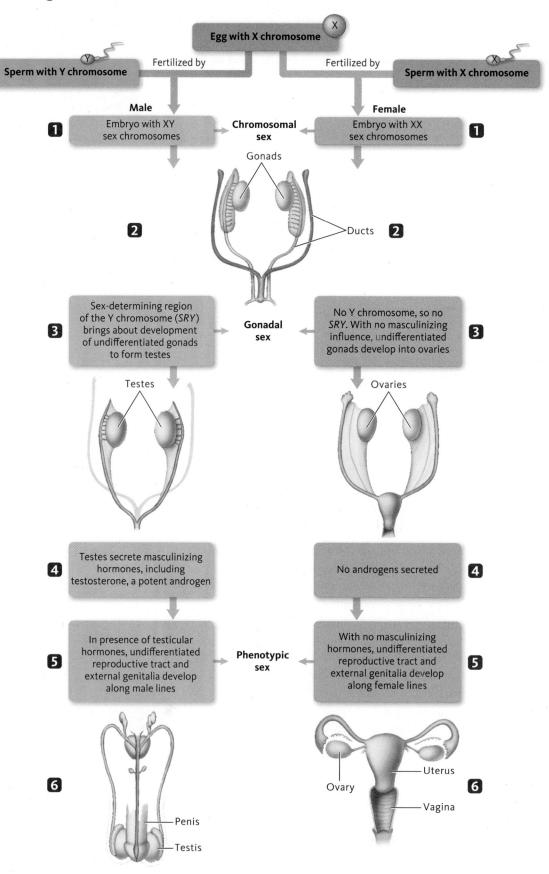

CASE B: Athlete Fails Sex Test

Sex testing was used in the Olympics and other world-class competitions following charges that some countries were allowing men to compete as women in events reserved for women.

Maria Patino was a first-rate athlete, a hurdler who represented Spain. At the World University Games in Japan in 1985, she was instructed to report to "Sex Control," where cells were scraped from the inside of her cheek to examine her sex chromosomes. That didn't bother her because she had already passed such a "sex check" at the Helsinki World Championships in 1983. But she had forgotten to bring her "certificate of femininity" (see Figure 1.13) to the Japanese games and had to be retested.

Figure 1.13 Certificate of Femininity
Maria Patino's certificate of femininity, issued before she competed in Japan. This states (in English and French) that she is a female and has passed the "sex test."

Maria Patino was a hurdler. Here she is finishing a race.

A few hours after the test, an official told her the results were abnormal. "What does that mean?" she wondered. She was asked to go to the hospital in the morning, before her race, to be retested again.

Maria was in and out of the hospital in a few minutes. The doctors spoke only Japanese, and no one said a word to her. When she was being driven to the stadium, just before the start of her race, Maria was told to fake an injury and withdraw from the games. Stunned, she did as she was asked. Later she learned that she had failed the sex test.

After returning home, Maria went to an endocrinologist, who diagnosed her condition as **complete androgen insensitivity (CAI)**, also called androgen insensitivity syndrome (AIS). This meant that although she was raised as a female and looked like a woman, her sex chromosomes were XY, not XX. Chromosomally, she was a man!

Some scientific, legal, and ethical questions come to mind when reading this case. Before addressing those questions, let's look a little further into the biology behind sex determination in humans.

What was Maria Patino's condition? Complete androgen insensitivity is caused by a mutation in a gene on the X chromosome. This gene is called the **androgen receptor (AR)** gene. People with this condition lack molecular sensors (called receptors) for testosterone on the surface of all their cells, making it impossible for the cells to respond to testosterone when it is present. This is especially important in those cells that form the internal and external sex organs. The presence of an XY chromosome set normally causes the nonspecific gonads to develop into testes and produce testosterone. But in CAI, cells in the gonad cannot respond to the presence of testosterone, and, therefore, development continues as if no testosterone were present, which means that the female duct system and female external genitals develop.

Individuals with this condition are chromosomal males (XY), but at birth, they appear to be females and are raised as girls (Figure 1.14). As adults, they do not menstruate, are infertile, and have

complete androgen insensitivity (CAI) a genetic condition that causes an XY karyotype child to appear female

androgen receptor (AR) gene a gene that encodes the protein that acts as a receptor for male hormones

Figure 1.14 Androgen Insensitivity Syndrome An individual with complete androgen insensitivity.

Figure 1.15 **Barr Bodies**

Inactive X chromosomes (Barr bodies) in a cell from a human female. XY males do not form Barr bodies.

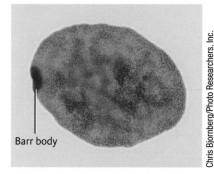

Barr body

Chris Bjornberg/Photo Researchers, Inc.

Barr body the inactivated X chromosome in human females (and other mammals)

well-developed breasts, very little pubic hair, and often undeveloped testes in their abdomen. This condition takes several forms with some variations in female development.

Years later, after finding out about her condition, Maria Patino was quoted as saying, "It was obviously devastating to me. I had devoted my life to sport. But it was never a question for me of my femininity. In the eyes of God and medicine I am a woman."

How did Maria Patino pass one sex test, but not another?

In human females, one X chromosome in all cells (except developing eggs) becomes tightly coiled and inactivated early in development. This coiled chromosome is referred to as a **Barr body** (see Figure 1.15) and, when stained, can be seen in a microscope. The inactivated X chromosome can come from the mother or the father. Males have only one X chromosome and do not form Barr bodies. Because females have two X chromosomes in each cell, and one or the other becomes inactivated, some cells express X chromosome genes only from the mother's X chromosome and some cells express genes only from the father's X chromosome.

The original sex test required of Maria Patino and all other female athletes used cells scraped from inside the cheek. The scrapings were placed on a microscope slide and the cells were stained with a dye that makes the Barr body visible. A number of cells are examined with a microscope to see if Barr bodies are present. If they are present, the athlete is certified as eligible to participate in competition as a female. Males

CASE B: QUESTIONS

Now that we have learned about the stages of sex determination and what can go wrong during development, let's look at the issues raised in Maria Patino's case.

1. Knowing what you do about sex determination, do you think Maria Patino is female? Why or why not?

2. In the first sex test, when cells from her cheek were examined, she passed and was certified as a female athlete. Should this have been enough to verify that she was a female? Why or why not?

3. Do you think Maria has a muscular advantage over other female runners because she is XY? If so, would this have made a significant difference?

4. When Olympic and world championship competition required sex testing, only females were tested. Is this fair? Why or why not?

5. Should all newborns have their sex chromosome status determined? (See Figure 1.16.) Would this avoid any problems that might arise later? Why or why not?

6. What problems might this testing cause for those with abnormal sex chromosomes, such as those we discussed in this chapter?

7. At a hearing, let's assume you are Maria's attorney.

Maria Patino was a hurdler. Here she is finishing a race.

Courtesy, Professor Maria José Patino

a. What argument would you make using the science related to sex determination? Be sure to make your argument simple enough for a juror to understand.

b. What argument would you make using the law?

c. What would you ask for as compensation?

8. At a hearing, you are the Olympic Committee's attorney.

a. What argument would you make using the science related to sex determination? How would you make it understandable?

b. What argument would you make to defend the actions of the committee?

c. What action would you recommend the committee take? Why?

9. Based on Maria's case, should any changes in sex testing be instituted?

10. Why didn't Maria find out about this condition earlier? Do you think the fact that some female athletes in strenuous sports do not menstruate may have complicated her diagnosis?

11. Could Maria sue her doctors for malpractice? Why or why not?

12. Who else might she sue?

Figure 1.16
A child with complete androgen insensitivity, the same condition as Maria Patino.

Courtesy Cynthia P. Stone

who are XY do not form Barr bodies, so this test was always thought to be a reliable way of preventing males from gaining an advantage by competing in events for female athletes.

Unfortunately, this sex test presents several problems. First, the test is not completely reliable. If Barr bodies do not stain or cannot be seen because cells overlap with one another, an XX female can fail the test. Second, XXY males form Barr bodies and can pass a chromosomal sex test designed to identify females. Third, some females have only one X chromosome (called an XO condition) and do not form Barr bodies. More importantly, for our case, Marina Patino was XY and her cells did not form Barr bodies. The fact that she passed one sex test but failed another is an indication of how unreliable the Barr body test is in detecting XX individuals and screening out XY individuals.

Eventually, in the face of evidence that the Barr body test was unreliable, organizers of the Olympic games and other competitions turned to a series of other tests in a continuing effort to screen out males who might attempt to compete as females, even though there is little evidence that this has ever occurred. Finally, after recognizing that testing caused emotional stress, stigmatization, and discrimination, sex testing was abolished before the 2000 Olympic games and has also been discontinued in most other athletic competitions.

1.3 Essentials

- **The most important developmental events occur in the first trimester.**
- **The internal sex organs begin developing in the seventh or eighth week of pregnancy.**
- **Development of the sex organs of a fetus is controlled by the hormones testosterone and estrogen.**
- **One X chromosome is inactivated in XX female cells and forms a Barr body.**
- **Disorders of sexual development such as CAI (complete androgen insensitivity) are genetically controlled.**

1.4 What Are the Legal and Ethical Issues Associated with Sex Determination?

As you can see from Maria Patino's situation, people with some genetic conditions are not always treated fairly. In fact, they are often discriminated against. When this occurs, it can be handled in a number of ways. One way is to file a lawsuit that forces employers and others to treat them equally. Another way is to pass laws that make it clear that society will not tolerate this type of behavior.

Every state of the United States has passed laws that protect people from being discriminated against because of their race, color, sex, origin, religion, or age. This type of discrimination usually occurs in critical areas such as housing, public accommodations, education, transportation, communication, recreation, health services, voting, and access to public services.

It seems as though Maria Patino was discriminated against because she was competing as a woman but could not pass the "sex test." This meant that her chromosomes showed that she was genetically a male, although she was physically a female.

People with complete androgen insensitivity fall into a medical category referred to as *intersexuality*. A medical definition of intersexuality is "a condition in which chromosomal sex of a person is not consistent with his or her phenotypic sex, or in which the phenotype is not classifiable as either male or female."

Some other conditions fall under this category. We will be discussing chromosomal abnormalities called Klinefelter syndrome (47,XXY), and Turner syndrome (45,X) in later chapters.

The Americans with Disabilities Act (ADA) addresses the problems faced by Americans in employment, insurance, and other areas. This federal act was signed into law in 1990 and protects the rights of people with disabilities. Many of these disabilities are caused by genetic conditions.

Under the ADA, a person with a disability is defined as an individual who:

a. has a physical or mental impairment that substantially limits one or more major life activities; or

b. has a record or history of such an impairment; or

c. is perceived or regarded as having such an impairment.

Do people with intersexuality fall under the ADA? This has been difficult to determine, but individuals with other genetic conditions are clearly protected by the ADA. One of these is achondroplasia (see Chapter 4), a form of dwarfism. Because of their height, people with this condition may be discriminated against in employment, education, or insurance. The ADA is designed to prevent this type of discrimination.

Table 1.1 lists some of the ethical and legal questions associated with discrimination related to sex determination.

Table 1.1 Some Legal and Ethical Questions about Sex Discrimination

Question	How Are These Questions Decided?	Related Case or Legal Issue
Can a person be denied a job because of his or her sex?	Laws require companies to hire without discrimination. Lawsuits are filed and decided by individual judges.	It is difficult to determine why a person is not hired unless it is specifically stated.
Was Maria Patino a victim of discrimination?	The rules of the committee and the Olympics did not allow someone to compete as a woman if she could not pass the chromosomal test.	Maria Patino did not file a lawsuit, but others have forced the removal of the "sex test" rules in international athletic competitions.

Spotlight on Law

Frye v. U.S.
293 F. 1013 (1923)

If Maria Patino had sued in a court of law, the court would have had to decide whether she is a woman or a man. As you saw when you answered the case questions, this situation does not always have a simple answer, and it involves an understanding of scientific principles. The following case was extremely important in helping courts determine what science is relevant as evidence in trials and what is not. Because *Frye v. U.S.* was the first case to address what science should be accepted in court, any lawsuit filed by Maria in the United States would have used it when considering the science related to sex determination and CAI. The questions formulated in this case were used for more than 70 years as the standard for accepting or rejecting scientific evidence in U.S. courts.

As we consider the legal cases throughout the book, we will be referring to *Frye* as it has been applied over the years to the principles of genetics and other areas of science in a court of law.

FACTS OF THE CASE

James Alphonso Frye was on trial for murder. He had maintained his innocence since his arrest. Early on, his attorney suggested that he take a test to determine whether he was telling the truth. The test, the systolic blood pressure deception test, was an early version of the lie detector test used today.

The court was asked to accept Frye's "truthful" results as evidence to support his plea of innocence. The technique was new, but his attorney had an expert witness. William Marston, an attorney and psychologist, had done a great deal of

research on how changes in the body's physiology correlate with lying. He claimed that the test measured whether the defendant was telling the truth by monitoring his systolic blood pressure, which goes up and down depending on how hard the ventricles of the heart are contracting. Marston testified that fear always produces a rise in systolic blood pressure, and lying, along with fear of detection, also raises blood pressure. However, the judge excluded Marston's testimony, stating that the science of deception testing was too new to be admitted.

Frye's lawyer took the case to the Court of Appeals of the District of Columbia, which ruled that the systolic blood pressure deception test had "not gained standing and scientific recognition," and therefore it upheld the lower court's decision not to admit Marston's testimony.

More importantly, the court set guidelines for the admission of scientific evidence. These guidelines stated that a scientific principle would have to gain "general acceptance" in the "particular field to which it belongs." The so-called general acceptance rule asked three questions:

1. To which scientific field does the evidence or testimony belong?

2. Is the evidence or testimony generally accepted in this field?

3. What constitutes general acceptance?

To be admissible, the scientific testimony must be scrutinized by a relevant scientific community and be significantly tested and confirmed.

If Maria's case had gone to court, the science behind the Barr body test to determine Maria's sex would have to have been addressed, and the acceptance of these tests by the scientific community would have to have been determined.

QUESTIONS

How much testing should be done before science is admitted into evidence?

What three questions would you ask an expert witness to determine whether he or she is an expert?

Should expert witnesses be paid? Why or why not?

All science begins as "new" or experimental. How long should it take before it becomes "real" science?

Fingerprints are now considered good evidence after about 50 years of use in trials. But they are not as ironclad as we might think. Often partial fingerprints are the only evidence available and can be difficult to identify. But the public likes fingerprints. Should public opinion have anything to do with admission in court? Why?

6. Today, the results of lie detector tests are not allowed as evidence. Why? Should they be?

7. In his book *The Truth Machine*, James Halperin envisions a future with a machine that can determine lying with no error. What effect do you think this would have on courtrooms?

8. Recently, the Supreme Court has given judges more control in determining what science should be allowed to be presented in trials. What effects might this freedom have on the standards for the admission of scientific evidence in court?

The Essential 10

1 Two methods of sex selection are sperm sorting and preimplantation genetic diagnosis (PGD). (Section 1.1)
Sperm sorting involves separation of X- and Y-bearing sperm.

2 Females have two X chromosomes; males have an X chromosome and a Y chromosome. (Section 1.1)
The X comes from either the mother or father, but the Y must come from the father.

3 Chromosomal sex is determined at fertilization. (Section 1.1)
The sperm that joins with the egg contains either an X or a Y.

4 Many developmental changes occur in the fetus between fertilization and birth. (Section 1.3)
The complex process of development is directed by chromosomes, the genes they carry, and both internal and external factors.

5 Sex is only one of the thousands of traits determined during development. (Section 1.2)
Many parts of sex development are hormonally controlled.

6 The internal sex organs begin developing in the seventh or eighth week of pregnancy. (Section 1.3)
Sexual development begins with the gonads.

7 The hormones testosterone and estrogen are secreted by the testes and ovaries. (Section 1.3)
Testosterone, estrogen, and their derivatives control development of the gonads, the duct system, and the genitals.

8 The most important developmental events occur in the first trimester. (Section 1.3)
Each trimester lasts about 3 months.

9 One X chromosome is inactivated in XX female cells and forms a Barr body. (Section 1.3)
The Barr body is formed when the X chromosome coils and attaches to the inside of the nuclear membrane.

10 Disorders of sexual development such as CAI (complete androgen insensitivity) are genetically controlled. (Section 1.3)
Mutations in many genes can affect formation of the sex organs.

Review Questions

1. Some physical traits specific to males or females usually appear during the teenage years. What are these attributes called?

2. If a child has a Y chromosome, what is his or her chromosomal sex? What is the child's chromosomal sex if he or she has two X chromosomes and a Y chromosome?

3. What is the sex ratio?

4. Can the sex of a baby be chosen during pregnancy? Why or why not?

5. At what point during pregnancy does each of the following structures develop?
   a. heart
   b. skeleton
   c. penis

6. Does the sex ratio in a population change over time? Explain.

7. What is PGD and why is it successful in sex selection?

8. How much does a fetus grow in the third trimester?

9. When sperm are formed, what percentage of them carry X chromosomes?

10. What percentage of the sperm carry Y chromosomes?

Application Questions

1. Do you think the chromosomes of every baby born should be examined? Why or why not?

2. Historically, men have often been furious with their wives for not giving them male children. Research this topic and find out when it became clear that women were not at fault.

3. When a baby is developing, doctors can operate on the fetus to correct certain errors. When the repair is finished, the fetus is put back into the uterus to finish developing. What do you think might be the biggest problem in this type of operation?

4. Suppose you are an attorney and a woman comes to you with the following problem. She had an ultrasound test early in her pregnancy, and her doctor said she was having a boy. But the woman gave birth to a girl, and now she wants to sue. How would you use *Frye v. U.S.* to explain that you cannot represent her?

5. Which do you think is the most dangerous time in the development of the fetus? Why?

6. There is about a 90% chance of producing a female using sperm sorting to select sperm carrying an X chromosome, and about a 73% chance of producing a male using sperm sorting to select sperm carrying a Y chromosome. Sex selection using PGD is 100% effective but produces extra embryos that must be stored for future use, donated, or destroyed. As a genetic counselor, you meet with a couple who wants to use sex selection to have a boy in order to balance a family with four girls. What questions would you ask them, and what options would you explore with them?

7. You are a physician. A couple with five girls comes to you for a second opinion. They want to have another child but, for personal

reasons, do not want to use sex selection. They consulted a reproductive expert who explained to them that the chances of having a boy are 50%, but since they have already had five girls, the odds are overwhelmingly in favor of their next child being a boy. They would like to know more about the odds of having a boy. What would you tell them? Would your answer change if you were an attorney? Why or why not?

8. Females carry two X chromosomes (XX) and males carry an X and a Y chromosome (XY). This means that females have two copies of all X chromosome genes and males have only one (in general, genes on the X are not found on the Y chromosome). If each gene produces one unit of product, does this mean that females make two units of product for X chromosome genes and males make one unit of all X chromosome gene products? Why or why not?

Online Resources

Preparing for an exam? Log on at www.cengagebrain.com for study tools to help you assess your understanding. If assigned by your instructor, the Case A and Spotlight on Law activities for this chapter, "Choice of Baby's Sex Now Available" and "*Frye v U.S.*," will also be available.

Learn by Writing

If you are interested in sharing your ideas about the questions raised in Chapter 1 or other issues related to what was presented here, we suggest you start a blog. Throughout the book we will give you some suggestions of discussion questions that you may want to include in your blog to share with members of your class or others.

Ask your instructor if your school has a blogging site or go to one of the free sites available on the internet (www.blogger.com, www.googleblog.com, www.tumblr.com, or others) and invite members of your class or others to contribute. This way you can write your opinions, share them with others, and get comments back.

Watch for other "Learn by Writing" suggestions in the Application Questions in various chapters, but don't wait to see it. Address interesting cases, questions, and ideas in every chapter . . . see what develops.

Here are some ideas to address in your blog from Chapter 1:

- The sex of a baby is a personal subject among families. It is very important in some societies to have a child of a certain sex. Does that mean we should give everyone the opportunity to pick the sex of his or her child?

- Intersex patients often say they are discriminated against if they tell others about their condition. Should this information remain private or be made public?

- A number of years ago, a child was undergoing a circumcision and a mistake was made. His penis was cut off. The doctors told the parents they would do surgery to make the child an external female and they should take him home and raise him as a girl. Do you think this would work? Maria Patino was raised as a girl successfully. How is she different from this boy?

WHAT WOULD YOU DO IF...?

. . . you were given the option to use sex selection to pick the sex of your baby?

. . . you were asked to write a rule that would govern sex testing in the Olympics?

. . . you were asked to vote on a law that would require sex chromosome testing of all newborns in your state?

. . . you were married to a woman who could not have a baby and the doctor told you that she carried an XY chromosome set?

. . . your daughter did not menstruate by age 15 and the doctor told you that her chromosomes are XY?

- Should the law actually be involved in any of these personal decisions?

- Address any other question or comment that came to mind as you read this chapter.

◼ In the Media

Dr. Phil (CBS), January 16, 2007

AIS Child on Dr. Phil Show

Kayleigh, whose parents brought her to the *Dr. Phil* show, has been diagnosed with androgen insensitivity syndrome (AIS) and is now 4 years old. Having the same syndrome as Maria Patino, Kayleigh looks female but has XY sex chromosomes.

Her mother said, "I was having to daily reinforce to Kayleigh that she was a girl, and if I talked to her about being a girl and not a boy, she would throw a temper tantrum. She cries and says she is a boy." Kayleigh also wanted to potty-train standing up.

To see highlights of this episode online, go to www.cengagebrain.com.

QUESTION

Do you think it is possible for a child this young to understand whether she is male or female?

USA Today, December 19, 2006

Female Runner Disqualified after Failing Gender Test

An Indian runner, Santhi Soundarajan, was stripped of her medal in the 800-meter run in the 2006 Asian games when her sex test came back abnormal. It was reported that she "appeared to have abnormal chromosomes." The official also said the test revealed more Y chromosomes than allowed. Athletes are not required to take a gender test but may be asked to do so.

To access this article online, go to www.cengagebrain.com.

Santhi Soundrarajan, an Indian runner who failed the sex test because of androgen insensitivity syndrome (AIS).

AP Photo/M. Lakshman

QUESTIONS

1. If this is the same condition (AIS) that Maria Patino has, should Santhi be allowed to continue as a female runner?

2. Research what has happened in Santhi's case since 2006.

AZ Capitol Times, February 12, 2011

House Outlaws Abortions for Sex and Race Selection
Caitlin Coakley

The Arizona House of Representatives passed HB 2442, a bill that outlaws abortion because of the sex or race of the fetus. Arguing on the floor, Rep. Steve Montenegro, R-Avondale, the bill's sponsor, was adamant that the bill had nothing to do with a woman's legal right to have an abortion and, instead, was a measure to prohibit bigotry and discrimination. But Democrats argued that these kinds of abortions do not occur with any frequency in their state. The bill now goes to the Senate for discussion.

To access this article online, go to www.cengagebrain.com.

QUESTION

Is a law stopping sex selection via abortion absolutely necessary? Why or why not?

CBC News, May 18, 2007

Canadians Buying 6-week Gender Determination Test

A new home test that can determine the sex of a fetus as early as 6 weeks has become quite popular among Canadians. This test, which requires a sample of the mother's blood, can determine whether male DNA is present.

The testing company, DNA Worldwide, said the results could come back in as few as four to six days. The company has studied the possibility of refusing to send the test to countries where couples practice sex selection and abortion.

To access this article online, go to www.cengagebrain.com.

QUESTION

If this test became available worldwide, do you think many people would use it to determine the sex of their child?

2

Assisted Reproductive Technology

REVIEW

Parts of a Cell (BB1.4)

How Does a Baby Develop from Fertilization
    to Birth? (1.2)

CENTRAL POINTS

- Sperm and eggs are formed in specialized organs.

- Males and females have different reproductive organs.

- Infertility is a serious problem in many parts of the world.

- Medical techniques are available to help people who are infertile.

- Legal cases are deciding the fate of frozen embryos.

CASE A: A Couple Considers Fertility Treatment

Brian and Laura have been married for 10 years. During the first 5 years of their marriage, they didn't really think about having children, mainly because both of them were very busy; Laura was an attorney, and Brian was finishing his master's degree.

But when Laura's sister became pregnant, it started them thinking. They decided to have a child and assumed that getting pregnant would be easy. But that was not the case. After a year of trying to conceive, they decided to talk to Laura's doctor. Shortly before the appointment with the gynecologist, Dr. Franco, the couple watched a television show that talked about new techniques to help infertile couples. Laura and Brian were amazed to learn that about 7 million American couples are infertile, and clinics all over the country specialize in helping them. The procedures were expensive, though, and they wondered what they would do if they couldn't get pregnant the "normal" way.

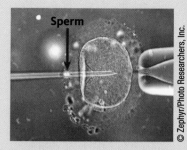

A human sperm being injected into an egg.

Brian thought they should adopt and not put themselves through the tests and procedures. Laura disagreed; she wanted a genetically related child.

Some questions come to mind when reading about Brian and Laura's situation. Before we address those questions, let's look at the biology behind fertility.

A Human Egg

2.1 How Are Sperm and Eggs Made?

As discussed in "Biology Basics: Cells and Cell Structure," all humans begin life as a single cell, the zygote, produced by the fusion of a sperm and an egg, a process called fertilization.

The **testes** (the male gonad) produce **sperm** and male sex hormones called **androgens**. One example of an androgen is testosterone. The **ovaries** (the female gonad) produce **eggs** and female sex hormones called **estrogens**. Early in gamete formation, sperm and egg cells undergo a form of cell division called **meiosis** (which will be discussed in Chapter 3), during which the chromosome number is reduced from 46 to 23. Fertilization of a sperm and egg to form a zygote restores the chromosome number to 46.

2.1 Essentials

- **Sperm are produced by meiosis in the testes of the male.**
- **Eggs are produced by meiosis in the ovaries of the female.**

testes the male organ where sperm are formed

sperm the male sex cell

androgens male sex hormones

ovaries female organs that make eggs

egg the female reproductive cell

estrogens female hormones

meiosis a form of cell division that creates egg and sperm

2.2 What Organs Make Up the Male Reproductive System?

Follow along with Figure 2.1 as we discuss the parts of the male reproductive system. Testes (1) form in the abdominal cavity during male embryonic development; before birth, they descend into the scrotum (2), a pouch of skin located outside the body. The male reproductive system also includes the following:

- the penis (3)
- a duct system that transports sperm out of the body (4, 5, 6)
- three sets of glands that secrete fluids to maintain sperm viability and motility (7, 8, 9)

The components of the male reproductive system are summarized in Table 2.1.

The testes (Figure 2.2) contain tightly coiled **seminiferous tubules**, where sperm are produced in a process called **spermatogenesis**. Spermatogenesis occurs most efficiently at a specific temperature (32–35° C) that is 2 to 5 degrees below body temperature. The scrotum holds the testes away from the body and, therefore, at the proper temperature for sperm production. After they are formed, sperm move through the male reproductive system in stages. In the first stage, they move to the **epididymis** (4) for storage.

In the second stage, when a male is sexually aroused, sperm move from the epididymis into a connecting duct called the **vas deferens** (5). The walls of the vas deferens are lined with muscles that contract rhythmically to move sperm forward. In the final stage, sperm are propelled by muscular contractions through the **urethra** (6) and expelled from the body during ejaculation.

As sperm are transported through the duct system, secretions are added from three sets of glands. The **seminal vesicles** (7) contribute **fructose**, a sugar that serves as an

seminiferous tubules the region of the testes where sperm production occurs

spermatogenesis formation of sperm

epididymis an area of the testes that stores sperm

vas deferens a connecting duct in the testes

urethra a tube present in males and females where urine is carried out of the body

seminal vesicles a set of male reproductive glands

fructose a sugar that supplies energy to the sperm

Figure 2.1 The Parts of the Male Reproductive System

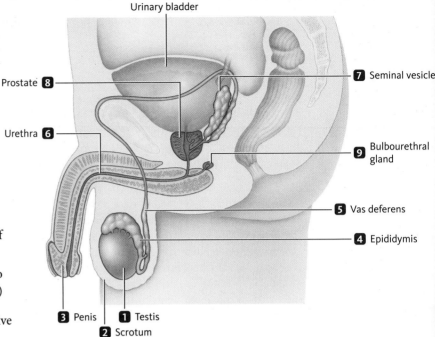

- Urinary bladder
- Prostate **8**
- Urethra **6**
- **7** Seminal vesicle
- **9** Bulbourethral gland
- **5** Vas deferens
- **4** Epididymis
- **3** Penis
- **1** Testis
- **2** Scrotum

Table 2.1 The Male Reproductive System

Component	Function
Testes	Produce sperm and male hormones
Epididymis	Stores sperm
Vas deferens	Conducts sperm to the urethra
Prostate gland	Produces seminal fluid that nourishes sperm
Urethra	Conducts sperm to the outside of the body
Penis	Organ of copulation
Scrotum	Holds testes away from the body to keep the temperature lower than body temperature

Figure 2.2 Cross-Section of a Testis
The tightly coiled seminiferous tubules are where sperm are produced.

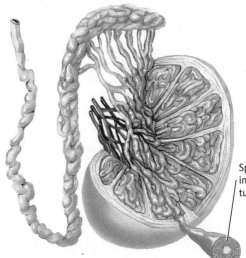

Sperm are formed in the seminiferous tubules

Figure 2.3 **Sperm Formation** A cross section of a seminiferous tubule showing the stages in sperm formation. Once formed, haploid spermatids differentiate to form mature sperm.

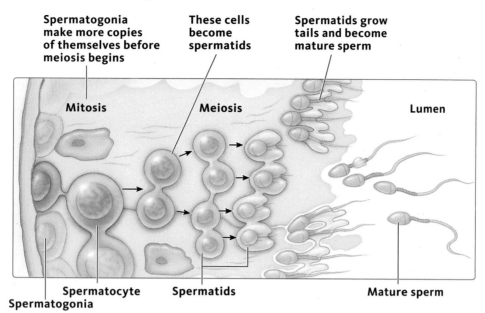

Spermatogonia make more copies of themselves before meiosis begins

These cells become spermatids

Spermatids grow tails and become mature sperm

Mitosis

Meiosis

Lumen

Spermatocyte

Spermatids

Mature sperm

Spermatogonia

energy source for the sperm, and **prostaglandins**, locally acting chemical messengers that stimulate contractions of the female reproductive system to assist in sperm movement. The **prostate gland** (8) secretes a milky, alkaline fluid that neutralizes acidic vaginal secretions and enhances sperm viability. The **bulbourethral glands** (9) secrete a mucus-like substance that provides lubrication for intercourse. Together, the sperm and these various glandular secretions make up **semen,** a mixture that is about 95% secretions and about 5% sperm.

How are sperm formed?
Follow along with the drawing of spermatogenesis in Figure 2.3. Sperm production occurs in the seminiferous tubules. Spermatogonia divide by mitosis to form **spermatocytes** (with 46 chromosomes). These cells undergo meiosis to produce intermediate cells called **spermatids**. Spermatids are haploid and have 23 chromosomes, but they are not yet sperm and do not have tails. Spermatids undergo structural changes to form mature sperm with tails.

Spermatogenesis begins at puberty and continues throughout life. If a **vasectomy** is performed (Figure 2.4), the vas deferens are cut and closed off. The three glands still produce their secretions, but the sperm are absent from the ejaculate. This surgery does not affect sperm production; sperm are still produced, but the body absorbs the sperm formed in the seminiferous tubules.

2.2 Essentials
- **The major parts of the male reproductive system are the testes, urethra, and penis.**
- **Sperm production takes place in the seminiferous tubules.**

2.3 What Organs Make Up the Female Reproductive System?

Follow along with the description of the female reproductive system in Figure 2.5, page 36. The ovaries are a pair of oval-shaped organs (1) about 1.5 inches (3 cm) long,

Figure 2.4 **In a vasectomy, a portion of the vas deferens is removed to prevent sperm from leaving the testis**
Often, this procedure can be reversed.

prostaglandins chemicals secreted into the semen that causes contractions of the vagina during intercourse

prostate gland a male organ that makes a fluid for sperm viability

bulbourethral glands male organs that create a fluid that lubricates intercourse

semen the liquid that sperm travel in

spermatocytes cells that divide by meiosis to form spermatids

spermatids early sperm

vasectomy a procedure that cauterizes the vas deferens

Figure 2.5 **The Parts of the Female Reproductive System**

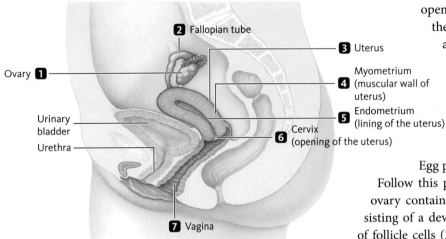

Ovary **1**

Urinary bladder

Urethra

2 Fallopian tube

3 Uterus

4 Myometrium (muscular wall of uterus)

5 Endometrium (lining of the uterus)

6 Cervix (opening of the uterus)

7 Vagina

then rebuilt during the next menstrual cycle. The lower neck of the uterus, the **cervix** (6), opens into the **vagina** (7). The vagina receives the penis during intercourse and also serves as the birth canal. It opens to the outside of the body behind the urethra. The components of the female reproductive system are summarized in Table 2.2.

How are eggs produced?

Egg production (oogenesis) occurs in the ovary. Follow this process as it is shown in Figure 2.6. Each ovary contains several hundred **follicles** (1), each consisting of a developing egg surrounded by an outer layer of follicle cells (2). Cells called oogonia (with 46 chromosomes) begin meiosis in the third month of a female's embryonic development, and then stop. In this state, they are called **oocytes** (3) and remain arrested in meiosis until just before they are released from the ovary. Meiosis is not completed until after they are fertilized. This means that at birth, a female infant has a lifetime supply of developing eggs in her ovary.

Beginning at puberty, a developing egg, called a secondary oocyte (4), is released (ovulated) from a follicle (5) during each menstrual cycle. Over a female's reproductive lifetime, between 400 and 500 oocytes will be released from the ovary and move into the fallopian tube.

If an egg is fertilized, the action of the sperm entering the egg triggers the completion of meiosis. When the sperm and egg nuclei fuse, a diploid cell, carrying 46 chromosomes and called a zygote, is formed.

Reread Section 1.2 in Chapter 1 to review how the zygote progresses from this point.

Table 2.2 **The Female Reproductive System**

Component	Function
Ovaries	Produce eggs and female hormones
Fallopian tubes	Transport sperm and egg
Uterus	Holds embryo and fetus
Vagina	Receptacle for sperm; birth canal

located in the abdominal cavity. Each ovary produces eggs in a process called **oogenesis**. Adjacent to each ovary are the **fallopian tubes** (2), which have fingerlike projections that partially surround the ovary. After the egg is released from the ovary, fertilization occurs in the fallopian tubes, and the fertilized egg (now called a zygote) moves into the uterus (3).

The fallopian tubes open into the **uterus**, a hollow, pear-shaped muscular organ about the size of your fist (about 3 inches [7.5 cm] long and 2 inches [5 cm] wide). The uterus consists of a thick, muscular outer layer called the **myometrium** (4) and an inner tissue layer called the **endometrium** (5).

The innermost layers of the endometrium build up and are shed at menstruation if fertilization has not occurred,

oogenesis development of eggs

fallopian tubes the structure that carries the embryo to the uterus

uterus an organ in the female where an embryo grows

myometrium the muscle of the uterus

endometrium lining of the uterus

cervix opening of the uterus

vagina the female organ that connects the uterus to the outside of the body

follicles a structure surrounding the developing egg

oocytes developing eggs

2.3 Essentials
- **The major parts of the female reproductive system are the ovaries, fallopian tubes, and uterus.**

2.4 Is Infertility a Common Problem?

Essentially, three elements are needed for a successful conception: First, the sperm and the egg must be produced and be healthy. Second, for fertilization to occur, the sperm and egg must interact. This usually occurs in the fallopian tube. Finally, the resulting embryo needs a place to grow (Figure 2.7).

If any one of these components is missing or not functioning properly, infertility may result. Usually a physician will make a diagnosis of infertility after a couple has been trying to conceive for about one year with no success. Although

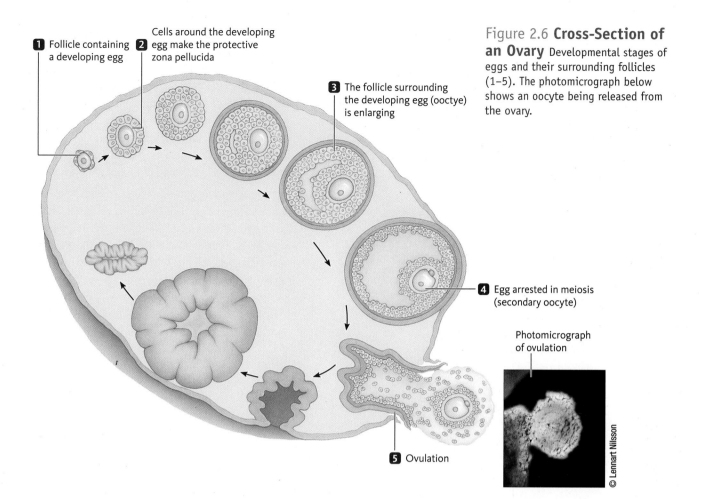

1 Follicle containing a developing egg

2 Cells around the developing egg make the protective zona pellucida

3 The follicle surrounding the developing egg (ooctye) is enlarging

4 Egg arrested in meiosis (secondary oocyte)

5 Ovulation

Figure 2.6 **Cross-Section of an Ovary** Developmental stages of eggs and their surrounding follicles (1–5). The photomicrograph below shows an oocyte being released from the ovary.

Photomicrograph of ovulation

© Lennart Nilsson

the exact problem in Brian and Laura's case is not clear, they seem to have an infertility problem. They are not alone. In the United States, the number of infertile couples has increased greatly over the last 10 years, to about one in six couples. In 40% of infertility cases the woman is infertile, in 40% the male is infertile, and in 20% the cause is unexplained.

Infertility is also related to increasing age; about 28% of couples in their late 30s are infertile. As a woman ages, her hormone levels fluctuate, and eventually her ovaries stop producing eggs. As this process, known as **menopause**, continues, it contributes to a reduction in fertility and eventually leads to the end of a woman's reproductive ability (Figure 2.8, page 38). As a woman reaches her 30s and 40s, a number of other things can contribute to infertility. First, the eggs in her ovary begin to age, and if they are ovulated and fertilized, they often do not implant in the uterus or are miscarried early in the pregnancy. The rate of miscarriage in women over 40 is approximately 75%.

The aging of eggs can make it difficult for the sperm to fertilize the egg. This is probably due to changes on the egg's surface. Finally, it seems that the problem with aging eggs may not involve the nucleus. When the nucleus of an older woman's egg is placed into the cytoplasm of a younger woman's egg, normal fertility and development can result. This process, nuclear transfer, is sometimes used as a treatment for infertility. In this case, the resulting fetus is the genetic child of the older woman who donated the nucleus. This is similar to the process used in cloning Dolly, the sheep (discussed in Chapter 15).

Despite the fact that many fertility problems are age-related, up to 20% of women in the United States delay having their first child until they are over the age of 35. Many of these pregnancies occur with the help of specialists who treat infertility. In addition, women over 35 years of age have a higher risk of having children with chromosomal abnormalities, a problem we will discuss in Chapter 3.

Figure 2.7 **A Cross-Section of a Normal Uterus** Here, an embryo implants and develops.

© Anatomical Travelogue/Photo Researchers, Inc.

Vagina Uterine cavity

menopause a condition where a woman no longer ovulates

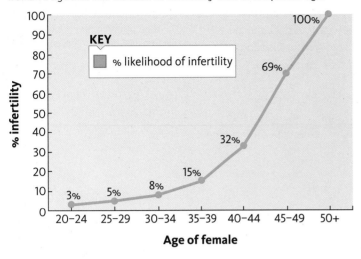

Figure 2.8 This graph shows the relationship between a woman's age and the increase in infertility that accompanies age.

There are two types of infertility: primary and secondary. If a couple has no children and cannot conceive, they are said to have primary infertility. But if a couple has had one child and has trouble conceiving a second, the problem is called secondary infertility. Some 3.3 million couples are thought to have secondary infertility. Its causes are unexplained, but age may be a factor in some cases. The discussion here will focus on primary infertility in both men and women.

What are the causes of infertility in women?

Three of the things needed to conceive are associated with the female reproductive process: the egg, the fallopian tube, and the uterus (a functional sperm is the fourth). As you can see in Figure 2.9, abnormalities in hormonal regulation of ovulation are the most common cause of infertility in women. Ovulation depends on levels of the female hormones, estrogen and progesterone, and their interactions.

Sperm can survive for several days in a woman's body, but an egg must be fertilized soon after ovulation. If a newly ovulated egg and a sperm are not present in the fallopian tube at the same time, conception will not occur. Therefore, ovulation and its timing are very important. Several problems can affect or prevent ovulation:

- **Hormone levels.** If there is no estrogen or too little estrogen, the ovary will not release an egg. Usually estrogen levels rise during the first part of a woman's cycle. When the estrogen level rises to a peak, it triggers ovulation and a rise in another hormone, called **luteinizing hormone (LH)** (home ovulation kits test for the presence of LH and can determine when a woman is ovulating). If estrogen levels are too low or are irregular, ovulation can be affected.

luteinizing hormone (LH) a hormone produced at the time of ovulation

secondary amenorrhea a condition of not menstruating

Hormonal problems occur in about 50% of all cases in which ovulation does not occur.

- **Absent or damaged ovaries.** Surgical removal of the ovaries, damage to the ovaries caused by other surgeries, or inflammation, radiation, or infection are causes of infertility. Ovarian cysts or infection can also prevent egg maturation or release. If there are no ovaries, then no eggs can be produced. Some genetic conditions cause a female child to be born without ovaries.

- **Premature menopause.** Although it is rare, some women stop menstruation and enter menopause at an early age. Some researchers think this happens more often in women who are extremely athletic and who have had low body weight for many years. This lack of menstruation in athletes might be why Maria Patino's condition was not detected earlier in her life (see Chapter 1). Recall that she would have been infertile because she had complete androgen insensitivity.

A related but reversible condition, **secondary amenorrhea**, is a lack of menstruation caused by several factors, including stress, low body weight, nutrition, and excessive physical conditioning.

Obesity can be a factor as well. Women who are obese and ovulate normally have trouble becoming pregnant. Studies show that extra body weight affects hormone levels as well as menstruation cycles, making pregnancy more difficult to achieve.

Fallopian tube blockage, another cause of infertility, occurs in about 15% of all cases. If these tubes are blocked, then the sperm and egg cannot interact.

Sexually transmitted bacterial and viral infections can cause inflammation, scarring, and closing of the fallopian tubes. Even seemingly unrelated conditions, such as appendicitis or a bowel problem called colitis, can cause inflammation in the abdomen that results in blocked fallopian tubes. Previous surgeries can cause formation of scar tissue, called adhesions, which can prevent eggs from moving through the fallopian tubes, contributing to infertility.

Figure 2.9 Factors That Contribute to Infertility in Women

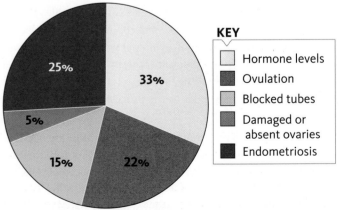

Also shown in Figure 2.9, uterine problems can be a cause of infertility. If a woman has no uterus because it was removed or she was born without one, she will not be able to carry a child. About 25% of infertility cases are caused by endometriosis, a problem with the inner lining of the uterus. In addition, if the innermost layers of the endometrium are not properly formed at each menstrual cycle, an embryo cannot implant, or the fetus detaches from the uterine wall. This problem causes many miscarriages.

What are the causes of infertility in men?
In the male reproductive system, problems with sperm formation and with the sperm themselves are the main causes of infertility. As shown in Figure 2.10, low sperm count is the most common cause of infertility in men.

Figure 2.10 **Factors That Contribute to Infertility in Men**

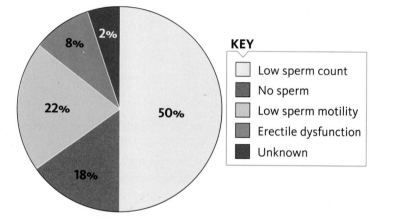

Male infertility can be caused by any number of structural and functional problems. Some of these are:

- **Low sperm count.** Too few sperm (less than 20 million per ejaculation) make it difficult for the necessary number of sperm to swim up the fallopian tubes and meet the egg. Low sperm count can be caused by exposure to chemicals or radiation in the environment, tight underwear, drugs (including marijuana), mumps and other diseases (such as diabetes), lead exposure, pesticide exposure, obesity, and alcohol consumption. In addition, injury to the testes and undescended testes can also be causes of infertility.

- **Low sperm motility.** When the sperm move too slowly, they cannot get to the egg in time for fertilization. Sometimes sperm have malformed or extra tails or do not swim toward the egg in an organized way. The results are similar to those associated with low sperm numbers.

- **Impotence.** If the penis cannot get enough blood supply, it cannot become erect, resulting in **erectile dysfunction (ED).** The causes of impotence can be emotional, physical (hormonal), or drug related. Certain conditions such as high blood pressure and diabetes can also cause erectile problems.

- **No sperm in the ejaculate.** The causes of **aspermia** can include surgery (vasectomy), injury, drugs, and birth defects such as undescended testes or being born with no testes.

What other factors influence fertility in men and women?
Both personal habits and environmental factors can affect a couple's chances of having a child. For example, smoking lowers sperm count in men and increases the risk of miscarriage and low-birth-weight babies in women. Overall, for each menstrual cycle, smoking by men or women reduces the chances of conceiving by one-third. Women who are significantly overweight or underweight often have difficulty becoming pregnant.

BY THE NUMBERS

11%
Percent of women (ages 15–44) who have received infertility treatment

7.3 million
The number of couples who are infertile in the United States

70%
The percentage of women age 30–35 using IVF who become pregnant

62%
The percentage of women age 40–44 using IVF who become pregnant

14%
The percentage of women older than age 44 using IVF who become pregnant

erectile dysfunction (ED) an inability of the penis to become erect

aspermia a condition in which no sperm are produced

Environmental factors, such as exposure to chemicals, pesticides, and radiation, also affect fertility. For example, in one study, women living on farms who had mixed or applied pesticides in the two years before trying to conceive were twenty-seven times more likely to be infertile than women who were not exposed to these chemicals. Other environmental agents such as lead, ethylene oxide (used to sterilize surgical instruments and medical supplies), and radiation have been shown to negatively affect fertility.

The relationship between age and infertility has taken on new importance as many people in the United States and other industrialized countries postpone marriage and/or pregnancy until educational or career goals are reached. Unfortunately, this delay means that a woman is older when she decides to have a child, and as we have seen, age is associated with decreased fertility. Increased maternal age is also associated with an increased risk of having a child with a chromosomal abnormality. It is estimated that once a woman has celebrated her 42nd birthday, she has a less than 10% chance of having a baby with her own eggs. Age is not a significant problem in male infertility.

During the "sexual revolution" of the 1960s and 1970s, there were higher incidences of infertility caused by untreated venereal diseases such as gonorrhea and chlamydia. This trend seems to have reversed since the appearance of AIDS (acquired immunodeficiency syndrome) in the 1980s; it brought common use of barrier methods of contraception (condoms), which prevent HIV (human immunodeficiency virus) infection and most venereal diseases.

Another social factor, the increasing difficulty many couples encounter in finding a child to adopt (a result of improved birth control and the availability of legal abortion), has increased the demand for medical answers to infertility regardless of their complexity and cost. We have also become more aware of our own genetics. Couples are more determined than ever to have their own genetic children.

2.4 Essentials

- **Infertility is a serious problem in the United States.**
- **If there are problems with egg or sperm formation or with the uterus, infertility may result.**
- **Either the male or the female of a couple can be infertile.**
- **Infertility in women increases with age.**
- **The most common form of female infertility is hormone imbalance.**

in vitro **fertilization (IVF)** a method of assisted reproduction done in a laboratory

2.5 How Do Assisted Reproductive Technologies Help with Infertility Problems?

Because of the increase in infertility over the last 10 years, a rapidly expanding medical specialty that treats infertile couples has been created. Many major medical centers now have fertility clinics, and there are independent programs in most cities. Many insurance companies cover such treatments, making them accessible to large numbers of couples.

This chapter will look closely at three of the most common methods used in assisted reproductive technologies (ART):

- donation of gametes
- *in vitro* fertilization
- surrogacy

Most fertility specialists begin by analyzing the hormone levels of both the woman and the man. If levels are low, then the physician may give the woman hormones (estrogen) to stimulate ovulation and/or give the man testosterone to stimulate sperm production.

If hormone treatments do not result in conception, more complex procedures may be used. One of these is gamete donation. If a couple is not producing sperm, eggs, or both, they may look to using a sperm and/or egg donor.

Sperm donation (Figure 2.11), one of the earliest methods of ART, has been widely used for almost 50 years. Sperm is provided by known or unknown donors and either used immediately or frozen in a cryotank (Figure 2.12) filled with liquid nitrogen for later use.

Figure 2.11 **An Ad Seeking to Recruit Sperm Donors**

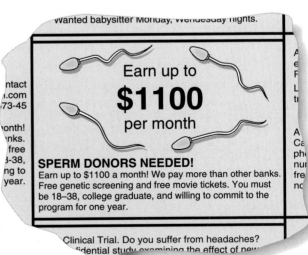

Figure 2.12 Sperm and embryos are stored at −180° C under liquid nitrogen in a cryogenic tank

© Hank Morgan/Photo Researchers, Inc.

Figure 2.13 In artificial insemination, sperm from a donor is placed in the uterus This laboratory technician is checking to make sure the sperm to be used in insemination is the correct sample.

© Mauro Fermariello/Photo Researchers, Inc.

Many sperm banks in the United States and other parts of the world offer sperm from qualified donors. Whether someone qualifies as a donor usually depends on his age, his health, his family history, and the viability of his sperm. Web sites offer couples, physicians, or women a chance to pick the traits they want in a donor and have the frozen sperm of their choice sent to them for artificial insemination (Figure 2.13). In this procedure, sperm are placed into the woman's uterus at ovulation, and if an egg is fertilized, a pregnancy may result. Sperm donation is an option for those with low sperm count, low sperm mobility, or aspermia. Single women who want to have children may also use sperm donors.

Some sperm banks, including the California Cryobank, one of the nation's largest, have developed a database of donors who resemble celebrities. A committee from the cryobank reviews donor profiles and adds a celebrity name to specific donor files.

As an alternative to sperm donation, for men with low sperm counts, sperm from several ejaculations can be pooled and concentrated to increase the chances of fertilization. For men with blocked ducts, sperm can be retrieved from the epididymis or the testis using microsurgery.

One of the most successful methods of ART is *in vitro* **fertilization (IVF)**. It was originally developed to find a way to bring the sperm and egg together to ensure fertilization. If the fallopian tubes are blocked, the sperm and egg do not have a place to interact, and fertilization will not occur. If, however, the sperm and egg are placed in a dish, they will be very close together, and the chances of fertilization will be increased, which is what happens in IVF (Figure 2.14).

Using sterile techniques, the sperm and egg are placed in a dish in the laboratory, and a technician watches the process of fertilization using a microscope. After fertilization, the dish containing the newly formed zygotes is placed in an incubator until the embryos are ready to place in the woman's uterus. Often, up to three or four embryos are transferred to the uterus to increase the chances of implantation.

Couples wanting their own genetic child often use IVF. The man contributes sperm and the woman an egg (read below for the process by which eggs are removed).

A woman who is not producing eggs can opt to use an egg donor (Figure 2.15, page 42). Egg donation is more difficult because the process of egg retrieval is time-sensitive and involves surgery. In addition, because of problems associated with freezing eggs, they are generally used for IVF immediately after retrieval. Although freezing eggs has been somewhat successful, it is not performed regularly. However, it is an option in

Figure 2.14 *In Vitro* Fertilization (IVF) and Implantation Stages (1–4) in the process of IVF. This procedure has resulted in thousands of births to otherwise infertile couples.

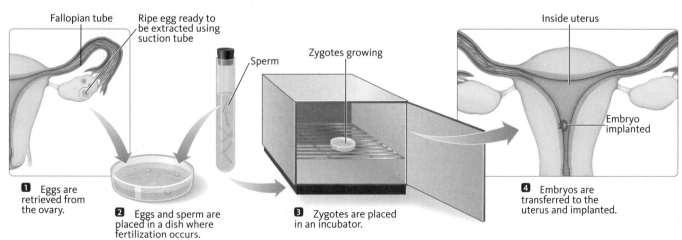

Fallopian tube

Ripe egg ready to be extracted using suction tube

Sperm

Zygotes growing

Inside uterus

Embryo implanted

❶ Eggs are retrieved from the ovary.

❷ Eggs and sperm are placed in a dish where fertilization occurs.

❸ Zygotes are placed in an incubator.

❹ Embryos are transferred to the uterus and implanted.

Figure 2.15 This Ad Ran in the Tufts University Daily Newspaper in November 1999 The price paid for egg donors has increased tremendously since then.

Wanted instructors for portrait drawing class

EGG DONOR NEEDED!

We are a loving infertile couple looking for a woman to help us have a baby. We are looking for a college student or graduate, age 21–33, with blue eyes and blonde or light brown hair. Compensation: $25,000 *plus expenses.*

...ee genetic screening, free medical evaluations ...me movie tickets. You must be 21-25, coll...

some cases. For example, women about to undergo chemotherapy who may want children in the future may freeze their own eggs for later use.

A donor who provides eggs to an infertile couple must undergo hormone shots to stimulate multiple ovulations and a surgical procedure (**laparoscopy**) to remove the eggs. Using a microscope, a technician sorts the eggs to remove those that are too young or too old to be used in fertilization. After retrieval, the donated eggs are used for *in vitro* fertilization. The egg may be fertilized with the sperm of the male partner or the sperm donor.

If a woman is using egg donation because her ovaries have stopped producing eggs, the infertility might be age related. When scientists analyzed eggs produced by older women, they found that the cytoplasm was the source of infertility. Scientists developed a method of injecting cytoplasm from a younger woman's egg into the egg of an older woman, and a number of babies have been born to older women using this technique.

If a woman cannot carry a fetus to term because she has no uterus or her uterus does not function correctly, a surrogate mother can be used. Because of the barrier set up by the placenta and uterine wall, any woman can carry a genetically unrelated fetus without an immune response causing damage to mother or fetus.

There are two types of **surrogacy**. In egg donor surrogacy, the surrogate's egg is fertilized by artificial insemination using the sperm of the infertile couple's male partner, and the

surrogate carries the embryo to term. In gestational surrogacy (Figure 2.16), the surrogate carries the embryo to term, but the sperm and egg are from two other people, who may be unrelated to the infertile couple.

Often a relative or friend volunteers to carry the baby for an infertile couple and give the baby to them after birth. In other instances, the surrogate is a paid stranger. In such cases, a contract is signed (see Section 2.6), and the woman gives the baby to the couple after birth with any stipulations made clear in the contract.

As you can see in Table 2.3, a number of combinations of ART can be used. In one case, a combination of techniques produced a girl with five parents. A contract was made among a sperm donor (anonymous), an egg donor (anonymous), and a surrogate (gestational) mother. The couple who contracted for the little girl, the egg donor, the sperm donor, and the surrogate mother are the five parents. This case went to court (*Buzzanca v. Buzzanca*) when the couple who contracted for the child decided to divorce.

2.5 Essentials

- *In vitro* **fertilization (IVF) is a commonly used infertility treatment in which sperm and eggs are combined in a laboratory dish and embryos result.**

What other methods of ART have been developed?
Recently, a number of new techniques have become popular for treatment of infertility. One of these, intra-cytoplasmic sperm injection (ICSI), can be used to inject a single sperm directly into an egg (Figure 2.17). Often

Figure 2.16 Teresa Anderson, a surrogate mother carrying quintuplets for the parents, Mr. and Mrs. Gonzales.

Dave Cruz/Arizona Republic

laparoscopy a procedure used to see into the abdomen

surrogacy a method of assisted reproduction where one woman carries a fetus for another

Table 2.3 New Ways to Make Babies

Artificial Insemination and Embryo Transfer	In Vitro Fertilization (IVF)
1. Father is infertile. Mother is inseminated by donor and carries child.	1. Mother is fertile but unable to conceive. Egg from mother and sperm from father are combined in laboratory. Embryo is placed in mother's uterus.
2. Mother is infertile but able to carry child. Donor egg is inseminated by father via IVF. Embryo is transferred and mother carries child.	2. Mother is infertile but able to carry child. Egg from donor is combined with sperm from father and implanted in mother.
3. Mother is infertile and unable to carry child. Donor of egg is inseminated by father and carries child.	3. Father is infertile and mother is fertile but unable to conceive. Egg from mother is combined with sperm from donor.
4. Both parents are infertile, but mother is able to carry child. Donor of egg is inseminated by sperm donor. Then embryo is transferred and mother carries child.	4. Both parents are infertile, but mother is able to carry child. Egg and sperm from donors are combined in laboratory (also see number 4, column at left).
	5. Mother is infertile and unable to carry child. Egg of donor is combined with sperm from father. Embryo is transferred to donor (also see number 2, column at left).
	6. Both parents are fertile, but mother is unable to carry child. Egg from mother and sperm from father are combined. Embryo is transferred to donor.
	7. Father is infertile. Mother is fertile but unable to carry child. Egg from mother is combined with sperm from donor. Embryo is transferred to surrogate mother.

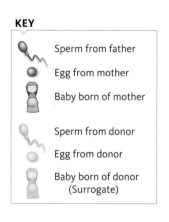

KEY

- Sperm from father
- Egg from mother
- Baby born of mother

- Sperm from donor
- Egg from donor
- Baby born of donor (Surrogate)

sperm are formed without tails and cannot swim to the egg. Even in these cases, ICSI has been successful.

Another technique mentioned previously, egg freezing, has now been improved. When first developed, the freezing of eggs was difficult because during thawing, ice crystals inside the cell would damage the egg, making it unviable. By chemically treating the eggs before freezing, eggs have been frozen successfully, and a number of babies have been born using previously frozen eggs.

If a woman does not have ovaries due to a birth defect, surgery, or injury, she would not be able to conceive. Recently, ovaries have been transplanted from one woman to another. In one case, the donor ovary was from a twin sister, but in another case, the donor was a stranger, and in still another case, the transplanted ovary came from a cadaver. All the women who received these transplants were able to conceive and bear children.

Figure 2.17 Intracytoplasmic Sperm Injection (ICSI) In ICSI, a single sperm is injected into an egg, where it will fuse with the egg nucleus, producing a diploid zygote.

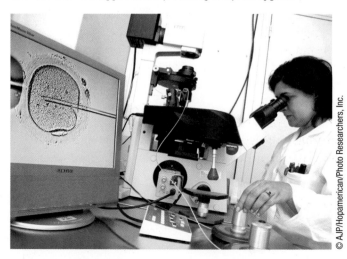

Figure 2.18 A cell being removed from an embryo for prenatal genetic diagnosis (PGD)

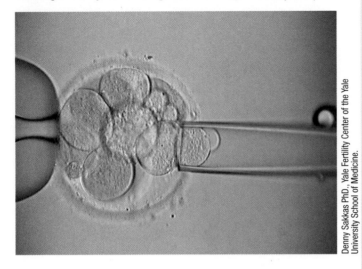

Denny Sakkas PhD., Yale Fertility Center of the Yale University School of Medicine.

In Chapter 1 we discussed the use of PGD (preimplantation genetic diagnosis) for sex selection. This procedure (Figure 2.18) has been used to select embryos for other reasons. Many parents who carry a genetic disorder that might affect their child use PGD and genetic testing of an embryonic cell to choose an embryo without this condition. However, in at least one case, parents with achondroplasia (a genetic form of dwarfism) chose an embryo *with* that trait.

A recent movie about PGD, *My Sister's Keeper,* is based on a novel by Jodi Picoult. The plot of both the book and the movie involves the use of PGD to choose an embryo with the same tissue type as a living child. Bone marrow from the new baby was used for a transplant to her sister, who suffered from leukemia.

In other ways to help infertile couples, physicians transfer gametes or zygotes directly into a fallopian tube to increase the chances of fertilization and implantation into the uterus. Transfer of gametes to the fallopian tube is called GIFT (gamete injection into the fallopian tube) and ZIFT (zygote injection into the fallopian tube). The success rates of some ART methods are summarized in Table 2.4.

CASE A: QUESTIONS

Now that we know more about infertility and its treatments, let's address the problem that Brian and Laura are facing.

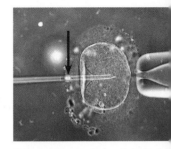

A human sperm being injected into an egg.

1. What tests should Dr. Franco perform on Brian and Laura?

2. List four things that Brian and Laura might do to solve their problem.

3. What do you think Brian and Laura should do? Why?

4. Give three reasons why Brian and Laura should use ART.

5. What are three credible reasons to adopt instead of using ART?

6. Do you know anyone who has experienced infertility? Do you know more than one couple with infertility problems? If so, how many? See if they will tell you their story; if so, write a paragraph about it.

7. Give two reasons why ART has become so important in the last 10 years.

What are some of the problems associated with the use of ART? As with many new technologies, unforeseen problems have arisen with the use of ART. During IVF, women are given hormone injections to stimulate the ovary to release many eggs. Some women hyperovulate and produce many more eggs than expected. This can be painful and possibly dangerous.

After IVF, the number of embryos implanted in the uterus can vary. In the past, two to four embryos were transferred to the uterus, but this often results in multiple births because some or all of the implanted embryos will develop. Multiple births often result in premature birth, low birthweight, and other complications.

Table 2.4 ART Success Rates in 2007

Type of ART	Resulted in Pregnancy (Single)	Resulted in Pregnancy (Multiple)	Pregnancies that Resulted in a Single Live Birth	Pregnancies that Resulted in Multiple Live Births
IVF with nondonor egg	22%	12%	82%	26%
IVF with frozen embryo	36%	21%	26%	7.8%
IVF with fresh embryos	43%	—	36%	7%
IVF with donor eggs	62%	—	53%	—

If a couple decides not to implant all the fertilized eggs, the remaining embryos can be frozen and kept for later implantations. Eventually, if not all the frozen embryos are implanted, many couples and physicians are reluctant to destroy them. In some cases, with permission of the genetic parents, embryos have been donated to other couples, a process known as embryo adoption. For many infertile couples, embryo adoption offers a less invasive and less expensive alternative to IVF.

2.6 What Are the Legal and Ethical Issues Associated with ART?

Many issues have come to light with the use of new ART procedures. A few of them will be discussed here.

During *in vitro* fertilization, a number of embryos can be produced. These embryos are placed in the woman's uterus or frozen for later use. Successful pregnancies have resulted from embryos frozen as long as 10 years. The big question is, what should be done with frozen embryos that no one wants?

When several embryos are placed in the uterus at one time, the chance of a multiple birth increases. The number of multiple births has increased dramatically in the last 10 years as a result of ART.

In situations involving sperm and egg donors, questions often arise about the paternity or maternity rights of the donors; a number of legal cases have addressed this issue.

Table 2.5 lists some of the ethical and legal questions raised by these procedures.

CASE B:
Single Mom Needs Dad

Jan Moppet wanted a child very badly but never found anyone she wanted to marry. So when she was about 35, her doctor told her that she had better think seriously about having a child before she got any older.

Dr. Ingram suggested that she have artificial insemination using sperm from a sperm bank. In this procedure, donor sperm are placed inside the woman; in a large percentage of women, pregnancy results. This sounded just like what Jan wanted.

She went through the procedure later that year and gave birth to a baby girl she called Alex. Lately, though, Jan has been thinking what the future might bring.

1. **Should Jan have chosen certain characteristics when picking her sperm donor?**
2. **Should Jan tell Alex about her father being a sperm donor?**
   a. **If yes, why and when?**
   b. **If no, why not?**
3. **When Alex reaches age 18, should she be allowed to find her father if she wants to do so?**
4. **You are Alex's attorney in her quest to find her father. Give five arguments you would make on her behalf.**
5. **If you were the sperm donor, list three reasons why you would not want Alex to find out your identity.**
6. **When Alex enters school, what might happen to her?**
7. **What is known about Alex's father's genetic makeup? Might she inherit a genetic disorder from him or carry a gene that she will pass on to her children?**

Spotlight on Law

Davis v. Davis
842 S.W. 2d 588, Tennessee Ct. of Appeals (1993)

During their marriage, Junior and Mary Sue Davis went to a fertility clinic to have *in vitro* fertilization. Mary Sue's eggs and Junior's sperm were fertilized in the laboratory, producing nine embryos. Two were implanted in Mary Sue's uterus, but she did not become pregnant. The other seven embryos were frozen. Before they could have a second try at IVF, they developed marital problems and decided to divorce. Junior and Mary Sue asked the divorce court (they lived in Tennessee) to make a judgment no other court had previously decided: each was asking for the embryos as part of the divorce settlement.

Junior wanted to destroy the embryos; Mary Sue wanted to implant the embryos and have children. In defense of her right to proceed with the implantation, she argued that the embryos were examples of potential human life, not typical property.

Junior's attorney argued that he should not be forced to be a parent; that this was his right under the Constitution; and that an embryo is not a person and therefore cannot be considered as a child.

The court needed to decide whether the embryos were property, children, or neither. If they were considered property, they would be split, as in most divorce law. If they were considered children, custody would have to be awarded.

The trial court awarded custody of the embryos to Mary Sue, calling them "children *in vitro*," and as such had a "right to life" and directed that Mary Sue be allowed to implant them.

Junior appealed the decision to the Tennessee Court of Appeals, which rejected the "right to life" rationale. It also stated that Junior had a right *not* to be a father and awarded control of the embryos to Junior and Mary Sue jointly. The court held that the Davises shared an interest in the frozen embryos.

After an appeal to the Tennessee State Supreme Court, it was concluded that frozen embryos are neither "persons" nor "property" but "occupy a position of special respect because of their potential for human life." It felt that the main issue is the individual's constitutional right to privacy. If a person has a right to procreate, then he or she also has a right not to procreate. If Mary Sue were allowed to implant the embryos, Junior would have no control over his parental status. Being an unwilling parent would place financial, emotional, and legal burdens on Junior. Therefore, the court awarded the embryos to Junior.

QUESTIONS

1. In your opinion, to whom should the court have awarded the embryos? Why?

2. If neither spouse had been awarded the embryos, what should have happened to them? Argue both sides.

3. Which attorney's argument do you agree with? Give three reasons.

4. In Australia, two frozen embryos were left when a couple died in a plane crash. The couple had millions of dollars, and many women wanted to be implanted with these embryos because they thought the resulting children would inherit the money. The Australian court decided that no one would get the embryos. Do you agree with this decision? Why or why not?

5. Is there a duty to protect human embryos from harm? If so, who should protect them? If not, why not?

6. What should be done with unclaimed frozen embryos?

7. What should be included in a contract for a couple going through IVF? List some possible choices.

8. After many tries to get pregnant, what could the Davises have tried besides IVF? Research possible options.

Table 2.5 **Some Legal and Ethical Issues Associated with ART**

Question	How Are These Questions Decided?	Related Case or Legal Issue
What should be done with leftover embryos?	This is usually decided by the fertility clinic and the parents of the embryos.	In the United Kingdom, a law requires that unclaimed embryos be destroyed after five years.
Should sperm and egg donors have parental rights to their children?	This is usually decided by courts if the donor wants to press the issue.	Individual cases have gone to court in the United States and the United Kingdom. For the most part, the parents who have raised the child retain the rights.
If a couple divorces, who should get custody of frozen embryos?	This is usually decided by courts if the couple wants to press the issue (see "Spotlight on Law: *Davis v. Davis*").	Most cases have used *Davis v. Davis* (see "Spotlight on Law") as a model and decided that no person can be forced to be a parent. Some have decided to destroy the embryos.
Should sperm and egg donors be paid for their contributions?	They are paid in the United States. Sperm donors can donate many times and are paid each time. Egg donors are given more compensation because of the complex process of egg retrieval.	No court case has developed from this issue, but each sperm bank and fertility clinic handles it differently.
Is a contract with a surrogate mother legal?	This legal question is dealt with on a state-by-state basis. One case in New Jersey, *In re Baby M*, decided the answer was no. New Jersey still does not accept these contracts, but other states do.	In a number of cases, surrogates have wanted to keep the babies they carried, but courts have upheld the contracts, especially in gestational surrogacy.
Should surrogates be compensated?	If the amount of money is indicated in the contract, this has been held to be legal.	Some feel that paying surrogates is the same as baby selling. Recently, however, the number of women being gestational surrogates has increased.

The Essential 10

1 Sperm are produced by meiosis in the testes of the male. (Section 2.1)

For every sperm-forming cell, four sperm are formed.

2 Eggs are produced by meiosis in the ovaries of the female. (Section 2.1)

For every egg-forming cell, one egg is formed.

3 The major parts of the male reproductive system are the testes, urethra, glands, and penis. (Section 2.2)

The testes and penis are external, while the glands and urethra are internal.

4 The major parts of the female reproductive system are the ovaries, fallopian tubes, and uterus. (Section 2.3)

All major parts of the female reproductive system are internal.

5 If there are problems with egg or sperm formation or with the uterus, infertility may result. (Section 2.4)

Eggs and sperm are needed to form an embryo; if either gamete is missing or defective, the man or woman cannot reproduce without medical help.

6 Infertility is a serious problem in the United States. (Section 2.4)

There are many causes of infertility, both environmental and genetic.

7 Either the male or the female of a couple can be infertile. (Section 2.4)

Many statistics exist about the number of infertile couples. They show that the causes can usually be determined by medical tests.

8 Infertility in women increases with age. (Section 2.4)

As a woman ages, her ovaries begin to shut down, and she has trouble producing viable eggs.

9 The most common form of female infertility is hormone imbalance. (Section 2.4)

Hormone levels in both females and males can cause problems in egg and sperm production.

10 *In vitro* fertilization (IVF) is a commonly used treatment for infertility. In IVF, sperm and eggs are combined in a laboratory dish to produce embryos. (Section 2.5)

In vitro fertilization was developed in the United Kingdom and has been practiced worldwide for over thirty years.

Review Questions

1. Fill in the following table by listing all infertility problems in women and the causes of each.

Infertility Problem	Cause 1	Cause 2

2. Fill in the following table by listing all of the infertility problems in men and the causes of each.

Infertility Problem	Cause 1	Cause 2

3. Look back at the tables in questions 1 and 2, and describe which assisted reproduction technique has been successfully used in treating one of the fertility problems you listed.

4. List the two types of surrogacy.

5. Do some research to explain how a vasectomy reversal is done.

6. What is it called when a man has few or no sperm?

7. Name a male hormone that doctors use to increase sperm production.

8. What are the three main things needed to have a baby?

9. In *Davis v. Davis*, summarize the major reason the Tennessee State Supreme Court gave for awarding the embryos to Junior Davis.

10. List three causes of infertility that are environmental.

Application Questions

1. Do you think that in the future, more doctors will specialize in assisted reproduction? Why or why not?

2. If the use of assisted reproduction increases in the future, what effect do you think this will have on adoptions? On abortions?

3. Of the two types of surrogacy, which might have more legal problems and why?

. . . you needed to pick someone to be your surrogate?

. . . a friend asked you to donate sperm for her baby?

. . . a friend asked you to donate an egg so that she and her husband could have a baby?

. . . you were a legislator asked to vote on a law to destroy all abandoned frozen embryos?

. . . you were a fertility specialist and were counseling a couple carrying quadruplets?

4. Research some other methods of assisted reproduction. Write a paragraph or two to describe them and discuss how often they are used.

5. Table 2.4 lists some statistics for conceptions and births. Research the numbers of patients who use sperm donation and their success rates.

6. Research has shown that the age of the egg is more important than the age of the female, and that older women can have children by implanting donated eggs fertilized by IVF or by freezing their own eggs for later use. This allows women to extend their child-bearing years, even after menopause. What problems, if any, do you foresee if older women (55 years of age and older) have children? List them.

7. Do some research about men who have fathered children after age 60. Why do you think it is more common for older men to have children than it is for women who are the same age? Why does our society accept this?

8. Ovarian transplants are one way to treat female infertility. As a judge, how would you handle a case in which a woman receives an ovary transplant, has a child, and is faced with a claim by the donor that she is the genetic mother of the child and wants custody or visitation rights?

9. Go back to Table 2.3 and reread the sections on artificial insemination and *in vitro* fertilization. Imagine that each of the numbered cases resulted in a legal custody hearing. Knowing what you do about the law and ART, who do you think would get custody in each instance?

Online Resources

Preparing for an exam? Log on at www.cengagebrain.com for study tools to help you assess your understanding. If assigned by your instructor, the Case A and Spotlight on Law activities for this chapter, "A Couple Considers Fertility Treatment" and "*Davis v Davis,*" will also be available.

In the Media

Jodie Foster's Baby
Steve Sailer

Hollywood actor Jodie Foster searched for the perfect sperm donor in 1998, and proudly announced that after a long hunt, she had been impregnated with the gametes of a tall, dark, handsome scientist with an IQ of 160. Delighted with the result, she went back to the same clinic for her second child in 2002.

To access this article online, go to www.cengagebrain.com.

QUESTIONS

1. How do you think being a celebrity affects a question such as sperm donation?

2. Research and find some other celebrities who have used sperm or egg donation.

UTISET, March 3, 2011

Identification Scares Off Sex Cell Donors
Unknown

The number of human egg and sperm donors has plummeted in Finland. Officials say the probable cause is a law that gives children conceived using donated sperm or eggs the right to know the identity of their biological parents.

To access this article online, go to www.cengagebrain.com.

QUESTION

When the United Kingdom first passed a similar law, donor numbers plummeted. Why do you think men stopped donating?

Thanh Nien News, March 3, 2011

Surrogate Moms to Give Birth in Thailand
Unknown

Last month, Thai police raided two houses in a Bangkok suburb and discovered 15 Vietnamese women, all of whom had been enlisted into an illegal surrogacy service. The women, including two who had already given birth, were taken to a detention center. Four of the women insist they were tricked into surrogacy and held against their will. The eleven others said they willingly traveled to Thailand to work as surrogate mothers last year.

The pregnant mothers wish to give birth. They said they agree to hand the babies to their biological parents only if the parents agree to pay compensation.

To access this article online, go to www.cengagebrain.com.

QUESTIONS

1. If the women understand what they are to do, is this wrong?

2. In many countries, such as Australia and Germany, surrogacy is illegal. Does that make services that allows couples to use surrogacy okay?

Pink or Blue: Is It a Boy or Girl?
Acu-Gen Biolab launches *Baby Gender Mentor* marketing and information site.

Acu-Gen, based in Boston, has developed a home test for the sex of a fetus using the mother's blood. A lancet and filter paper are included in the kit; the woman pricks her skin and the blood is sucked into the filter paper, which is sent back to the company for analysis. Acu-Gen claims a 99.9% accuracy rate. Parents can pay a small fee to find out the sex of their child without waiting for the ultrasound.

To learn more about Acu-Gen's home test product online, go to www.cengagebrain.com.

QUESTIONS

1. Do you think many people will use this as a form of sex selection? Why or why not?

2. This is probably the first of many home tests for women who want to find out information about their fetuses. What other things might an expectant mother want to test for? What might be the result of this on a large scale?

3

Changes in Chromosome Number

CENTRAL POINTS

- Normally, humans have 46 chromosomes, but chromosome numbers can vary among individuals.

- Several different tests can be used to determine the chromosome number in a fetus.

- Changes in chromosome number or structure can cause genetic disorders.

- Lawsuits can result from problems with genetic testing.

CASE A: Results Worry Pregnant Woman

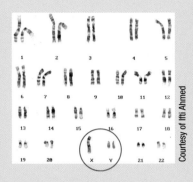

Martha Lawrence was nervous about going to the human genetics unit at the hospital. Her physician referred her because she was unexpectedly pregnant at age 41, and at that age she was at risk for having a child with a chromosomal abnormality, especially Down syndrome. She was 18 weeks pregnant, which would not leave much time if she wanted to have an abortion, considering it takes 7–14 days to get the results of an amniocentesis procedure. Women older than age 35 have a much higher risk of having a child with a chromosomal abnormality than younger women.

Martha previously had two normal pregnancies; her children are now 13 and 17. The genetic counselor, Dr. Gould, suggested that Martha have amniocentesis (removal of fluid and cells from around the fetus) to examine the fetal chromosomes.

In discussing the results of the test, Dr. Gould said that the fetus did not have an extra copy of chromosome 21 and therefore did not have Down syndrome. But the analysis did show the presence of an extra Y chromosome (XYY instead of XY), a condition called Jacobs syndrome. This condition is fairly common; each day in the United States, five to ten boys are born with an XYY set of chromosomes.

Some questions come to mind when reading this case. Before we can address those questions, let's look at how mistakes in cell division can lead to chromosomal abnormalities.

A Girl with Down Syndrome

3.1 What Is a Chromosome?

As discussed in "Biology Basics: Cells and Cell Structure," chromosomes are thread-like structures in the nucleus of a cell that carry genetic information (Figure 3.1, page 52).

In humans, chromosomes exist in pairs. Members of a pair are called homologous chromosomes; one member of the pair is inherited from each parent, for a total of 46 chromosomes. The homologous chromosomes carry identical genes at specific places called loci. The genes on each homologue are identical but may be represented by different forms of the gene, called **alleles**. Remember from Chapter 1 that one pair of chromosomes are called the **sex chromosomes**. There are two types of sex chromosomes, the X chromosome and the Y chromosome. All other chromosomes (numbers 1–22) are called **autosomes**.

As shown in Figure 3.1a, chromosomes have four important parts: (1) the centromere, (2) the short arm (p arm), (3) the long arm (q arm), and (4) the **telomere**. Centromeres are constricted areas on the chromosome, located at different places on different chromosomes (Figure 3.1b). Telomeres are sections of DNA at the ends of chromosomes that keep chromosome ends from sticking to each other (Figure 3.1c). After each cell division, the telomeres shorten. Scientists are studying this phenomenon as it applies to aging.

alleles different forms of the same gene

sex chromosomes chromosomes involved in sex determination; in humans, the X and Y chromosomes are sex chromosomes

autosomes chromosomes other than the sex chromosomes; in humans, chromosomes 1–22 are autosomes

telomeres a section on the ends of chromosomes that is related to aging

Table 3.1 **Comparison between Homologous Chromosomes and Sister Chromatids**

	Homologous Chromosomes	Sister Chromatids
Centromere	Different	Same
Size	Same	Same
Genes	Same	Same
Alleles	Same or different	Same

When chromosomes become visible during cell division, they have already replicated, as you can see in Figure 3.1. Each replicated chromosome consists of two structures, called chromatids, held together by a centromere. Chromatids joined together by a common centromere are called sister chromatids. Table 3.1 compares the properties of homologous chromosomes and sister chromatids.

3.1 Essentials

- **The normal diploid (2*n*) chromosome number in humans is 46.**

Figure 3.1 **Chromosome Structure** (a) A replicated chromosome consists of several parts, including the two chromatids, a centromere, and telomeres, structures at the ends of each chromatid. (b) The short arm is called the p arm and the long arm is the q arm. The length of each arm depends on the location of the centromere. (c) Telomeres can be visualized by staining.

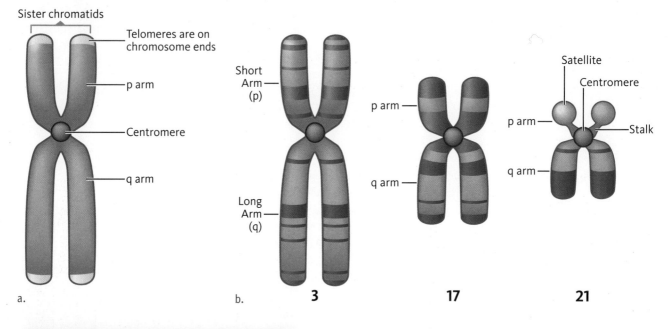

Sister chromatids
Telomeres are on chromosome ends
p arm
Centromere
q arm
a.

Short Arm (p)
Long Arm (q)
b. **3**

p arm
q arm
17

Satellite
Centromere
p arm
Stalk
q arm
21

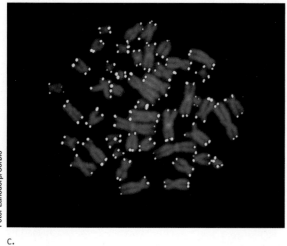

Peter Lansdorp/Corbis

c.

3.2 What Can Cause Changes in the Number of Chromosomes?

Eggs and sperm (gametes) are produced in the ovaries and testes by a form of cell division called **meiosis**. At the beginning of meiosis, diploid cells (2*n*) each contain two copies of all chromosomes. During meiosis, members of all chromosome pairs separate from each other to produce haploid (*n*) cells, each of which contains 23 chromosomes. This reduction in chromosome number is accomplished by two rounds of cell division, called meiosis I and meiosis II (Figure 3.2). As the chromosomes become visible at the beginning of meiosis I, they have already replicated and contain sister chromatids held together at the centromere. The cell division at meiosis I

Figure 3.2 Summary of Chromosome Movements in Meiosis

Replicated homologous chromosomes appear and pair with each other in prophase I. At metaphase I, pairs of homologous chromosomes are aligned at the center of the cell and separate from each other in anaphase I, producing two cells. In meiosis II, the chromosomes migrate to the cell center, the centromeres split, and the sister chromatids are converted into chromosomes. Each of the resulting haploid cells contains one copy of each chromosome.

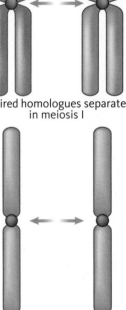

Members of chromosome pair
Sister chromatids   Sister chromatids

Each chromosome pairs with its homologue

Paired homologues separate in meiosis I

Sister chromatids separate and become individual chromosomes in meiosis II

separates members of a **homologous chromosome pair** (but not the sister chromatids). Then during meiosis II, the centromeres split, converting the sister chromatids into separate chromosomes.

In females, division of the cytoplasm (cytokinesis) at meiosis I is unequal. One cell, destined to become the egg, receives about 95% of the cytoplasm. The other cell, which will not become an egg, is called a **polar body**. Similarly, in meiosis II, the egg receives most of the cytoplasm, and another polar body forms. Therefore, meiosis results in one egg in females. In males, meiosis results in the formation of four sperm.

Table 3.2 summarizes the events of meiosis, and Figure 3.3, page 54, summarizes the movement of chromosomes in meiosis.

Normally, meiosis results in gametes that contain 23 chromosomes. However, in a small percentage of cases, chromosomes fail to separate properly during one of these two divisions (Figure 3.4, page 55). This event, called **nondisjunction**, results in some gametes with two copies of a given chromosome, and some gametes with no copies of that same chromosome. If gametes with an abnormal number of chromosomes are involved in fertilization, the resulting zygote will have an abnormal number of chromosomes. Variations in chromosome number that involve one or a small number of chromosomes are called **aneuploidy**. Depending on which chromosomes are involved, the effects of aneuploidy can range from non-life-threatening to devastating and lethal. An error during meiosis that resulted in a sperm carrying two copies of a Y chromosome is probably what happened in Martha Lawrence's case. Aneuploidy is a major cause of early miscarriages; about 5% of all pregnancies are aneuploid. In addition, aneuploidy is the leading cause of mental retardation and developmental disabilities and places a significant social, emotional, and financial burden on millions of families.

3.2 Essentials

- **Meiosis is involved in formation of the sperm and egg.**
- **In meiosis, members of a chromosome pair separate from each other, forming gametes that contain the haploid chromosome number.**
- **Most aneuploidy is the result of mistakes in meiosis.**

meiosis cell division that results in sperm or egg with half the number of chromosomes

homologous chromosome pair members of a chromosome pair

polar body two small cells produced in meiosis I and meiosis II that do not become gametes and are ultimately discarded

nondisjunction a failure of chromosomes to separate properly during meiosis, resulting in the incorrect number in sperm or egg

aneuploidy an abnormal number of chromosomes

Figure 3.3 **The Stages of Meiosis**
In meiosis I, chromosomes with sister chromatids become visible early in prophase (a). During metaphase I, the chromosomes migrate to the middle of the cell, and spindle fibers in the cytoplasm attach to the centromeres (b). At anaphase I, members of a chromosome pair separate from each other and move toward opposite sides of the cell (c). The result is the formation of two nuclei in telophase I (d), and two new cells form by division of the cytoplasm. In prophase of meiosis II,

Meiosis I

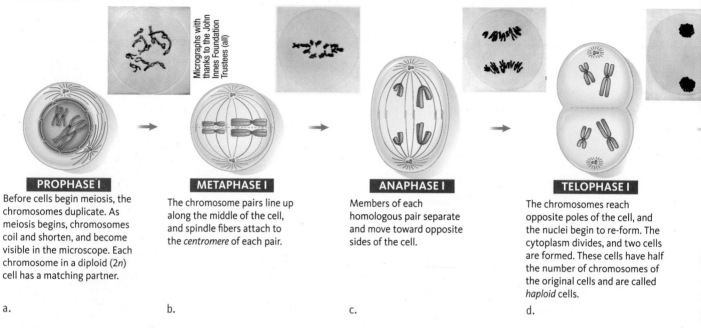

PROPHASE I

Before cells begin meiosis, the chromosomes duplicate. As meiosis begins, chromosomes coil and shorten, and become visible in the microscope. Each chromosome in a diploid (2n) cell has a matching partner.

a.

METAPHASE I

The chromosome pairs line up along the middle of the cell, and spindle fibers attach to the *centromere* of each pair.

b.

ANAPHASE I

Members of each homologous pair separate and move toward opposite sides of the cell.

c.

TELOPHASE I

The chromosomes reach opposite poles of the cell, and the nuclei begin to re-form. The cytoplasm divides, and two cells are formed. These cells have half the number of chromosomes of the original cells and are called *haploid* cells.

d.

Table 3.2 **Meiosis**

Stage	Characteristics
Interphase I	Chromosome replication.
Prophase I	Chromosomes become visible, homologous chromosomes pair, chromatids undergo crossing over (recombination). Nuclear envelope breaks down.
Metaphase I	Paired chromosomes align at equator of cell.
Anaphase I	Members of each chromosome pair move to opposite poles of the cell.
Telophase I	Chromosomes uncoil, nuclei form.
Cytokinesis (division of the cytoplasm)	Cytoplasm divides, forming two cells. In females: unequal division of cytoplasm forms egg and polar body. In males: equal division forms two spermatocytes.
Prophase II	Chromosomes re-coil.
Metaphase II	Unpaired chromosomes align at the equator of the cell.
Anaphase II	Centromeres split, daughter chromosomes pull apart.
Telophase II	Chromosomes uncoil, nuclear envelope forms, meiosis ends.
Cytokinesis (division of the cytoplasm)	Cytoplasm divides, forming four cells. In females: unequal division of cytoplasm forms egg and polar bodies. In males: equal division forms four spermatids.

the chromosomes become coiled and begin to move toward the center of the cell (e). During metaphase II (f) the chromosomes are aligned at the center of the cell and become attached to spindle fibers. During anaphase II, the centromeres split, converting the sister chromatids into chromosomes, which move to opposite sides of the cell (g). Finally, after the chromosomes uncoil, nuclear envelopes re-form, and the cytoplasm divides; the result is four cells, each with the haploid number of chromosomes (h). Meiosis is now complete.

Meiosis II

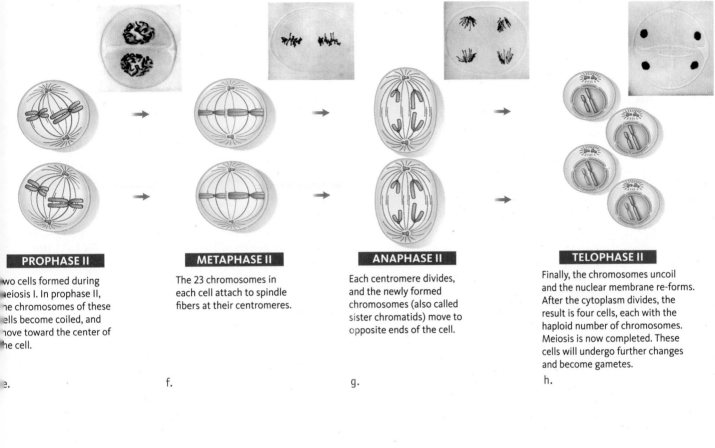

PROPHASE II

wo cells formed during eiosis I. In prophase II, he chromosomes of these ells become coiled, and ove toward the center of he cell.

e.

METAPHASE II

The 23 chromosomes in each cell attach to spindle fibers at their centromeres.

f.

ANAPHASE II

Each centromere divides, and the newly formed chromosomes (also called sister chromatids) move to opposite ends of the cell.

g.

TELOPHASE II

Finally, the chromosomes uncoil and the nuclear membrane re-forms. After the cytoplasm divides, the result is four cells, each with the haploid number of chromosomes. Meiosis is now completed. These cells will undergo further changes and become gametes.

h.

Figure 3.4 Nondisjunction The failure of chromosomes to properly separate at either the first or the second division in meiosis is called nondisjunction. The result is an abnormal number of chromosomes in sperm or eggs. If these abnormal gametes are involved in fertilization, the result can be a child with an abnormal number of chromosomes.

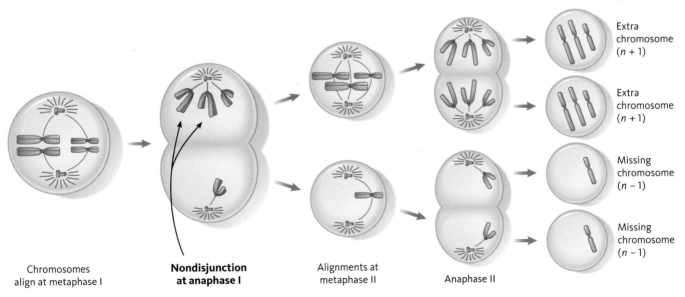

Chromosome number in gametes:

Extra chromosome ($n + 1$)

Extra chromosome ($n + 1$)

Missing chromosome ($n - 1$)

Missing chromosome ($n - 1$)

Chromosomes align at metaphase I

Nondisjunction at anaphase I

Alignments at metaphase II

Anaphase II

3.3 Can We Identify Chromosomal Abnormalities in a Fetus?

Because of Martha Lawrence's age, her genetic counselor suggested that she have a procedure called **amniocentesis** to analyze the fetal chromosomes and detect any abnormalities. Figure 3.5 shows a developing fetus in the uterus and how amniocentesis is used to obtain a sample of cells for chromosome analysis.

How does amniocentesis work? Amniocentesis is used routinely to collect fetal cells and amniotic fluid for analysis in a cytogenetics lab. Aneuploidy, other chromosomal abnormalities, and biochemical disorders can be detected by amniocentesis, and the sex of the fetus can also be determined.

During amniocentesis, the fetus and placenta are first located by ultrasound. Ultrasound machines use sound waves to create an image of the fetus so it can be located in the uterus. A needle (Figure 3.5) is inserted through the abdominal and uterine walls (avoiding the placenta and fetus) and into the sac surrounding the fetus. Some of the fluid surrounding the fetus, called **amniotic fluid**, is collected. This fluid consists of fetal urine, water, and cells shed from the fetus. The fluid and the cells it contains therefore reflect the genetic makeup of the fetus.

After the cells are recovered from the fluid, they are grown in the laboratory for several days. Then the cells are broken open, and the chromosomes are spread onto a microscope slide and stained to produce the banding patterns that identify specific chromosomes. The slide is examined under a microscope, and an attached camera transmits the chromosome images to a computer (Figure 3.6), where they are digitized, processed, and assembled to make a **karyotype**. In a karyotype, as shown in Figure 3.7, chromosomes 1–22 (the autosomes) are arranged in pairs according to size and centromere location. The sex chromosomes (X and Y) are usually placed on the karyotype separately from the autosomes.

Amniocentesis is usually performed at or after the 16th week of pregnancy. Processing and analyzing the results usually takes about 7–14 days. As with all medical procedures, some risks are associated with amniocentesis. Because there is a 0.2–0.3% (1 in 300 to 1 in 500) risk of miscarriage associated with amniocentesis, the procedure is normally only used under certain conditions:

Figure 3.5 Amniocentesis In amniocentesis, a syringe needle, guided by ultrasound, is inserted into the uterus. The needle passes through the fetal membranes to collect a small sample of amniotic fluid that contains cells. The cells can be collected and grown in the laboratory for chromosome analysis.

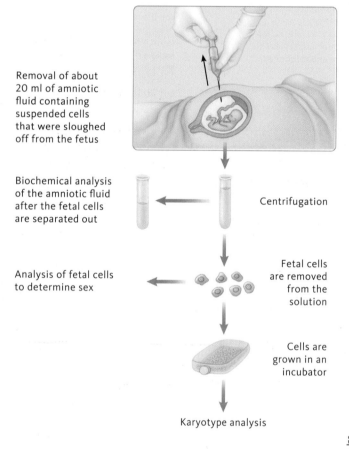

Removal of about 20 ml of amniotic fluid containing suspended cells that were sloughed off from the fetus

Biochemical analysis of the amniotic fluid after the fetal cells are separated out

Centrifugation

Analysis of fetal cells to determine sex

Fetal cells are removed from the solution

Cells are grown in an incubator

Karyotype analysis

Figure 3.6 A Cytogenetics Lab The chromosomes on the microscope slide are displayed on the screen. This image is converted into a karyotype using software.

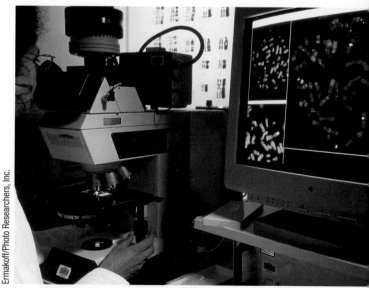

Ermakoff/Photo Researchers, Inc.

amniocentesis a prenatal test for genetic conditions done by analyzing cells in the amniotic fluid

amniotic fluid a fluid that surrounds the fetus and offers protection

karyotype a way of displaying human chromosomes

Figure 3.7 A Karyotype In chromosome analysis, the results are displayed in a diagram called a karyotype. The chromosomes are arranged in pairs, by size, and by centromere location. The sex chromosomes are often shown at the bottom middle. In this normal karyotype, there are two copies of each autosome, and an X and a Y chromosome, indicating that the cells for this analysis were taken from a male.

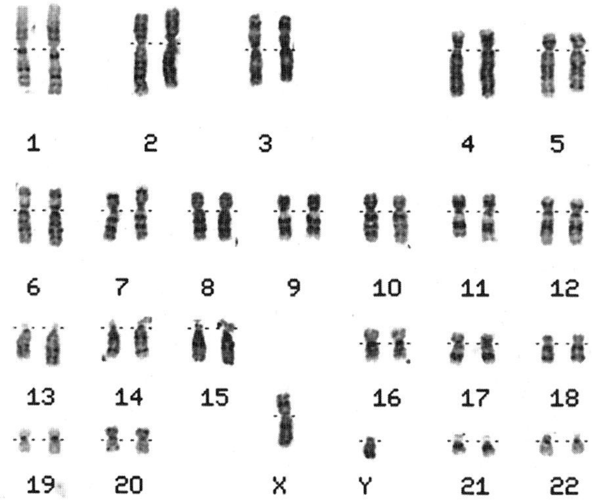

Courtesy of Ifti Ahmed

Figure 3.8 Chorionic Villus Sampling In CVS, a catheter inserted into the uterus through the vagina is used to collect cells from the chorion, a membrane produced by the fetus. Cells from the chorion are used for chromosome analysis and karyotype preparation.

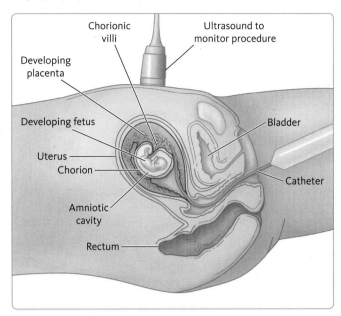

- When the mother is older than age 35. The risk for having children with chromosomal abnormalities increases dramatically for women in this age group. At age 41, Martha Lawrence is a candidate for this test.

- When the mother has already had a child with a chromosomal aberration. The recurrence risk in such cases is 1–2%.

- When either parent has one or more structurally abnormal chromosomes. This condition may cause an abnormal number of chromosomes in the fetus.

- When the mother carries a mutant allele on the X chromosome that causes a genetic disorder.

- When the parents have experienced unexplained infertility or a number of previous miscarriages.

Are there other tests for aneuploidy?

Another method of prenatal chromosome analysis, called **chorionic villus sampling** (CVS), can be performed earlier in pregnancy (at 10–12 weeks, compared with 16 weeks for amniocentesis). Because the cells removed by CVS are rapidly dividing, test results are available within a few hours or a few days. With amniocentesis, fetal cells must be grown in the laboratory for several days before test results can be read.

In CVS (Figure 3.8), a flexible catheter is inserted through the vagina or abdomen into the uterus, guided by ultrasound images. Cells of the chorionic villi (a fetal tissue that forms part of the placenta) are removed by suction and analyzed immediately. This gives CVS a distinct benefit over amniocentesis, because the results can be determined in the first trimester of pregnancy. Both CVS and amniocentesis can be coupled with recombinant DNA technology for prenatal diagnosis. The use of recombinant DNA techniques for prenatal diagnosis of genetic disorders is discussed in Chapter 8.

As with amniocentesis, some risk is associated with CVS. Women who have had this procedure may experience cramps, infection, and miscarriage. Studies indicate that the risk for CVS-related miscarriage when the test is done in a major medical center is about the same as that for amniocentesis (1 in 300 to 1 in 500).

Are there less invasive tests for aneuploidy?

Less invasive methods of prenatal diagnosis are being developed. Researchers are using two different approaches for obtaining fetal DNA for testing: (1) fetal cells isolated from the mother's blood, and (2) free fetal DNA (ffDNA) that also floats in the mother's blood.

Fetal cells can cross the placenta and enter the mother's bloodstream. However, due to the small number of fetal cells that cross the placenta, more study is being done on the use of ffDNA as a method of prenatal diagnosis.

The amniotic fluid surrounding the fetus also contains ffDNA produced when fetal cells break down and the chromosomes become fragmented. Chromosomes from the father and the mother contribute to the ffDNA in this fluid (Figure 3.9a, page 58).

chorionic villus sampling (CVS) a prenatal test that removes cells from the chorion that surrounds the embryo

This ffDNA can also cross the placenta, so that the mother's blood carries ffDNA. However, free DNA from her own cells breaking down is also in her blood (Figure 3.9b). Because the DNA the fetus inherits from the mother is identical to her own, only DNA from the father can be easily identified in her blood. At this time ffDNA is primarily used to identify fetal sex. The use of this genetic technology (see Chapter 17) may eventually replace amniocentesis and chorionic villus sampling as methods for prenatal diagnosis.

Are there other chromosomal variations?

In addition to the XYY condition present in Martha Lawrence's fetus, other birth defects are caused by changes in chromosome number or structure. These changes may involve additional copies of all the chromosomes in a cell (a cell with 69 or 92 chromosomes instead of 46), a condition called **polyploidy**; the gain or loss of individual chromosomes (aneuploidy); or structural changes within individual chromosomes.

Figure 3.10 shows some of the structural changes found in chromosomes.

The diploid number ($2n$ or 46) of chromosomes in body cells and the haploid number (n or 23) in gametes are the normal condition. The term aneuploidy means without the normal condition (a– means "without"). The simplest form of aneuploidy involves the gain or loss of a single chromosome. The gain of one chromosome is known as **trisomy** (it is written as $2n + 1$, or 47 chromosomes); the loss of one chromosome is known as **monosomy** (written as $2n - 1$, or 45 chromosomes).

The most common cause of both trisomy and monosomy in humans is nondisjunction, the failure of chromosomes to separate properly during meiosis. There are two cell divisions in meiosis, and nondisjunction can occur in either the first or the second division.

3.3 Essentials
- **Amniocentesis and chorionic villus sampling can detect aneuploidy during pregnancy.**
- **An extra chromosome, a condition called trisomy, can be fatal to the fetus.**
- **A missing chromosome is a condition called monosomy.**

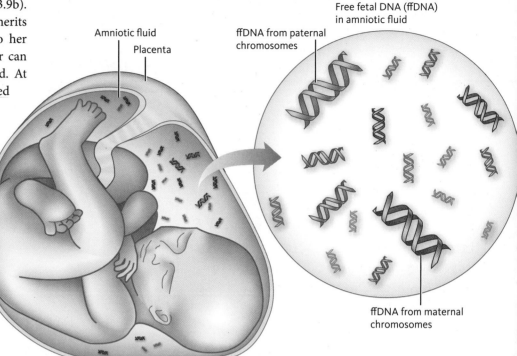

Figure 3.9 **Amniotic Fluid Contains DNA from the Fetus** Fetal DNA can cross the placenta and enter the mother's bloodstream, which contains DNA from the mother. Isolation and analysis of fetal DNA in the maternal blood is an alternative to amniocentesis and chorionic villus sampling.

Amniotic fluid
Placenta
ffDNA from paternal chromosomes
Free fetal DNA (ffDNA) in amniotic fluid
ffDNA from maternal chromosomes

Figure 3.10 **Structural Changes in Chromosomes** Some of the common structural abnormalities seen in chromosomes include deletions (a) where parts of the chromosome are lost, and duplications (b) of a chromosome segment.

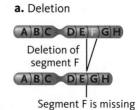

a. Deletion

Deletion of segment F

Segment F is missing

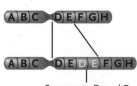

b. Duplication

Segments D and E are duplicated

polyploidy having extra chromosome sets

trisomy a condition where an organism is diploid except for one chromosome that is present in three copies, making a total of 47 chromosomes

monosomy having one less than the diploid number of chromosomes

3.4 What Are the Effects of Monosomy and Trisomy?

In the following sections, we will examine some of the symptoms of monosomy and trisomy involving the autosomes, chromosomes 1–22. Then we will consider monosomy and trisomy involving the sex chromosomes (X and Y).

Monosomy involving any autosome is fatal and results in miscarriage early in pregnancy. This happens because when a chromosome is missing, all the genes on that chromosome are also missing. Trisomies involving the majority of autosomes also cause the death of the fetus early in development. Only a few autosomal trisomies result in live births (trisomy 8, 13, 18, and 21). Trisomy 21 (Down syndrome) is the only autosomal trisomy in which survival into adulthood is possible (Figure 3.11).

What is the leading risk factor for trisomy?

The only known risk factor for Down syndrome and other forms of autosomal aneuploidy is maternal age (Figure 3.12). Young mothers have a lower probability of having a child with Down syndrome than do older mothers. However, the risk increases rapidly after age 35, which is why the genetic counselor recommended amniocentesis for Martha Lawrence. At age 20, a mother's risk of having a child with Down syndrome is 1 in 2000 (0.05%); by age 35, the risk has climbed to 1 in 111 (0.9%); and at age 45, 1 in 33 (3%) of all newborns have trisomy 21. Paternal age has also been proposed as a factor in autosomal trisomy, but no clear link has been found.

Why is maternal age a risk factor?

No one knows for certain why the age of the mother increases the risk for aneuploidy. One idea is that older eggs are more prone to nondisjunction as they complete meiosis after ovulation or fertilization. The eggs of a female fetus begin meiosis well before the child is born. These eggs enter meiosis I and then stop, and remain stopped until just before ovulation. Under hormonal stimulation that begins at puberty, one egg per month completes meiosis I as it is released from the ovary and then enters meiosis II. The final steps in meiosis II are not completed unless the egg is fertilized. Younger women have a low risk of having an aneuploid child; their eggs have been in meiosis I for only 15–25 years. Eggs ovulated at age 40 have been in meiosis I for more than 40 years. During this time, events in the cell or the action of environmental agents might damage the egg, making nondisjunction more likely.

Another explanation focuses on a process called maternal selection. According to this idea, interactions between the uterus and an implanting embryo normally lead to spontaneous abortion of embryos with an abnormal number of chromosomes. As women age, maternal selection may become less effective, allowing more chromosomally abnormal embryos to implant and develop. More research is needed to clarify the underlying mechanisms. *young women's bodies abort embryos w/ incorrect # of chromosome*

The symptoms of Down syndrome and some other autosomal trisomies are summarized in Figure 3.13, page 60.

Are there other causes of Down syndrome?

Down syndrome is one of the most common birth defects and occurs once in about every 800 births. About 95% of the time, the condition is caused by the presence of three individual copies of chromosome 21 (trisomy 21). A rare second form of Down syndrome (about 5% of all cases) is caused by a translocation involving chromosome 21. In these cases, most of

Figure 3.11 **Adult with Down Syndrome**

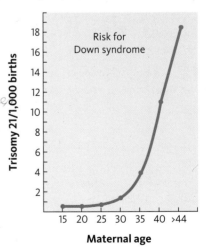

Figure 3.12 **Relationship between Maternal Age and Down Syndrome** The frequency of Down syndrome births increases dramatically when the mother is older than age 35.

Figure 3.13 Some of the Common Trisomy Conditions Found in Humans

Autosomal Trisomies

Karyotype

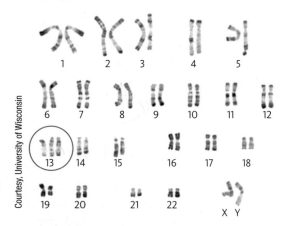

Courtesy, University of Wisconsin

Name
Trisomy 13: Patau syndrome (47,+13)

Number of Chromosomes	How Common?	Survival
47, extra #13	1 in 15,000 live births	1–2 months

Symptoms
Trisomy 13 involves facial malformations, eye defects, extra fingers or toes, and feet with large protruding heels. Internally, there are usually severe malformations of the brain and nervous system, as well as congenital heart defects.

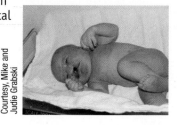

Courtesy, Mike and Judie Grabski

Name
Trisomy 18: Edwards syndrome (47,+18)

Number of Chromosomes	How Common?	Survival
47, extra #18	1 in 11,000 live births; 80% are females	2–4 months

Courtesy of Ifti Ahmed

Symptoms
Infants are small at birth, grow very slowly, and have mental disabilities. Clenched fists, with the second and fifth fingers overlapping the third and fourth fingers, and malformed feet are also characteristic. Heart malformations are almost always present, and heart failure or pneumonia usually causes death.

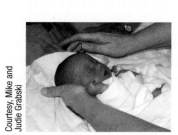

Courtesy, Mike and Judie Grabski

Name
Trisomy 21: Down syndrome (47,+21)

Number of Chromosomes	How Common?	Survival
47, extra #21	1 in 800 live births (changes with age of mother)	Up to age 50

Hironao NUMABE, M.D., D.M.Sc., Tokyo Medical University

Symptoms
Down syndrome is a leading cause of childhood mental retardation and heart defects in the United States. Affected individuals usually have a wide, flat skull; folds in the corner of the eyelids; and spots on the iris. They may have furrowed, large tongues that cause the mouth to remain partially open. Physical growth, behavior, and mental development are retarded, and approximately 40% of all children with Down syndrome have congenital heart defects. In spite of these handicaps, many individuals with Down syndrome lead rich, productive lives.

Lauren Shear

chromosome 21 becomes attached (translocated) to another chromosome, often 14 or 15, an event called a Robertsonian translocation (Figure 3.14a).

If the translocated chromosome is inherited along with two normal copies of chromosome 21, the result is Down syndrome (Figure 3.14b). However, it is possible for a person to inherit the translocated chromosome and not have Down syndrome. These individuals carry the translocated chromosome and one copy of chromosome 21. However, because the translocation can be passed on to the next generation, carriers are at great risk for having a child with Down syndrome. It is important for all children with Down syndrome and their parents to have karyotypes prepared to determine which form of Down syndrome is present in the family.

What other accidents of cell division cause genetic disorders?
Sometimes, instead of receiving one chromosome from each parent, an embryo receives two copies of the same chromosome from one parent and no copies of that chromosome from the other parent. This accident, **uniparental disomy** (UPD), can occur during the formation of the egg or sperm or during development after fertilization. UPD is associated with some genetic disorders.

Let's briefly look at two disorders caused by UPD involving chromosome 15 (Figure 3.16, page 62). One of these is called Prader-Willi syndrome (PWS) and results when a child inherits both copies of chromosome 15 from the father. Affected children can be mentally retarded and have uncontrolled appetites that result in obesity. The second disorder is called Angelman syndrome (AS) and occurs when a child inherits both copies of chromosome 15 from the mother. Children with AS are also mentally retarded and have speech problems. UPD is related to imprinting, which will be described in Chapter 6.

Does aneuploidy affect the sex chromosomes?
Aneuploidy involving the X and Y chromosomes is more common than autosomal aneuploidy. Monosomy involving autosomes is always fatal, but this is not necessarily true for the sex chromosomes. Monosomy involving the X chromosome, a condition called Turner syndrome (45,X), occurs in about 1 in 10,000 female births (Figure 3.15, page 62). Although 1 in 10,000 female births are 45,X, many more 45,X embryos are lost through miscarriage.

On the other hand, monosomy for the Y chromosome (45,Y) always results in miscarriage; no live births for this condition have been reported.

uniparental disomy a condition in which both copies of a chromosome are inherited from one parent

Figure 3.14 **Robertsonian Translocations and Down Syndrome** (a) These translocations can produce a chromosome that contains the long arms of chromosomes 14 and 21. (b) Gametes containing the translocated chromosome when fertilized by a normal gamete produce three types of surviving individuals: those with a normal karyotype, those who carry the translocation and are at risk for having a Down syndrome child, and those with Down syndrome. Other combinations of chromosomes result in zygotes that do not survive.

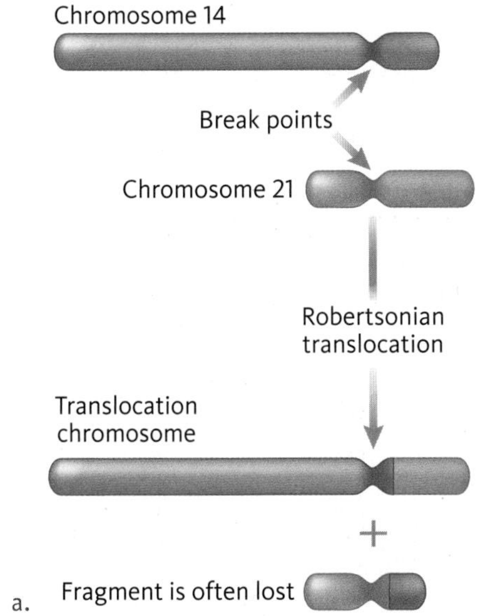

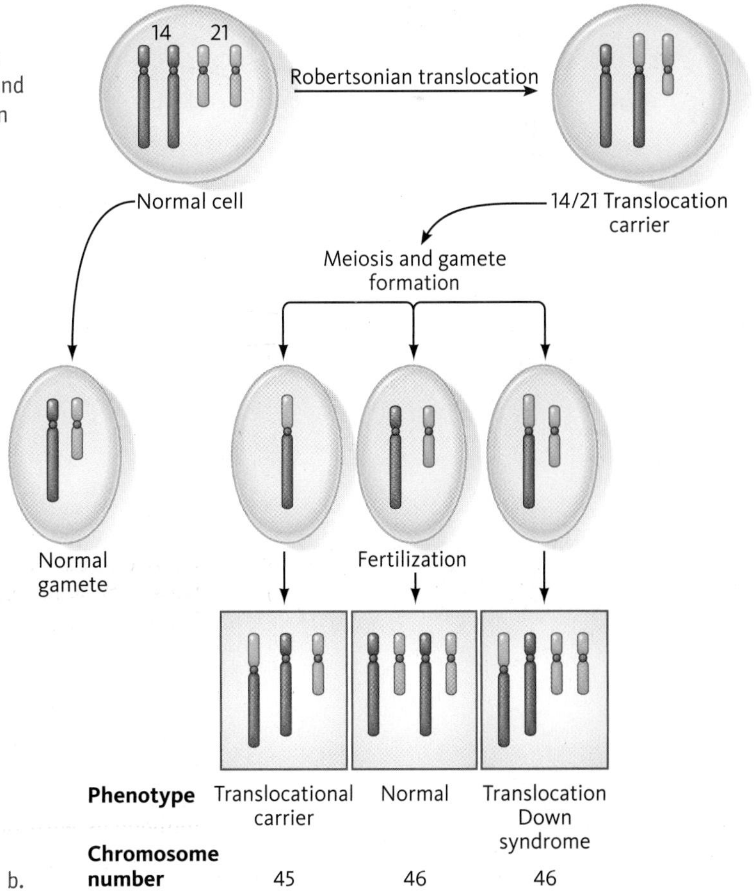

Chromosome 14

Break points

Chromosome 21

Robertsonian translocation

Translocation chromosome

+

a. Fragment is often lost

14   21

Robertsonian translocation

Normal cell

14/21 Translocation carrier

Meiosis and gamete formation

Normal gamete

Fertilization

Phenotype	Translocational carrier	Normal	Translocation Down syndrome
Chromosome number	45	46	46

b.

5–7%
Percentage of all childhood
deaths due to aneuploidy

**0.3–1 oz
(10–30 ml)**
Amount of fluid removed in an
amniocentesis

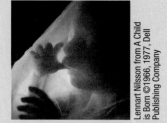

Lennart Nilsson from A Child
is Born ©1966, 1977, Dell
Publishing Company

40%
Percentage of children with Down
syndrome who have heart defects

1 in 2000
The risk of having a child with
Down syndrome if the mother is
younger than age 20

1 in 33
The risk of having a child with
Down syndrome if the mother
is age 45

Other aneuploid conditions that involve the sex chromosomes include Klinefelter syndrome (47,XXY), and Jacobs syndrome (47,XYY), the condition present in Martha Lawrence's fetus. These conditions are summarized in Figure 3.17, page 63).

3.4 Essentials

- **Individuals with Down syndrome (trisomy 21) can survive to adulthood.**

3.5 Is There a Way to Evaluate Genetic Risks?

Understanding the results of a genetic test and what it means for a family and its present and future children can be an overwhelming responsibility. Fortunately, a health professional called a **genetic counselor** can help individuals or families understand the results of a genetic test. The information provided by a genetic counselor may include the following:

- medical facts, including the diagnosis, progression, management, and any available treatment for a genetic disorder
- how heredity contributes to the disorder and the risk of having children with this disorder
- how to adjust to the disorder in an affected family member
- alternatives for dealing with the risk of recurrence

Genetic counselors achieve these goals in a nondirective way. They provide all the information available to individuals or family members, so the person or family can make the decisions best suited to them based on their own cultural, religious, and moral beliefs.

Who are genetic counselors? Genetic counselors are health care professionals with specialized training in medical genetics, psychology, and counseling. They

genetic counselor a person who works with patients to explain genetic testing and its results

Figure 3.15 **Adult with Turner Syndrome**

Lew Stamp/Akron Beacon Journal

Figure 3.16 **Uniparental Disomy** Inheritance of both copies of a chromosome from one parent can result in genetic disorders such as Prader-Willi syndrome and Angelman syndrome.

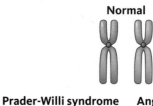

Normal

Prader-Willi syndrome

Two copies of mother's
chromosome 15

Angelman syndrome

Two copies of father's
chromosome 15

Figure 3.17 Some of the Common Aneuploidies Involving the Sex Chromosomes in Humans

Sex Chromosome Trisomies

Karyotype

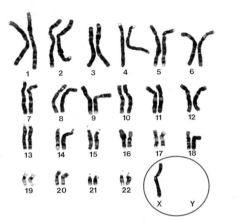

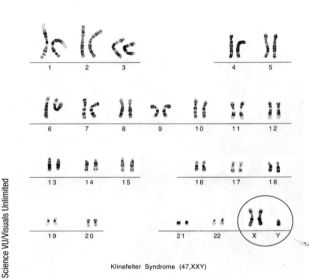

Klinefelter Syndrome (47,XXY)

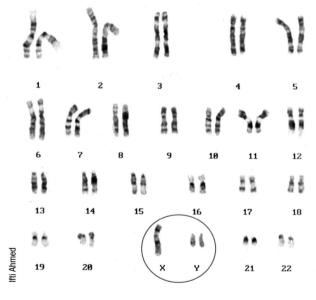

Name
Turner syndrome (45,X)

Number of Chromosomes
45, missing one X

Survival
To adulthood

Symptoms
Babies born with Turner syndrome are always female and live to adulthood. They are short and wide-chested with undeveloped ovaries. At birth, puffiness of the hands and feet is prominent, but this disappears in infancy. Many also have narrowing of the aorta. They have normal intelligence.

How Common?
It is estimated that 1% of all conceptions are 45,X and that 95–99% of all 45,X embryos die before birth. Turner syndrome occurs with a frequency of 1 in 10,000 female births.

Name
Klinefelter syndrome (47,XXY)

Number of Chromosomes
47, an extra X

Survival
To adulthood

Symptoms
The features of this syndrome do not develop until puberty. Affected individuals are male but are usually sterile. Some men with Klinefelter syndrome have learning disabilities or mild retardation. Many men with Klinefelter are mosaics. This means some cells have the XY chromosome combination and others have XXY. In these men, the symptoms are less severe.

How Common?
1 in 1000 males

Name
XYY syndrome (47,XYY)

Number of Chromosomes
47, with an extra Y

Survival
To adulthood

Symptoms
These individuals are above average in height and thin. Some affected individuals may have personality disorders, behavioral problems, and/or some form of retardation. Others have learning disabilities; some have very mild symptoms. They often experience severe acne during adolescence.

How Common?
1 in 1000 male births (about 0.1% of the males in the general population)

usually work as part of a health care team, providing information and support to people and families who have genetic disorders or may be at risk for an inherited disorder. Genetic counselors identify families at risk, investigate the condition in the family, interpret information about the disorder, analyze inheritance patterns and risks of recurrence, and review available options with the family.

Why do people go to genetic counselors?

People ask to see a genetic counselor if they have a family history of a genetic disorder, cancer, a birth defect, or a developmental disability. Women older than age 35 and members of specific ethnic groups in which particular genetic conditions occur more frequently are often seen by genetic counselors, who teach them about their increased risk for genetic or chromosomal disorders and the genetic tests available to them.

Counseling is especially recommended for several groups of people:

- women who are pregnant, or are planning to become pregnant, after age 35
- couples who already have a child with mental retardation, a genetic disorder, or a birth defect
- couples who would like testing or information about genetic disorders that occur more frequently in their ethnic group
- couples who are first cousins or other close blood relatives
- individuals who are concerned that their job, lifestyle, or medical history may pose a risk to a pregnancy, including exposure to radiation, medications, chemicals, infection, or drugs
- women who have had two or more miscarriages or babies who died in infancy
- couples whose infant has a genetic disease diagnosed by routine newborn screening
- people who have, or are concerned that they might have, an inherited disorder or birth defect

How does genetic counseling work?

Most people see a genetic counselor after a prenatal test or after the birth of a child with a genetic disorder. Individuals may also seek genetic counseling to determine their risk of being affected by a condition they may already have, or their risk of being a carrier. The counselor usually begins by constructing a detailed family and medical history and a pedigree. A pedigree shows the family's genetic history as a diagram using symbols to represent family members (pedigrees will be discussed in Chapter 4). Then he or she explains the basic concepts of genetics and how the condition is inherited (Figure 3.18).

When couples seek information before they have a child, blood tests of the couple and their families can be used along with pedigree analysis to help determine what, if any, risks are present. If a trait is found to be genetically determined, the counselor constructs a risk assessment profile that explains the risk of having a child who has the condition.

To help both couples and individuals understand how genes and gene products are related to a genetic disorder, genetic counselors use a number of teaching techniques. They share information that allows an individual or a couple to make informed decisions.

What is the future of genetic counseling?

As information from the Human Genome Project (HGP) changes medical care, the number of genetic disorders that can be diagnosed by genetic testing is increasing dramatically, and the role of the genetic counselor will become more important. In addition, with the rise of genetic testing, its availability to the general public, and a greater understanding of genetics, more genetic counselors will be needed in the future. A number of universities and colleges are adding genetic counseling training programs for students who are interested. Results from the HGP are already changing the

CASE A: QUESTIONS

Now that we understand some basic information about chromosomal abnormalities and how they arise, let's look at some of the issues raised in Martha Lawrence's situation. The uncertainty surrounding the impact of an XYY child illustrates Dr. Gould's dilemma about Martha's case.

Courtesy of Itti Ahmed

1. Based on the karyotype that Dr. Gould obtained by amniocentesis, list five options available to Martha.

2. Which do you think is the best option in this situation?

3. Give three reasons why someone might disagree with your choice.

4. Knowing what you do about the different aneuploid conditions, if the diagnosis were Down syndrome, would your decision change? Why or why not?

While analyzing the results of Martha's amniocentesis, Dr. Gould had some questions herself:

5. Keeping in mind that Martha was primarily concerned about her age and the chances of having a child with trisomy 21, should Dr. Gould tell her about the abnormal XYY karyotype? Why or why not?

6. Dr. Gould considered telling Martha's husband about the test results before she told Martha. Should she have?

Figure 3.18 Genetic counseling sessions offer an opportunity for family members to learn about genetic disorders and the risk of having affected children.

Copyright © Robin Nelson/Photo Edit

focus of genetic counseling from reproductive risks to adult-onset conditions such as cancer, diabetes, and cardiovascular diseases. The information provided by the counselor allows at-risk individuals to adopt lifestyles that may reduce the impact of the disorder, to make decisions about whether to have children, and to plan for medical care that they or their affected children may require later in life.

3.5 Essentials

- **As a woman ages, the chances that she will have a child with aneuploidy increase, and she can get help from genetic counselors.**

3.6 What Are the Legal and Ethical Issues Associated with Chromosomal Abnormalities?

In the U.S. legal system, Martha Lawrence would have an option to sue Dr. Gould if she were not offered amniocentesis and later gave birth to a child with a genetic disorder. There are two possible legal scenarios: a **wrongful-birth suit** or a **wrongful-life suit**. Both types of lawsuits are based on the idea that the birth or the life of the child is wrong because of either the action or the inaction of the physician or genetic counselor.

CASE B:
Test Results Worry Physician

Martha Lawrence's case really had Dr. Gould thinking. She remembered a patient back in 1996 who taught her quite a lot about the legal aspects of her practice. This patient was named Donna Slotin. She gave birth to a son named A.J. At the time, Donna was 31 and was not considered a high-risk pregnancy; however, A.J. was born with a genetic condition that causes serious mental retardation and other physical disabilities. After the birth and diagnosis, an attorney called Dr. Gould and said that the Slotins were going to sue her because Donna was not given an amniocentesis.

A physician evaluating a karyotype.

James King-Holmes/Photo Researchers, Inc.

1. **What is the first thing Dr. Gould should have done?**
2. **What would the Slotins' attorney have argued?**
3. **What would Dr. Gould's attorney have argued?**
4. **What do you think could have been done to avoid a lawsuit?**

Table 3.3 explains the differences between the two legal actions.

The outcomes in these types of cases are based on two questions:

1. Could a diagnosis of this condition have been made in time to have an abortion?
2. Was the condition serious enough that a reasonable person would have had an abortion?

Most courts of law allow wrongful-birth suits for two reasons: *Roe v. Wade* gave a woman an alternative to birth, and physicians have extensive medical malpractice insurance, so financial compensation is often not a problem. However, only five states allow wrongful-life suits, because courts are uncomfortable declaring that someone should never have

wrongful-birth suit a lawsuit that is brought by parents against a physician or lab who did not correctly identify a birth defect

wrongful-life suit a lawsuit that is brought by the child against a physician or lab who did not correctly identify a birth defect

Table 3.3 Some Legal and Ethical Issues Associated with Wrongful Births and Wrongful Life

In a Wrongful-Birth Case	In a Wrongful-Life Case
The parents sue the physician.	The child sues the physician.
The parents argue that because the physician's inaction did not make it possible for them to take a prenatal test, they could not make an informed decision as to whether they should have the baby.	The child argues that because of inadequate advice by the physician, he or she was born into a life of pain and suffering.
This type of case relies on *Roe v. Wade*, the case that legally allowed abortion in the United States, because the mother must say that if she had known that the child had this condition, she would have aborted it.	This type of case also relies on *Roe v. Wade* for similar reasons.
Reasons for such suits include the following: The physician neglected to tell the parents of the test or the risk of having such a child; the physician or lab mixed up the results, did the test incorrectly, or did the wrong test; or the physician deliberately chose not to tell the parents.	Reasons for such suits are the same as for wrongful-birth suits.

been born. An example of a wrongful-life suit is examined in "Spotlight on Law: *Becker v. Schwartz*."

Wrongful-life and wrongful-birth suits are different from malpractice suits, because in these cases the physician has done nothing to cause the injury (the birth defect). In malpractice suits, the physician is accused of malpractice ("bad" practice), because he or she caused the damage to the patient.

Some other types of cases have evolved from these types of lawsuits:

- Wrongful conception: when birth control methods (surgical or chemical) don't work
- Wrongful pregnancy: when an abortion isn't completed and the mother has medical problems

Martha Lawrence may want to read about some of the interesting legal aspects of the XYY karyotype. Early investigations found that people with the XYY karyotype have a tendency toward aggressive behavior associated with the presence of an extra Y chromosome. In effect, this may mean that some forms of violent behavior are genetically determined. In fact, the XYY karyotype has been used on several occasions as

a legal defense (unsuccessfully, so far) in criminal trials (see, for example, *People v. Tanner*, 13 Cal.App. 3d [1970]).

The question asked in these trials and the subsequent studies of children with the XYY karyotype is: Can we find a direct link between the XYY condition and criminal behavior? Using today's science, there doesn't seem to be strong evidence to support such a link. In fact, most males with the XYY karyotype lead socially normal lives.

In the United States, long-term studies of the relationship between antisocial behavior and the XYY karyotype were discontinued. Researchers feared that identifying children with potential behavioral problems might lead parents to treat them differently and result in behavioral problems as a self-fulfilling prophecy. The uncertainty surrounding the XYY condition illustrates Dr. Gould's dilemma about Martha Lawrence's case and may make Martha's decision all the more difficult.

3.6 Essentials

- **Wrongful-life and wrongful-birth lawsuits can result from incorrect prenatal diagnosis.**

Spotlight on Law

Becker v. Schwartz
386 N.E.2d 807 (NY 1978)

The following case is an example of the type of lawsuit Ms. Lawrence's child might file if Dr. Gould fails to tell Ms. Lawrence about the XYY condition: a wrongful-life case.

Delores Becker got pregnant at age 37. Her physician, who treated her for her entire pregnancy, did not tell her about amniocentesis and her increased risk of having a child with Down syndrome. Her son was born with Down syndrome. Even though it was 1978, amniocentesis was being used.

The Beckers filed a wrongful-life suit on behalf of their son, and the resulting case, *Becker v. Schwartz*, was eventually heard by the New York Court of Appeals. Delores Becker testified that if she had been informed of her son's condition, she would have had an abortion. As a result of her physician's failures, she did not learn of her son's mental disability until after he was born.

RESULTS

The court dismissed the complaint, holding that no one has the right to be born free of disease. A $2500 settlement was reached, and the baby was given up for adoption. Some attorneys analyzing the case later thought that the Beckers could have received more money if they had gone to an even higher court.

QUESTIONS

1. What was the basis for the suit?

2. Do you think that the court found correctly?

3. How much money do you think the jury might have given for the care and "loss" of their son?

4. Knowing what you do about Down syndrome, do you think the parents should have raised the child and not sued? Why or why not?

The Essential 10

1. **The normal chromosome number in humans is 46. (Section 3.1)**
   There are 23 pairs of chromosomes, including the sex chromosomes (XX or XY).

2. **Meiosis is involved in formation of the sperm and egg. (Section 3.1)**
   The egg and sperm have the haploid number of chromosomes.

3. **In meiosis, members of a chromosome pair separate from each other, forming gametes that contain 23 chromosomes. (Section 3.2)**
   If the chromosomes don't separate correctly, aneuploidy can occur.

4. **Most aneuploidy is the result of mistakes in meiosis. (Section 3.2)**
   Aneuploidy can result in monosomy or trisomy.

5. **Two tests that can detect aneuploidy during pregnancy are amniocentesis and chorionic villus sampling. (Section 3.3)**
   Each one of these tests has associated risks and benefits.

6. **An extra chromosome, a condition called trisomy, can be fatal to the fetus. (Section 3.3)**
   Risk to the fetus depends on which chromosome is involved.

7. **A missing chromosome is a condition called monosomy. (Section 3.3)**
   When a chromosome is missing, all genes on that chromosome are also missing.

8. **Individuals with Down syndrome (trisomy 21) can survive to adulthood. (Section 3.4)**
   People with Down syndrome face many serious challenges.

9. **As a woman ages, the chances that she will have a child with aneuploidy increase, and she can get help from genetic counselors. (Section 3.5)**
   It is recommended that women age 35 and over have amniocentesis when pregnant.

10. **Wrongful-life and wrongful-birth lawsuits can result from incorrect prenatal testing. (Section 3.6)**
    Testing problems can be due to problems in laboratories.

Review Questions

1. At which stages of meiosis is a chromosomal aberration most likely to occur?

2. State the purpose of meiosis.

3. List two situations that would require karyotype analysis of a newborn.

4. Explain why trisomy involving a large chromosome can be more serious than one involving a small chromosome.

5. Do some research on how often trisomies in chromosomes 1–22 occur and report your findings.

6. What are the stages of meiosis?

7. List two reasons nondisjunction occurs.

8. What are the risk factors in amniocentesis?

9. List three types of trisomy.

10. State two types of patient who might need a genetic counselor.

11. How many chromosomes are in a diploid cell? A haploid cell?

12. Draw a chromosome and label its parts.

13. What is the function of a telomere?

14. Define a polar body.

15. What name is assigned to chromosomes 1–22?

16. What might be one reason a woman would refuse an amniocentesis?

17. What is ffDNA?

18. Explain how uniparental disomy can occur.

19. How are Prader-Willi syndrome and Angelman syndrome different?

20. List the steps to making a karyotype in the cytogenetics lab.

Application Questions

1. If a woman is younger than age 35, physicians don't usually recommend prenatal testing, because the maternal risk factor for genetic disorders is low. Research this topic, and list two ways this practice may change in the future.

2. During amniocentesis, a very long needle is used to extract the amniotic fluid. Find a photo of an amniocentesis needle, find out how long it is, and discuss how the test can be discussed with a woman without frightening her.

3. Occasionally a baby is harmed during prenatal diagnosis. Does a physician need to discuss this possibility with the patient? Why or why not?

4. Obviously, some chromosomal aberrations are more serious than others. List two differences between monosomy of the X chromosome and monosomy of any other chromosome.

5. It seems pretty clear that a physician who diagnoses a chromosomal aberration did not cause the condition. Yet some are sued by the parents of these children. Is this ethical?

6. There is a chromosomal abnormality called triploidy that occurs when the fertilized egg contains three copies of each chromosome (69 chromosomes). Where do you think the extra set of chromosomes comes from? Hint: Many triploid embryos have an XYY sex chromosome set.

7. Amniocentesis is not performed until about the 16th week of pregnancy. One reason physicians wait is to make sure that the cells recovered are from the fetus and not from the mother. Clearly, if a karyotype of the recovered cells shows a Y chromosome, the cells are from the fetus. Why?

8. A pregnant mother comes to you and insists on having the procedure at week 12. The result shows cells that have two X chromosomes. How could you test to find out if the recovered cells belong to the mother or the fetus?

9. State one reason someone might want an amniocentesis as early as 12 weeks.

10. Continuing the "By the Numbers" chart, research and fill in the numbers below:

How many amniocenteses are done every year?	
How many Down syndrome babies were born in one recent year?	
How many Down syndrome babies were born 10 years ago?	

Online Resources

Preparing for an exam? Log on at www.cengagebrain.com for study tools to help you assess your understanding. If assigned by your instructor, the Case A and Spotlight on Law activities for this chapter, "Results Worry Pregnant Woman" and *Becker v. Schwartz*," will also be available.

In the Media

Newsomewellness.com, March 2011

Simple Blood Test May Soon Be Used for Down Syndrome

A simple blood test during pregnancy may give hope in detecting Down syndrome early on, according to a new study. An experiment conducted using blood samples has identified all normal and all Down syndrome pregnancies with 100 percent accuracy.

Cypriot scientists took blood samples from pregnant women and mothers of babies with and without Down syndrome. The study authors said that in each case, the test quickly identified the chromosomal variation, pointing out 14 Down syndrome cases and 26 normal fetuses. The new test eliminates the risk of miscarriage and can identify Down syndrome in the 11th week of pregnancy.

To access this article online, go to www.cengagebrain.com.

QUESTIONS

1. If more women are tested in this way, what problems do you foresee?

2. We discussed this type of testing in this chapter and Chapter 7. List five reasons it will be done more in the future.

WHAT WOULD YOU DO IF...?

. . . you were Martha Lawrence? (See Case A.)

. . . your daughter, who has Down syndrome, wanted to have a child?

. . . you were a lawyer who had to defend a physician accused of wrongful birth? What arguments would you use?

. . . you were Dr. Gould and thought that Martha Lawrence was mentally unstable? (See Case A.)

. . . you were a member of a legislature that was voting on a bill that required every pregnant woman over the age of 35 to be tested by amniocentesis?

. . . you were a physician whose patient refused amniocentesis even though she was 42 years old?

. . . you were a pregnant woman considering amniocentesis, and you knew that the chance of miscarriage with an amniocentesis is 0.2%?

. . . you were a physician considering whether to tell your patients that the chance of miscarriage with an amniocentesis is 0.2%?

Science, January 12, 1973

Behavioral Implications of the XYY Genotype

One genetic study was carried out by Harvard child psychiatrist Stanley Walzer and Harvard Medical School geneticist Park Gerald. By 1968, they were screening all newborn males at Boston Hospital for Women and following up by studying the development of those with abnormal karyotypes such as XYY or XXY. The research was funded by a grant from the Centers for Studies of Crime and Delinquency of the National Institute for Mental Health.

Source: Ernest B. Cook, "Behavioral Implications of the XYY Genotype," Science, vol. 179, pp. 139–150, January 12, 1973.

QUESTIONS

1. What problems do you see with this study?

2. Do some research on this study, and find out why it was stopped.

Law & Order: SVU (NBC)

"Competence," May 10, 2002
"Clock," September 26, 2006

Law and Order Raises Questions

In the first of these episodes, a woman with Down syndrome who was raped wants to keep her baby, but her mother objects. In the second, a woman with Turner syndrome who looks young for her age fakes a kidnapping to run away with her boyfriend.

To access episode summaries and downloads online, go to www .cengagebrain.com.

QUESTIONS

1. Should these topics be covered in television shows?

2. Both of these episodes are about competency of individuals with genetic conditions. Who should determine if a person is competent?

How Genes Are Transmitted from Generation to Generation

CASE A: A Family's Dilemma

Alan Franklin, 17, has only vague memories of his mother; she died before he was 5 years old. But his older brother, John, told him about her. She had been sick for a very long time; she stumbled around and walked with jerky movements, and finally was in bed all the time. John said she died of **Huntington disease (HD)**.

Huntington disease is caused by a mutant form of a gene. A person who carries just one copy of this mutation will eventually have the disease. Most people with HD begin to show symptoms between ages 30 and 50. Muscle control and mental capacity slowly disintegrate until the person is bedridden and dependent on others for care. The disease is fatal, but its progression can take as long as 10 to 20 years.

Alan is a good student and wants to be a pilot in the Air Force; he wonders if he is at risk for developing HD. Recently in his biology class he learned that HD is inherited and that a test is available for people at risk for this disorder.

There are three children in the Franklin family. Mary Franklin, Alan's older sister, is engaged to be married to Bob Wilson next June. John's wife, Emily (Alan's sister-in-law), is pregnant, and they already have a 2-year-old son, Matthew.

Alan asked his biology teacher for more information about the test, and she gave him the name of Marta Wright, a genetic counselor at a nearby medical center. She suggested that he ask his father's permission to visit the counselor.

A family discussion about a genetic disorder.

Alan's father encouraged him to see the genetic counselor. During their first appointment, Ms. Wright diagrammed Alan's personal family history:

She explains that the top of this diagram contains a square, symbolizing his father, connected by a line to a circle symbolizing his mother. Her symbol is filled, because she had Huntington disease. Below these symbols is a square representing Alan. The counselor tells Alan that because his mother had HD, there is a 50% chance that he also carries the HD mutation. If he does, he will probably begin to show symptoms in middle age and may pass the gene on to some of his children.

Ms. Wright also tells Alan that if he takes the test for HD and the results show that he will develop HD, he probably will not get into the Air Force, and his dreams of being a pilot will not come true.

Some questions come to mind when reading about Alan's case. Before we address those questions, let's look at how genetic traits, including Huntington disease, are passed from generation to generation.

A Father with His Albino Son

4.1 How Are Genes Transmitted from Parents to Offspring?

Thanks to the work of Gregor Mendel, a European monk who lived in the mid-1800s, we understand how genes are transmitted from parents to offspring in plants and animals, including humans. In his experiments with pea plants, Mendel chose plants that each had a different, distinguishing characteristic, called a *trait*. These traits included plant height and flower color. Let's look at one of his experiments to see what he did and how he explained the results.

In an early experiment, Mendel crossed tall pea plants with short pea plants to see how height was passed from parents to offspring. Geneticists use letters as symbols to identify each trait: in this case, we will use *T* for tall, and *t* for short. Mendel discovered specific patterns in the way traits are passed from one generation to the next. Some traits, such as shortness, disappeared in the first generation of offspring but reappeared in the next generation in about one-quarter of the offspring (a 3:1 ratio of tall to short). He called the trait that was present in the first-generation offspring a **dominant** trait, and the trait that was absent but reappeared in the next generation a **recessive** trait.

We now know that Mendel's conclusions about how traits are inherited are correct and apply to humans as well as pea plants. From the results of his experiments he theorized that traits are passed from generation to generation by "factors" (we call them *genes*) transmitted from parent to offspring. He concluded that each parent carries a pair of genes for a given trait but contributes only one of these genes to its offspring. We now know that this separation of members of a gene pair occurs during meiosis when members of a chromosome pair separate from each other, with each gamete receiving only one copy of each chromosome.

Members of a gene pair can differ. In our example, each parent carries a pair of genes specifying plant height, but this gene has two versions; one specifies tall plants (Figure 4.1a), the other short plants (Figure 4.1b). We call these variations of a gene **alleles**. Organisms that carry two identical alleles of a gene (*TT* or *tt*) are **homozygous**; those that carry nonidentical alleles (*Tt*) are **heterozygous**.

So, using letters to symbolize alleles in Mendel's experiment, the first cross looked like this:

TT (tall) × *tt* (short)
(in this cross, both parents are homozygous)

Tt (tall)
(in this generation, all the offspring are heterozygous)

In addition, Mendel concluded that the appearance of an organism (tall or short) can be different from its genetic makeup. What an organism looks like, or what we can observe about it, is called its **phenotype**, and the genetic makeup of an organism is called its **genotype**. Therefore, *TT*, *Tt*, and *tt* are genotypes, and *tall* and *short* are phenotypes. Organisms can have identical phenotypes but different genotypes. For example, because *tall* is a dominant trait, pea plants with the tall phenotype can have two different genotypes: homozygous (*TT*) or heterozygous (*Tt*).

dominant a trait expressed in the homozygous or heterozygous condition

recessive a trait expressed only in the homozygous condition

alleles different forms of a gene

homozygous having two identical alleles of the same gene

heterozygous having two different alleles of the same gene

phenotype the observable appearance of a genetic trait

genotype the genetic constitution of an individual

Figure 4.1 **Tall and Short Pea Plants** These two pea plants are similar to those Mendel used in his experiments. In his work, the tall plant (a) was about 8 feet tall and the short plant (b) was about 18 inches tall.

a.
b.

A diagram called a Punnett square can be used to visualize crosses and keep track of the possible combinations of alleles and phenotypes in the offspring (Figure 4.2). In this example, we will begin by crossing the heterozygous (*Tt*) offspring of the cross diagrammed above. In this way, we will be able to predict the possible genotypes and phenotypes of the offspring.

In constructing the Punnett square, the alleles carried by the gametes of one parent (male) are placed across the top of the square, and those carried by gametes of the other parent (female) are placed along the left side of the square. By combining alleles from the male gametes with alleles from the female gametes in the squares, all the possible genotypes and phenotypes from this cross can be predicted.

In addition, the Punnett square allows us to predict the chance that any of the offspring will be homozygous or heterozygous, and whether they will be tall or short. As you can see, there is a 25% chance that the offspring will be homozygous tall (*TT*), a 50% chance the offspring will be heterozygous tall (*Tt*), and a 25% chance that the offspring will be homozygous short (*tt*). It is clear from looking at the phenotypes that there is a 75% chance that the offspring will be tall and a 25% chance that the offspring will be short.

Figure 4.2 Allele Combinations in Offspring When Mendel crossed first-generation offspring of a cross between tall and short plants, each heterozygous plant produced two types of gametes, those carrying the dominant (*T*) allele, and those carrying the recessive (*t*) allele. Combinations of alleles produced by fertilization resulted in about three tall plants for each short one. Where did the short plants come from?

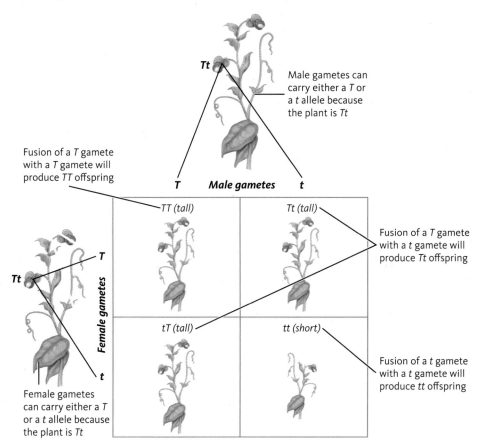

Male gametes can carry either a *T* or a *t* allele because the plant is *Tt*

Fusion of a *T* gamete with a *T* gamete will produce *TT* offspring

Fusion of a *T* gamete with a *t* gamete will produce *Tt* offspring

Fusion of a *t* gamete with a *t* gamete will produce *tt* offspring

Female gametes can carry either a *T* or a *t* allele because the plant is *Tt*

Figure 4.3 Genes on Chromosomes Genes occur in pairs on homologous chromosomes. Different versions of the same gene, called alleles, may be present on each homologue.

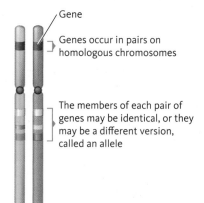

Gene

Genes occur in pairs on homologous chromosomes

The members of each pair of genes may be identical, or they may be a different version, called an allele

Figure 4.4 Gene and Chromosome Segregation During meiosis, homologous chromosomes separate from each other and end up in different gametes. Genes carried on these chromosomes also separate or segregate from each other so that each gamete carries only one member of a gene pair.

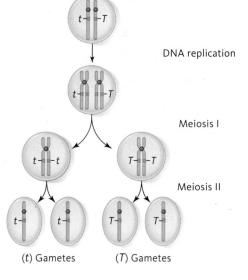

DNA replication

Meiosis I

Meiosis II

(*t*) Gametes   (*T*) Gametes

4.1 Essentials

- **Members of a gene pair segregate into gametes independently of other gene pairs so that all gametes have different combinations of genes.**
- **Gregor Mendel, a monk who lived in the 1800s, was the father of genetics.**
- **Mendel did his experiments with plants, but his important laws of genetics are still used today to study human genetics.**

4.2 Mendel, Chromosomes, Genes, and DNA

Mendel is still remembered because he established how traits and the genes that control them are passed from generation to generation. He did his work before it was known how chromosomes behave in meiosis, or that genes encoded in DNA are part of chromosomes. It is now known that the pairs of genes Mendel described are carried on members of a chromosome pair (Figure 4.3).

Two of his conclusions became the foundation of genetics:

1. Two copies of each gene separate from each other during the formation of egg and sperm (which takes place after meiosis). As a result, only one copy of each gene is present in the sperm or egg and is contributed to the offspring. This idea is called the **law of segregation** (Figure 4.4).

law of segregation separation of homologues into different cells in meiosis

Figure 4.5 **Independent Assortment of Chromosomes and Genes**

During meiosis, chromosomes and the genes on those chromosomes assort into gametes independently of other chromosomes and genes. The result is that gametes contain all combinations of assorting genes.

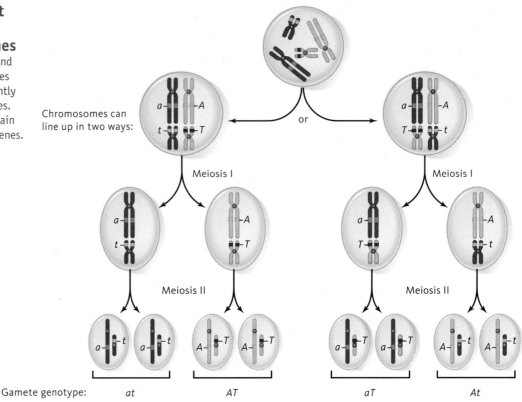

Chromosomes can line up in two ways:

or

Meiosis I · Meiosis I

Meiosis II · Meiosis II

Gamete genotype: *at* · *AT* · *aT* · *At*

2. Members of a gene pair segregate into gametes independently of other gene pairs so that gametes can have different combinations of parental genes. This idea is called the **law of independent assortment** (Figure 4.5).

Mendel's methods are still used to study the genetics of experimental organisms such as fruit flies, rats, and plants. These same laws discovered by Mendel are applied every day by genetic counselors such as Ms. Wright to help their clients understand their family's genetics.

How do Mendel's findings apply to human traits?

To illustrate that Mendel's two laws apply to human traits, let's follow the inheritance of a trait called *albinism*. Figure 4.6 shows a person with albinism.

We can use the letters *A* and *a* to represent the alleles in this case: *A* is the dominant allele for **pigment** formation (color), and *a* is the recessive allele for lack of pigment formation. People with albinism carry two copies of the recessive allele (*aa*) and cannot make pigment; they have pale white skin, white hair, and colorless eyes (Figure 4.6). The frequency of albinism varies in different populations, with the Hopi Indians having one of the highest rates—1 in 22 individuals affected.

Anyone carrying at least one dominant allele (*Aa* or *AA*) can make enough pigment to have colored skin, hair, and eyes.

We'll start our discussion of how Mendel's laws apply to humans with parents who are heterozygotes (*Aa*) with normal pigmentation (Figure 4.7).

Aa × Aa

During meiosis, the dominant (*A*) and recessive (*a*) alleles carried by each parent separate from each other and end up in different gametes (Mendel's law of segregation). Because each parent can produce two different types of gametes (one with *A* and another with *a*), the random union of gametes at fertilization can produce four possible combinations of alleles in the offspring:

AA, Aa, aA, aa

Figure 4.6 **Albinism in Humans**

The woman in this photograph has albinism and lacks pigment in her hair, eyes, and skin.

Rick Guidotti/Positive Images

law of independent assortment random distribution of homologues into cells during meiosis

pigment a colored substance

pedigree a diagram showing inheritance of familial traits

Figure 4.7 **Segregation of the Allele for Albinism in Humans** In each heterozygous parent, about half the gametes carry the dominant normal allele (*A*), and half carry the recessive allele (*a*) for albinism. The combination of alleles at fertilization means there is a 75% chance that a child will have normal pigmentation, and a 25% chance the child will be an albino.

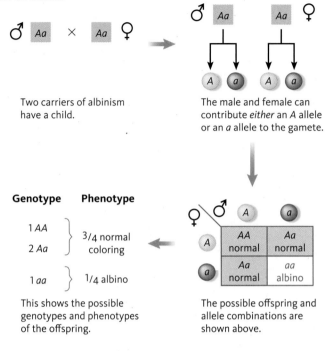

Two carriers of albinism have a child.

The male and female can contribute *either* an *A* allele or an *a* allele to the gamete.

This shows the possible genotypes and phenotypes of the offspring.

The possible offspring and allele combinations are shown above.

As Mendel experimented with plants, he did many different crosses over a short period of time and examined the phenotypes of hundreds of offspring. Obviously, we cannot do this with humans. However, if the parents in this family have enough children (say an improbable number such as 30–50), we may see something close to the predicted ratio of three pigmented to one albino that Mendel's laws predict.

Important: This means that if both parents are heterozygotes (*Aa*), each child has a 75% chance of having normal pigmentation and a 25% chance of being albino (a 3:1 ratio).

How is inheritance of traits studied in human genetics?

To study the inheritance of human traits, genetic counselors take a detailed family history. This history is used to construct a diagram called a **pedigree**, which shows all family members and identifies those affected with the genetic disorder in question as well as unaffected family members. In addition to interviews, counselors can also use letters, diaries, photographs, family records, and medical records to assemble the family history and construct the pedigree.

Pedigrees are made using a standard set of symbols and lines. Some of these are shown in Figure 4.8.

The pedigree in Figure 4.9 shows how these symbols and lines are used to make a sample pedigree covering two generations of a family.

To simplify the identification of generations and individuals in a pedigree, a numbering system uses Roman

Figure 4.8 **Symbols Used in Pedigree Construction** Some symbols used in pedigree construction are shown below:

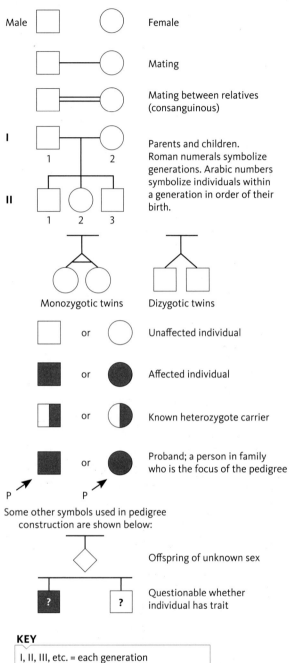

Some other symbols used in pedigree construction are shown below:

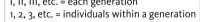

Offspring of unknown sex

Questionable whether individual has trait

KEY

I, II, III, etc. = each generation
1, 2, 3, etc. = individuals within a generation

Figure 4.9 **Sample Pedigree 1** This sample pedigree shows two generations of a family. The vertical line connects the parents (above) to the offspring (below). The children are connected to a horizontal sibship line in order of their birth. In this case, the first two children are girls, and the last two are boys.

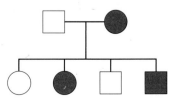

Figure 4.10 Numbering in Pedigrees In pedigrees, Roman numerals (I, II, III, etc.) are used to indicate generations. Within each generation, Arabic numbers (1, 2, 3, etc.) are used to identify individuals. Thus, everyone in the pedigree has a unique numerical identifier (e.g., II-2, III-1).

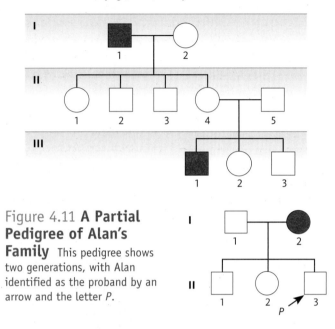

Figure 4.11 A Partial Pedigree of Alan's Family This pedigree shows two generations, with Alan identified as the proband by an arrow and the letter *P*.

numerals for generations and Arabic numbers for individuals (Figure 4.10).

Pedigrees are often constructed after a family member who has a genetic disorder has been identified, or when a family member requests counseling. In Alan's case, he requested that Ms. Wright construct the pedigree. When the family pedigree is constructed, Alan will be identified as the **proband**, the person who is the focus of the pedigree (Figure 4.11). He will be indicated on the pedigree by an arrow and the letter *P*. The proband may not be affected with the trait but may be interested in knowing his or her risk for carrying the trait and/or passing it on to future generations.

Collection of pedigree information is not always easy because knowledge about distant relatives is often incomplete, details about medical conditions can be forgotten over time, and family members are sometimes reluctant to discuss relatives who had abnormalities.

Look at the two pedigrees below. One shows a simple two-generation pedigree of Alan and his parents (made by Ms. Wright, the genetic counselor); the other is a family with albinism. What differences do you see? As we will see, when more information is available, a more complete pedigree can be prepared so that more analysis will be possible.

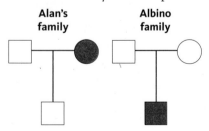

4.2 Essentials

- **Mendel discovered that only one member of a gene pair is present in a sperm or egg (law of segregation).**
- **Members of a gene pair separate into sperm and eggs independently of other gene pairs (law of independent assortment).**
- **Mendel's findings apply to human traits.**

4.3 What Can We Learn from Examining Human Pedigrees?

Analysis of a pedigree can determine whether a trait present in a family has a dominant or recessive inheritance pattern. Pedigrees can also be used to predict genetic risk in several situations, including the following:

- the risk to a fetus
- the risk of having an adult-onset disorder
- the risks for affected future offspring

Based on Mendel's conclusions, it is known that traits in humans can be inherited in several ways, each with a distinctive pattern of inheritance. To appreciate the differences in inheritance and the resulting differences in genetic risk, we will first discuss three possible patterns of inheritance: **autosomal recessive**, **autosomal dominant**, and **X-linked recessive**. Autosomal traits are carried on chromosomes 1–22, and X-linked traits are carried on the X chromosome.

How are autosomal recessive traits inherited? A pedigree showing affected and unaffected members over several generations can determine whether a trait has a recessive pattern of inheritance. The pedigrees of recessive traits, such as albinism, have several distinguishing characteristics:

- Unaffected parents can have affected children:

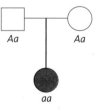

- All children of affected parents are affected:

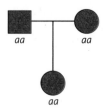

- For two heterozygotes (carriers), the risk of having an unaffected child is 75% and the risk of having an affected child is 25%.

- Male and female children are affected in roughly equal numbers. In addition, both the male and female parent must transmit the mutant allele for a child to be affected.

- In some cases, unaffected (heterozygous) parents of an affected (homozygous) individual may be related to each other. This relationship is called **consanguinity** and, in pedigrees, is indicated by a double line between the parents.

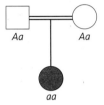

A pedigree illustrating the inheritance of an autosomal recessive trait is shown in Figure 4.12, and Table 4.1 lists several autosomal recessive genetic disorders.

Some recessive traits have relatively mild phenotypic effects that show up in traits such as hair color and eye color. Others can be life-threatening or even fatal disorders, including cystic fibrosis (CF) and sickle-cell anemia (discussed shortly). Many genetic disorders, including those listed below, occur when an important protein in the body is produced incorrectly or is missing altogether. In albinism, for example, the affected protein is involved in making the pigment melanin.

1. **Albinism** (*A* = normal coloring; *a* = albinism): Albinism is a group of genetic conditions associated with a lack of pigment (melanin) in the skin, hair, and/or eyes. In normal individuals, melanin is found in pigment granules inside

cells (called melanocytes) present in the skin, hair, and eyes. In people with albinism, melanocytes are present but contain no pigment because the individual cannot make melanin.

One common type of albinism is called *oculocutaneous albinism type I* (OCA1). Affected individuals have no pigment in their hair, skin, and eyes; they also have reduced eyesight, sensitivity to bright lights, and involuntary eye movements. This disorder is caused by defects in the way melanin is synthesized.

2. **Cystic fibrosis** (*C* = normal condition; *c* = cystic fibrosis): Cystic fibrosis is another recessive trait. CF affects the glands that produce mucus and digestive enzymes; it has far-reaching effects because these glands perform a number of vital functions. CF causes the production of thick mucus in the lungs that blocks airways (Figure 4.13, page 80); most people with cystic fibrosis develop obstructive

proband the individual who is the focus of a pedigree

autosomal recessive inheritance in which a trait does not appear in every generation

autosomal dominant inheritance in which a trait usually appears in every generation

X-linked recessive a recessive trait carried on the X chromosome

consanguinity a close family relationship between two individuals

albinism a genetic condition with a lack of the pigment melanin

cystic fibrosis a recessively inherited genetic disorder with abnormal mucus production

Figure 4.12 **Autosomal Recessive Pedigree** This pedigree shows how an autosomal recessive trait is transmitted through four generations of a family.

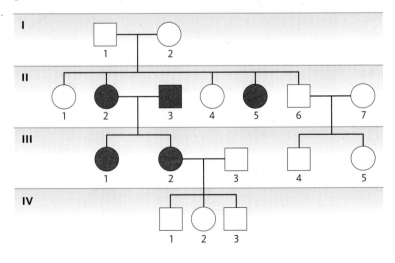

Table 4.1 **Examples of Autosomal Recessive Genetic Disorders**

Trait	Phenotype
Cystic fibrosis (CF)	Mucus production that blocks ducts of certain glands and lung passages
Phenylketonuria (PKU)	Excess accumulation of phenylalanine in blood, mental retardation
Sickle-cell anemia (SCA)	Abnormal hemoglobin, blood vessel blockage
Xeroderma pigmentosum (XP)	Lack of DNA repair enzymes, sensitivity to light, skin cancer
Tay-Sachs disease (TSD)	Improper metabolism affecting cells of the nervous system, death in childhood

Figure 4.13 **The Lungs of an Individual with Cystic Fibrosis** The air passages are clogged with mucus, making it difficult to breathe.

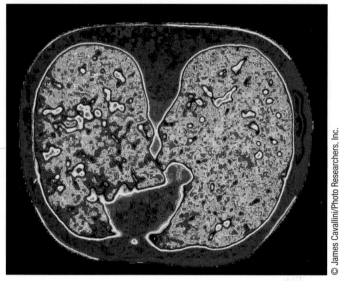

© James Cavallini/Photo Researchers, Inc.

Figure 4.14 **A Map of Geographic Regions Affected by Sickle-Cell Anemia** The percentage of the population carrying the sickle cell allele is shown by different colors. In some regions, more than 14% of the population are heterozygotes.

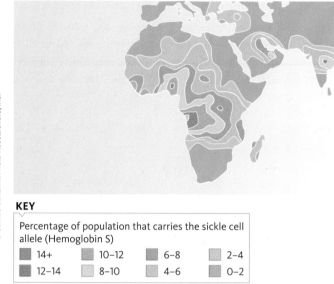

KEY

Percentage of population that carries the sickle cell allele (Hemoglobin S)

■ 14+	■ 10–12	■ 6–8	■ 2–4
■ 12–14	■ 8–10	■ 4–6	■ 0–2

lung diseases and infections that lead to premature death. Like many other genetic disorders, the frequency of cystic fibrosis is different in various populations. Among individuals of Northern European descent, about 1 in every 2000 children is affected. In other populations, the frequency is much lower: about 1 in 17,000 African American births and about 1 in 31,000 Asian American births.

The CF gene and its protein product, called cystic fibrosis transmembrane conductance regulator (**CFTR**), have been identified and studied in detail and used to develop new methods of treatment, including gene therapy, a procedure that will be discussed in Chapter 10.

3. **Sickle-cell anemia** (*S* = normal red blood cells; *s* = sickle cell): People with ancestors from parts of West Africa, the lowlands around the Mediterranean Sea, or some regions of the Indian subcontinent (Figure 4.14) have a high frequency of a genetic disorder called sickle-cell anemia (SCA). SCA is inherited as a recessive disorder that causes production of abnormal hemoglobin, a protein found in red blood cells. Hemoglobin transports oxygen from the lungs to the tissues of the body. In SCA, abnormal hemoglobin molecules stick together and form fibers that cause red blood cells to become crescent- or sickle-shaped (Figure 4.15).

CFTR cystic fibrosis transmembrane conductance regulator, the protein encoded by the cystic fibrosis gene

sickle-cell anemia a genetic disorder with altered red blood cell shape

Figure 4.15 **Normal and Sickled Red Blood Cells** A normal red blood cell (a) and a sickled red blood cell (b). Sickled cells often cannot pass through capillaries, and as they accumulate, they slow or stop blood flow to cells.

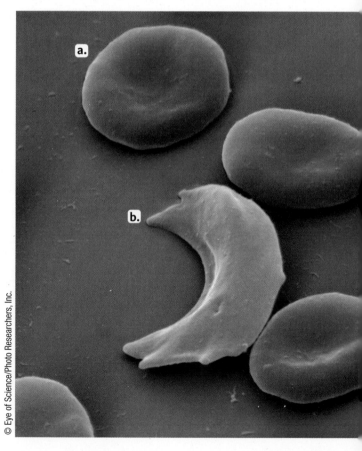

© Eye of Science/Photo Researchers, Inc.

These deformed cells are fragile and break open as they circulate through the body. New red blood cells are not produced fast enough to replace those that are lost, causing anemia. This also reduces the oxygen-carrying capacity of the blood.

People with sickle-cell anemia tire easily and can develop heart failure caused by an increased load on the circulatory system. The deformed blood cells clog small blood vessels and capillaries, further reducing oxygen transport. As oxygen levels fall in the body, more red blood cells become deformed, causing intense pain as blood vessels are blocked. In some affected areas, ulcers and sores appear on the skin. Blockage of blood vessels in the brain can also cause strokes and paralysis.

Sickle-cell anemia may be a serious disease, but the mutant allele protects those who carry it against malaria, an infectious disease that kills millions each year. The relationship between sickle-cell anemia and malaria will be explored in Chapter 15.

How are autosomal dominant traits inherited?
In disorders inherited in a dominant pattern, anyone who carries one copy of a mutant allele is a heterozygote (*Aa*) and, therefore, has the condition. Only in rare cases are dominant genetic disorders present in a homozygous condition (*AA*). Because it results in a double dose of the incorrect protein or none of the protein being produced, being homozygous for the mutant alleles is often fatal early in life. Unaffected individuals, on the other hand, carry two recessive alleles (*aa*) and have a normal phenotype.

Pedigrees of dominantly inherited traits have several distinctive characteristics:

- Unless a new mutation is involved, every affected individual should have at least one affected parent:

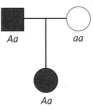

- Because most affected individuals are heterozygotes (*Aa*) and usually have a homozygous recessive (unaffected) partner (*aa*), each child has a 50% chance of being affected:

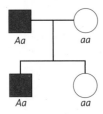

- The numbers of affected males and females are roughly equal.

- Because most affected individuals are heterozygous, two affected individuals may have unaffected children:

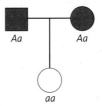

A pedigree showing the inheritance of an autosomal dominant trait is depicted in Figure 4.16.

Table 4.2 lists several autosomal dominant genetic disorders.

Figure 4.16 **Autosomal Dominant Pedigree** This pedigree shows four generations of a family with an autosomal dominant trait. Symbols with diagonal slashes represent deceased family members.

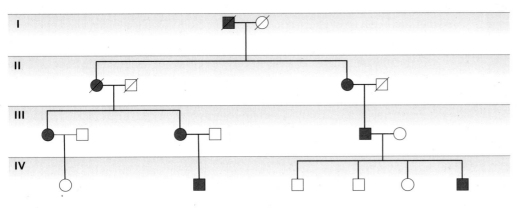

Figure 4.17 Tumors Associated with Neurofibromatosis (NF-1)

This woman has neurofibromatosis, a genetic disorder inherited as an autosomal dominant trait. In this case, tumors originating in cells of the nervous system have grown on her skin.

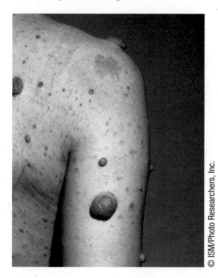

© ISM/Photo Researchers, Inc.

Figure 4.18 Effects of Huntington Disease

This woman has Huntington disease and uses a wheelchair because she has difficulty controlling movement of her limbs.

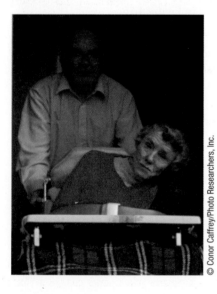

© Conor Caffrey/Photo Researchers, Inc.

neurofibromatosis an autosomal dominant condition with benign tumors

Huntington disease (HD) an autosomal dominant neurodegenerative disease

huntingtin the protein that is defective in Huntington disease

Table 4.2 Examples of Autosomal Dominant Genetic Disorders

Trait	Phenotype
Achondroplasia	Dwarfism associated with abnormalities in growth
Brachydactyly	Malformed hands with shortened fingers
Familial hypercholesterolemia	Elevated cholesterol levels, cardiovascular disease
Marfan syndrome	Connective tissue disorder, possible aortic aneurysm
Porphyria	Inability to metabolize porphyrin, episodes of mental disturbance, such as hallucinations or paranoia

Genetic conditions such as those listed below are inherited as dominant traits.

1. **Neurofibromatosis** (*N* = Neurofibromatosis 1; *n* = normal): A common autosomal dominant trait, affecting from 1 in 2500 to 1 in 3300 people, is type I neurofibromatosis (NF). The mutant NF allele can produce several different phenotypes. Some affected individuals (*Nn*) have only pigmented spots on their skin, called *café-au-lait spots*. Others have noncancerous tumors in the nervous system (Figure 4.17). In some cases, these tumors can be large and press on nerves, causing blindness or paralysis in body parts. In a small number of cases, tumors can cause deformities of the face or other body parts.

   Although NF is an autosomal dominant condition, it often appears in the children of unaffected parents. This happens because the gene for neurofibromatosis, *NF1*, mutates spontaneously from the normal to the mutant allele. The *NF1* gene has a very high mutation rate. When this happens, one parent's normal allele is changed, causing NF in the child. The causes of the high mutation rate in this gene are under study, but it is probably related to the large size of this gene (see Chapter 6 for a discussion of gene size and mutation rate).

2. **Huntington disease** (*H* = Huntington disease; *h* = normal): Huntington disease (HD) causes destruction of cells in certain areas of the brain. Symptoms begin slowly, usually between ages 30 and 50. Because HD occurs in midlife, affected individuals like Alan's mother may have already had children, who in turn may inherit this disorder. In the United States, HD affects about 1 in 10,000 people (Figure 4.18).

   Early signs of HD include irritability or apathy. Later, uncontrollable rapid arm and leg movements and a lack of coordination become more frequent. The disorder becomes progressively worse, and individuals with HD develop slurred speech and have difficulty eating or swallowing. Soon they are mentally and physically unable to care for themselves. There is no treatment for HD, and within 10 to 15 years after symptoms begin most affected individuals die from complications such as pneumonia, falls, or choking.

   As the disease progresses, cells in certain regions of the brain die because large amounts of a defective form of the protein called **huntingtin** accumulate in cells (Figure 4.19). The mutant huntingtin protein is present in all cells of the brain, but a second protein, found only in certain regions of the brain, helps make the mutant huntingtin protein lethal, leading to progressive cell death in these cells. More research is needed before drugs can be developed to treat this disorder.

   Because HD is inherited as a dominant trait, only one copy of the mutant allele is needed to cause this disorder (*H* is the Huntington allele; *h* is the normal allele). Most affected individuals are heterozygous (*Hh*), and each of their children has a 50% chance of having the disease.

Figure 4.19 Brain Cells in Normal and Huntington Disease Brains
Microscopic photographs of human brain cells. (a) Cells from a normal brain, showing dense clusters of blue-stained nuclei. (b) Cells from the brain of a person with Huntington disease. Notice there are fewer cells, as indicated by the smaller number of blue-stained nuclei. Cells in this brain have died because of the accumulation of the defective protein huntingtin.

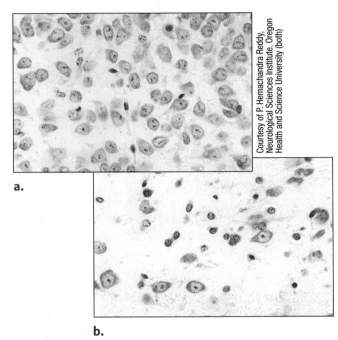

Courtesy of P. Hemachandra Reddy, Neurological Sciences Institute, Oregon Health and Science University (both)

a.

b.

Pedigree analysis and adult-onset disorders
Many of the genetic disorders we've discussed in this chapter have phenotypes expressed at or shortly after birth. However, some disorders, such as Huntington disease, develop only later in life and are called adult-onset disorders. In HD, symptoms appear between the ages of 30 and 50 years. Affected heterozygous individuals may already have children, who will each have a 50/50 chance of inheriting the condition.

Other dominantly inherited genetic disorders also develop later in adult life. One of these is adult polycystic kidney disease (ADPKD), a very common disease found in 1 in 600 to 1 in 1000 individuals. ADPKD is inherited as an autosomal dominant trait like HD, and symptoms first appear in midlife. Cysts begin forming in the kidney at this time, but at first there are very few symptoms, often just headaches and elevated blood pressure. The growth of cysts leads to kidney failure, and death may occur by age 60.

Adult-onset disorders can present a problem in pedigree analysis. At the time a pedigree is constructed and analyzed, some family members that carry the mutant allele may be too young to show any symptoms. If that is the case, they are placed on the pedigree as normal individuals. However, if they do not have symptoms but do carry the mutant allele, they will develop symptoms later in life. In order to calculate the risk to their children and other family members, genetic tests may be needed.

CASE A: QUESTIONS

Now that we understand how Huntington disease and some other disorders are inherited, let's look at some of the issues raised in the case of Alan and the rest of the Franklin family.

Catchlight Visual Services/Alamy

A family discussion about a genetic disorder.

1. Now that you know about Huntington disease, is Alan at a high enough risk to be tested?

2. If Alan is tested, must the genetic counselor or the doctor tell his siblings the results?

3. Even though Alan is 17, should he be able to have the test without his father's consent? Give three reasons pro or con.

4. What might happen if Alan's father says that Alan cannot be tested?

5. Now that you know when the symptoms of HD first appear, would the results of a test affect your decision whether Alan should be admitted to the Air Force Academy? Why or why not?

4.3 Essentials

- **Pedigrees are used to study family history and the distinctive patterns of inheritance seen in genetic disorders.**
- **From a pedigree, one can often determine how a trait is inherited.**
- **In recessive traits such as albinism, two unaffected parents may have a child with the condition. Two copies of the mutant allele are required for the condition to be expressed.**
- **In dominant traits, a heterozygote (*Aa*), who carries one copy of a mutant dominant allele, has the condition.**

4.4 How Are X-Linked Recessive Traits Inherited?

Females have two X chromosomes; therefore, they can be heterozygous (*Aa*) or homozygous (*aa*) for any gene carried on the X chromosome. Males, on the other hand, carry only one copy of the X chromosome and also carry a Y chromosome. The Y chromosome does not contain the same genes as the X chromosome and carries very few genes of its own. Genes on the X chromosome are called **X-linked**, and genes on the Y chromosome are called **Y-linked**.

X-linked genes carried on the X chromosome

Y-linked genes carried on the Y chromosome

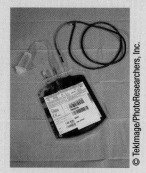
factor VIII a blood-clotting protein that is abnormal in hemophilia

hemizygous genes on the X chromosomes in males that have no alleles on the Y chromosome

muscular dystrophy a degenerative disease of muscles carried on the X chromosome

dystrophin the protein that is abnormal in muscular dystrophy

hemophilia a genetic disorder of blood clotting

Because males carry one X chromosome and one Y chromosome, any mutation in an X chromosome gene that causes a recessive genetic disorder will be expressed because there is no normal, dominant allele of the gene present on the Y. In other words, males cannot be heterozygous for any gene on the X chromosome. Because X-linked recessive genes are always expressed in males, they are affected by X-linked recessive genetic disorders far more often than females.

XX^*: females   Genotype: A^*a

X^*Y: males   Genotype: $^*aY$

X^*: recessive mutant allele

Because males cannot be homozygous or heterozygous for genes on the X chromosome, they are said to be **hemizygous** for all genes on the X chromosome.

Males give an X chromosome to their daughters and a Y chromosome to their sons. They never give an X to their sons. Females give an X chromosome to each of their children (review the inheritance of X and Y chromosomes in Chapter 1).

Because of the way the X chromosome is passed from parents to children, a distinctive pattern of inheritance for X-linked recessive traits shows up in pedigrees:

- Hemizygous (X^*Y) males and homozygous (X^*X^*) females are affected.

- These conditions are much more common in males than in females. In the case of rare traits, males are almost exclusively affected.

- Affected males who get the mutant allele from their mothers transmit it to all of their daughters but not to any sons:

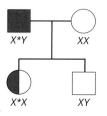

- Daughters of affected males are heterozygous, or carriers, and therefore unaffected. However, sons of heterozygous females each have a 50% chance of receiving the recessive gene and being affected:

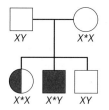

Geneticists have identified more than 850 X-linked recessive traits to date, including color blindness, muscular dystrophy, and hemophilia. Table 4.3 lists several X-linked recessive genetic disorders. The pedigree in Figure 4.20 shows the inheritance of an X-linked recessive disorder.

Genetic conditions such as the ones listed below are inherited as X-linked traits.

1. **Muscular dystrophy** (X^M = normal; X^m = muscular dystrophy): The most common form of muscular dystrophy is Duchenne muscular dystrophy (DMD), which affects about 1 in 3500 males. Infant boys with DMD appear healthy at birth but begin to develop symptoms between ages 1 and 6. Progressive muscle weakness is one of the first signs of DMD. The disease progresses rapidly, and by age 12 affected individuals usually must use a wheelchair because of muscle degeneration (Figure 4.21). Death usually occurs by age 20 because of respiratory infection or cardiac failure.

The DMD gene is located near one end of the X chromosome and encodes a protein called **dystrophin**. Dystrophin works inside the muscle cell to prevent the plasma membrane from breaking during muscle contraction (Figure 4.22a, b). When dystrophin is absent or defective, these cells are torn apart and eventually die.

There are two forms of X-linked muscular dystrophy. In DMD, which is the more serious form, no detectable amounts of dystrophin are present in muscle tissue. However, in the second, less serious form, called *Becker muscular dystrophy (BMD)*, a shortened, partially functional form of dystrophin is made. As a result, boys with BMD develop symptoms at a later age, have milder symptoms, and live longer than those with DMD. These two diseases are caused by different mutations in the same gene.

Table 4.3 **Examples of X-Linked Recessive Genetic Disorders**

Trait	Phenotype
Red-green colorblindness	Inability to see green and red
Hemophilia A	Inability to form blood clots
Lesch-Nyhan syndrome	Lack of HGPRT protein, mental retardation, self-mutilation
Muscular dystrophy	Duchenne-type, progressive condition with muscle wasting

2. **Hemophilia** (X^H = normal; X^h = hemophilia): One X-linked genetic disorder associated with lack of blood clotting is a recessively inherited disorder known as *hemophilia A*. In hemophilia A, a clotting factor called **factor VIII** is absent or present in reduced amounts. This genetic disorder causes a failure to form blood clots. As a result, affected individuals are in danger of bleeding to death from minor cuts or from hemorrhages caused by a bruise. They often require hospitalization to treat bleeding.

Figure 4.20 **X-Linked Recessive Pedigree** X-linked recessive traits are passed from heterozygous mothers (II-6 and III-5) to their sons, who each have a 50% chance of being affected. Affected males (I-1) cannot pass the trait to their sons, because they give them only a Y chromosome.

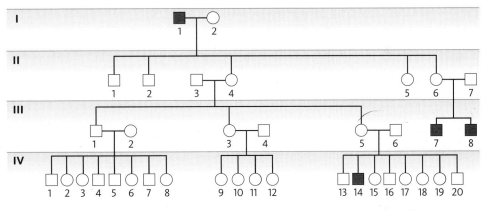

Figure 4.21 **Duchenne Muscular Dystrophy (DMD)** The boy in the wheelchair has DMD and is unable to walk because of muscle cell death in his legs.

Figure 4.22 **Distribution of Dystrophin** The location of dystrophin, the protein coded by the DMD gene in muscle cells. (a) In normal muscle cells, all the dystrophin is located in the cell membrane, and the cytoplasm looks dark. (b) In muscle cells of a boy with DMD, there is no dystrophin in the cell membranes (so the membranes are invisible), and defective copies of the dystrophin molecules accumulate in the cytoplasm, which stains a light blue. When dystrophin is not present in the cell membrane, muscle contractions eventually tear the membrane, and the cell dies.

a.   **b.**

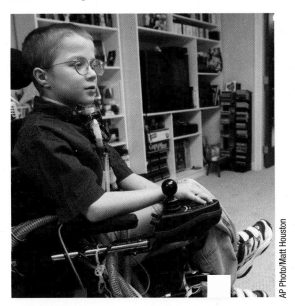

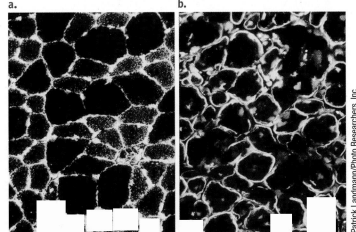

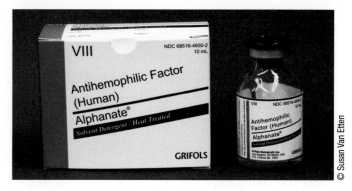

© Susan Van Etten

Hemophilia A is the most common form of X-linked hemophilia. Females are affected only if they are homozygous for this recessive gene ($X^h X^h$)*. Very few females have hemophilia A because it can occur only when a heterozygous female and a male with hemophilia have a girl. Even then, the chance she will have hemophilia A is only 50%. Until recently, very few males with hemophilia survived to reproductive age, a fact that further lowers the chances that a female will have hemophilia.

Treatment for hemophilia formerly involved periodic injections of concentrated clotting factor made from blood serum collected from a large number of donors. In the early to mid-1980s, a substantial number of clotting factor preparations were contaminated with HIV, the virus associated with AIDS. As a result, about half of all those (males) with hemophilia became infected with HIV. Many developed AIDS and died as a result of their treatment with contaminated blood products.

Today, recombinant DNA technology is used to make clotting factors that are free from contamination with viruses and other disease agents (Figure 4.23).

4.4 Essentials
- **In X-linked traits such as hemophilia, the mutant allele is carried on the X chromosome, and males are more often affected than females.**
- **Many traits are inherited through dominant, recessive, and X-linked genetics.**

---

*X^h = hemophilia gene.

Figure 4.24 **A Pedigree of Mitochondrial Disease** In mitochondrial inheritance, an affected mother passes the trait to all her children, who are all affected. Males cannot transmit the trait to any of their children.

Do mitochondrial genes follow Mendel's rules of inheritance?
Recall from "Biology Basics: Cells and Cell Structure" that mitochondria, organelles found in the cytoplasm of cells, are involved in the production of energy-containing molecules that are used to power many cell functions. Because mitochondria are descendants of free-living bacteria, they carry a set of their own genes on a circular DNA molecule. Because mitochondria and their genes do not segregate during meiosis, they do not follow Mendelian rules of inheritance.

Mitochondria are transmitted from mothers to all their children through the cytoplasm of the egg (sperm do not contribute any mitochondria to the fertilized egg). As a result, genetic disorders caused by mutations in mitochondrial genes have a distinct pattern of inheritance:

- Mitochondrial disorders are transmitted by affected mothers to all their children, who are all affected with the disorder.
- An affected female will transmit the disorder to all her children, but an affected male cannot transmit the disorder to any of his children.

A pedigree showing the inheritance of a mitochondrial trait is depicted in Figure 4.24. Mutations in mitochondrial genes usually reduce the amount of energy produced, and organ systems with high energy demand are most affected, including the central nervous system, muscles, and the liver. In addition, some genetic disorders associated with mitochondria are listed in Table 4.4.

Table 4.4 **Some Genetic Disorders of Mitochondria**

Trait	Phenotype
Kearns-Sayre syndrome	Short stature; retinal degeneration
Leber optic atrophy (LHON)	Loss of vision in central field; adult onset
Leigh syndrome	Degradation of motor skills
MELAS syndrome	Episodes of vomiting, seizures, stroke-like episodes
Progressive external ophthalmoplegia (PEO)	Paralysis of eye muscles

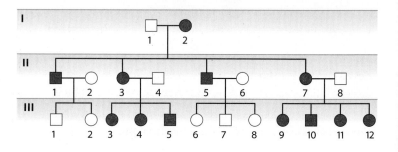

4.5 What Are the Legal and Ethical Issues Associated with Dominant and Recessive Genetic Conditions?

One of the questions a person might ask after reading Alan's family's dilemma is, if the family takes the Huntington test, who can see the test results? Alan wants to have a test to find out his genetic status, but could his future employers or the U.S. Air Force know his status?

Legally, medical records and test results are confidential, and doctor/patient confidentiality is one of the legal principles that allow patients to trust doctors with information. Because of this confidentiality, patients feel safe telling their doctors private medical concerns. A law was passed in 1996 that mandated such privacy rights. This law, called the Health Insurance Portability and Accountability Act, or HIPAA, created national standards to protect individuals' medical records and other personal health information.

HIPAA:

- gives patients the ability to control their health information
- allows only certain uses of health records
- protects the privacy of health information
- establishes civil and criminal penalties if patients' privacy rights are violated
- allows for certain exceptions—for example, to protect public health

If Alan discusses his test results with his physician, the family's insurance company might be able to see the results. When applying for health insurance, patients are asked to sign a waiver allowing the insurance company to see their medical records. In our case, Alan's family must have signed such a waiver.

The reasons that insurance companies and employers can access medical records vary, but some are:

- to keep statistics on certain conditions
- to get information for medical studies that will help the medical population as a whole
- to find information that will protect employees from illness
- to find information that might lower health care costs
- to decide whether to employ or cover certain individuals
- to decide the cost of insurance for an individual or company

On some levels, this seems to be the opposite of confidentiality, but insurance companies need to see patient's medical conditions before, during, and after they work with them.

Table 4.5 addresses some of the ethical and legal questions regarding privacy of medical information.

CASE B:
The Franklins Find Out More

Alan is now very concerned about his future. Because the genetic counselor has told him that there is a test for HD, he wonders whether he will need written permission from his father to be tested.

A session with a genetic counselor.

Alan decides to talk to his older siblings. He talks to Mary first. She says she wants to discuss testing with her fiancé. Mary's fiancé, Bob, thinks the test is a good idea but suggests that if the results are positive, she sign a prenuptial agreement stating that they will not have any children.

Alan's older brother, John, says he would take the test but wants his wife's fetus tested too. He believes that the baby should be aborted if it carries the Huntington gene—he still remembers his mom. His wife, Emily, agrees.

The genetic counselor then completes the family's pedigree:

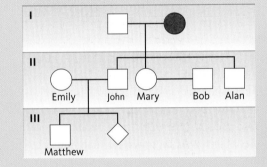

Alan wonders if he should even have the HD test.

1. **Do you think Alan should have the test? Give three reasons for your answer.**

2. **What might happen if Alan tests positive? List a few possibilities.**

3. **If Alan tests positive, will this affect John's or Mary's chances of having the mutant allele? Why or why not?**

4. **Now that you know how serious HD is, how do you think this might affect employment for members of this family?**

5. **Remember Martha Lawrence's case from Chapter 3? Her doctor had questions about whether she should tell Martha the results of her amniocentesis. Here the test results affect not only Alan but also his entire family. If the test results were given to Alan but not the rest of the family, what might happen? List three issues that might arise from this situation.**

Table 4.5 Legal and Ethical Questions about Medical Information Privacy

Question	How Are These Questions Decided?	Related Case or Legal Issue
Is a minor protected under the HIPAA laws?	State laws set guidelines for the age that allows coverage under HIPAA. However, teens in many states are allowed privacy.	In many states, a pregnant woman, no matter what her age, is protected under HIPAA and can make her own medical decisions.
Can employees sue if their medical records are released without their permission?	Yes, lawsuits were common before the passage of the HIPAA laws, but now there are fewer. Often, these questions are settled by arbitration with the employer and a mediator.	Some states limit the amount of money awarded in such lawsuits.
Can an employer mandate that employees or potential employees be tested for a condition such as Huntington disease?	If the condition is tied directly to one's employment, a test could be mandated. In the case of Huntington disease, this would probably not be applicable.	In *EEOC v. Burlington N. Santa Fe Railway Co.*, discussed in Chapter 5, the company tried to test its employees for an inherited form of *carpal tunnel syndrome*. Read "Spotlight on Law," Chapter 5.
Can an insurance company drop your insurance if you have a positive genetic test for a condition such as Huntington disease?	The Kennedy/Castlebaum Law specifically states that this could not happen. In addition, in 2008, the Genetic Information Nondiscrimination Act (GINA) was passed by Congress. This bill will prevent discrimination.	On May 21, 2008, President George W. Bush signed GINA into law. This should result in greater privacy in genetic testing.

Spotlight on Law

Tarasoff v. Regents of the University of California
551 P.2d 334 (Cal. 1976)

The question of whether medical records are confidential is a serious issue in the legal profession. We sign privacy agreements based on HIPAA laws every time we go to the doctor. The following case directly applies to this question.

Dr. Lawrence Moore, a psychologist employed by the University of California hospitals, was treating Prosenjit Poddar. During a session in 1969, Poddar confided that he intended to kill his ex-girlfriend, Tatiana Tarasoff, when she returned from her summer vacation.

Dr. Moore wasn't sure that Poddar would actually kill Tatiana, but he wanted to be safe, so he notified the campus police. The police detained Poddar but released him because he seemed rational and denied that he was going to kill his ex-girlfriend. Soon after, he stopped his treatment with Dr. Moore. Three months later, Poddar killed Tatiana.

Tatiana's parents sued the University of California for malpractice, claiming that Dr. Moore had a "duty to warn" Tatiana and that it could have saved her life.

RESULTS

The University of California (defendant) won the case, with the court stating that no duty to warn exists; the Tarasoff family appealed the decision. The Supreme Court of California reversed the decision, stating that therapists have a duty to take actions to protect third parties from violent patients, even if doing so breaches confidentiality. The court said that Dr. Moore could not escape liability just because Tatiana was not his patient; he still should have warned her.

In addition, when a therapist determines that a patient presents a serious danger to another, he or she has an obligation to use reasonable care to protect the intended victim from danger. However, the therapist must have a special relationship with either the person who might cause harm or the potential victim before he or she can breach confidentiality. The potential victim must be identifiable, and the harm must be foreseeable and serious. The court stated, "Privilege ends where public peril begins."

LEGISLATION

Some states have enacted laws based on the *Tarasoff* decision that require disclosure when certain dangers exist. They sta

: the therapist or doctor has an obligation to use reason-
: care to protect a potential victim.

QUESTIONS

Do you agree with the California Supreme Court's decision? Why or why not?

The Tarasoffs were asking for monetary damages to compensate them for the loss of their daughter. Do you think this is a good argument? Why or why not?

Did Dr. Moore do enough to protect Tatiana?

If not, what more should he have done?

What could this mean for the future of medicine?

6. This decision has been cited in cases in which a physician was trying to decide whether to tell a spouse that his or her partner had HIV. Does *Tarasoff* apply in this situation?

7. Does the *Tarasoff* case apply in Alan's situation? How?

8. If you were the attorney for the Tarasoffs, would you want a jury trial or a judge-only trial? Why?

9. A woman has a genetic test and finds that she carries the gene for cystic fibrosis, a recessive trait. Should her physician tell the patient's fiancé? Why or why not?

10. How could the decisions in the *Tarasoff* case be used in the argument in the case in question 9?

Review Questions

1. When Mendel studied his pea plants, he knew nothing about genes or chromosomes. How did he figure out that these traits were passed on? Do physicians today use this technique? When?

2. What are dominant traits, and how do they differ from recessive traits?

3. The law of segregation describes which cellular process?

4. Define independent assortment.

5. The same rules for recessive traits apply to humans as well as to Mendel's pea plants. Does this make it simpler or harder for parents of an affected child to understand a genetic counselor's explanation?

6. Because offspring of a parent with a dominant trait have a 50% chance of inheriting the mutant allele, are dominant traits more or less common than recessive traits? Explain.

7. Choose four conditions discussed in this chapter and complete the following chart.

Condition	Is the Condition Dominant, Recessive, or X-Linked?	On What Chromosome Is the Gene?	What Group Tends to Have the Condition?	Is There a Test for the Condition?

8. Explain the meaning of hemizygous.

9. Why don't males with X-linked traits pass them on to their sons?

10. Why are so few females affected with hemophilia?

11. How does dystrophin work in the muscles?

12. What goes wrong with the dystrophin gene to cause Duchenne muscular dystrophy?

13. Explain the difference between Duchenne and Becker forms of muscular dystrophy.

14. In sex-linked traits, explain why more males are affected than females.

15. Explain why hemophiliacs were affected by HIV in the 1980s.

The Essential 10

1 Only one copy of each chromosome is present in the sperm or egg and is contributed to the offspring. (Section 4.1)

When the sperm and egg combine, the number of chromosomes is 46.

2 Members of a gene pair segregate into gametes independently of other gene pairs, so that gametes have different combinations of genes. (Section 4.1)

Only one member of each gene pair goes into an egg or sperm.

3 Gregor Mendel, a monk who lived in the 1800s, was the father of genetics. (Section 4.1)

Mendel studied traits of pea plants such as tall and short stems.

4 Mendel did his experiments with plants, but his important laws of genetics are still used today to study human genetics. (Section 4.1)

Human traits are inherited in the same way as traits in pea plants.

5 Pedigrees are used to study family history and the distinctive patterns of inheritance seen in certain genetic conditions. (Section 4.2)

Anyone can construct a pedigree of his or her family.

6 From a pedigree, one can determine how a trait is inherited. (Section 4.2)

Using a pedigree, counselors can determine a couple's risk of their children inheriting certain conditions.

7 In recessive traits such as albinism, two unaffected parents may have a child with the condition. (Section 4.2)

Two copies of the mutant allele are required for the condition to be expressed.

8 In dominant traits, a heterozygote (*Aa*), who carries one copy of a mutant dominant allele, has the condition. (Section 4.2)

If a person has a dominant allele, his or her offspring have a 50% chance of inheriting it.

9 In X-linked traits the mutant allele is carried on the X chromosome, and males are more often affected than females. (Section 4.3)

One trait that is inherited this way is hemophilia.

10 Many traits are inherited through dominant, recessive, and X-linked genetics. (Section 4.3)

These traits run in families and can be tracked by pedigrees.

Application Questions

1. Mendel is mentioned in this chapter, but other scientists were also important in the early study of genetics. Research two of them, and write a short paragraph about their contributions.

2. Research the problems a person with albinism may have throughout his or her life, and write a short report.

3. Look up the pedigrees of the royal houses of Europe, and draw one. Why did certain traits (hemophilia and others) show up so frequently?

4. Make your own family pedigree, and trace one trait through a few generations.

5. Research three inherited conditions not discussed in this chapter, and list their symptoms.

6. Summarize how geneticists are studying Duchenne muscular dystrophy in the Amish population.

7. When you look at "By the Numbers," you can see how many carriers of CF there are in certain populations. Can you use this information to lower the numbers of babies born with CF?

8. Now that you know how height is inherited in pea plants, do you think height is inherited the same way in humans? Explain your answer.

9. In the pedigree below, determine how the trait is inherited. Is it dominant, recessive, or X-linked recessive? How can you tell?

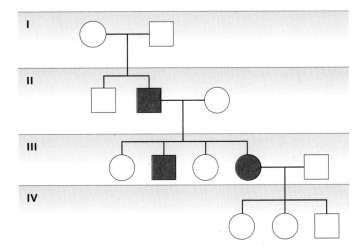

10. Draw the following pedigree: A man and a woman have three children in the following order: two sons and a daughter. The daughter marries and has monozygotic twins and later has a son who has cystic fibrosis.

11. The gene for neurofibromatosis, *NF1*, has a high rate of mutation. It is also a very large gene. Other genes with high mutation rates are also very large. What might explain this relationship between gene size and mutation rate?

Read the following before answering questions 12–15.

Carol Carr had three sons from her marriage with Hoyt Scott. Years after the children were born, Hoyt began showing signs of Huntington disease. Until his death in 1995, Carol devoted over two decades of her life caring for Hoyt. By the time of his death, two of their sons, Randy and Andy, had developed Huntington disease,

. . . you were Alan? (See Case A and Case B.)

. . . you were Mary? (See Case B.)

. . . you were on a committee deciding whether members of this family (Case A and Case B) could obtain health insurance?

. . . you were the genetic counselor advising Emily (Case B) about her unborn fetus?

. . . you were a member of a legislature voting on a bill that made it mandatory for all couples applying for a marriage license to be tested as carriers of cystic fibrosis? Why?

. . . you were asked to counsel a person whose genetic test was positive for HD?

and eventually both were placed in a nursing home. By June of 2002, both men were in the late stages of the disease when Carol shot both of them to death in their beds. She was arrested and charged with murder but pled guilty to two charges of assisting at a suicide. She was sentenced to 5 years in prison but was paroled after serving 21 months. As part of the plea agreement, Carol was prohibited from living with her third son, James, or being his care-taker. James also developed Huntington disease.

12. After having two children with Huntington disease, what were the chances that Carol and Hoyt would have a third child with this disease?

13. Visit the OMIM (Online Mendelian Inheritance in Man) site on the Web and search for Huntington disease. On that page, look up the average age of onset when the disease is inherited from the father or the mother. Are the symptoms more severe if HD is inherited from the father?

14. Draw up a list of reasons why euthanasia should be allowed in such cases, and a list of reasons why euthanasia should not be allowed in such cases.

15. Some states, including Oregon and Washington, allow assisted suicide for those suffering with certain fatal conditions. After researching the requirements for one of these states, determine if Randy and Andy would have qualified for assisted suicide for Huntington disease. People in the last stages of HD are often unable to communicate and express their wishes. Would such individuals qualify for assisted suicide? Under what circumstances would they qualify?

Online Resources

Preparing for an exam? Log on at www.cengagebrain.com for study tools to help you assess your understanding. If assigned by your instructor, the Case A and Spotlight on Law activities for this chapter, "A Family's Dilemma" and "*Tarasoff v. University of California*," will also be available.

Learn by Writing

In Chapter 1 we suggested that you start a blog with members of your class or others who are interested in human genetics. Now is a good time to revisit your blog and consider some of the issues surrounding testing for sickle-cell anemia. E-mail others you think might be interested and invite them to contribute.

Here are some ideas to address in your blog:

- Is a single gene defect such as sickle-cell anemia more serious than an extra or missing chromosome?

- Ms. Lawrence's case involves early testing. Alan's family might have benefited from these procedures. How?

- Should the law actually get involved in any of these personal decisions?

- Address any other question or comment that came to mind as you read this chapter.

- Does your state have a law based on *Tarasoff* that requires disclosure of dangers even if it involves a breach of confidentiality? If so, are any specific conditions spelled out?

In the Media

New York Times, June 8, 2008

Albinos, Long Shunned, Face Deadly Threat in Tanzania

Jeffrey Gettleman

Throughout sub-Saharan Africa albinos have been discriminated against for many years. Recently, however, in Tanzania, this has become much more deadly.

Nineteen albinos, adults and children, have been killed this year and their body parts sold.

One in 3000 people in Tanzania is albino, and witch doctors are selling the skin, bone, and internal organs to make into potions that are supposed to help people become rich.

In addition, some Africans believe that albinos have magical powers, and their hair has been sold on the market for high prices. Fishermen weave the hair into their fishnets in the belief that it will increase the catch.

To access this article and related video online, go to www.cengagebrain.com.

QUESTIONS

1. How might you try to protect albinos who are in fear for their lives?

2. What ideas do you have to educate the population and make them understand how albinism is inherited?

Inter Press Service News Agency, July 2, 2006

Striking Down the Taboos about Albinism

The Nigerian government plans to establish an agency to coordinate the affairs of its citizens with albinism, who number about 600,000. The people with albinism in this country are having similar problems as those in Namibia and other sub-Saharan countries.

Faced with many challenges, this agency for people with albinism is conducting a variety of initiatives in a bid to improve matters.

It will visit church leaders, tribal leaders, and councilors to appeal to them to help people with albinism with hats, creams, and sunglasses. Joseph Ndinomupya, president of the Namibia Albinism Association Trust, a private company, thinks this will work.

To access this article online, go to www.cengagebrain.com.

QUESTIONS

1. Is this type of action enough to change people's minds?

2. Does discrimination against people with albinism occur in the United States and other countries?

3. How does this story show progress in the protection of albinos?

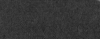

An Artist's Interpretation of the Core of a DNA Molecule

Robert Brocksmith/Photo Researchers

BIOLOGY BASICS 2

DNA and RNA

BB2.1 What Is the Evidence That DNA Carries Genetic Information?

In the early part of the 20th century, biologists identified chromosomes as the carriers of genetic information. Where did this idea come from? When observed in the microscope, the behavior of chromosomes as they were passed from cell to cell in meiosis paralleled the movement of genes from generation to generation as described by Mendel. Later, researchers identified the main chemical components of chromosomes: proteins and DNA (deoxyribonucleic acid). The question that remained was, which of these chemicals carries genetic information? Because it was known that proteins contained 20 different subunits in the form of amino acids, and DNA contained only four subunits (nucleotides), most scientists believed that only proteins were complex enough to carry genetic information.

Figure BB2.1 Colorized Electron Micrograph Showing Cells of *Streptococcus pneumoniae*

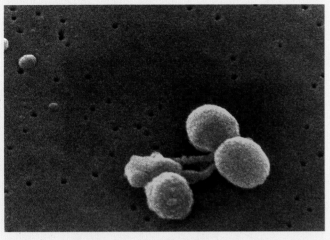

Dr. Richard Facklam/CDC Photo Library

As it turns out, it is actually the DNA in chromosomes that carries genetic information. The evidence for DNA as a carrier of genetic information came from an unexpected source: the study of an infectious disease. By the 1920s, it was known that one form of pneumonia was caused by a bacterium called *Streptococcus pneumoniae* (Figure BB2.1).

In the 1920s, Frederick Griffith studied two different strains of this bacterium. One strain (the S strain) caused pneumonia when injected into mice. Using a microscope, he saw that a capsule surrounded the cells of this strain. The other strain (the R strain) was not surrounded by a capsule and did not cause pneumonia when injected into mice. Griffith wondered if the capsule somehow caused the pneumonia. To test this idea, he devised a clever experiment. He killed S cells by heating them and mixed them with the living R cells. When he injected this mixture into the mice, they developed pneumonia and died (Figure BB2.2). The *Streptococcus* cells recovered from the dead mice had capsules surrounding them.

To explain how injection of a small number of dead cells with capsules and a small number of live cells without capsules could end up as a large number of cells with capsules that caused pneumonia, Griffith concluded that after injection into the mice, the living R cells acquired the ability to form a capsule like the S cells. To do this, he surmised that genetic information was somehow transferred from the dead S cells into the living R cells, allowing them to make a capsule and, therefore, cause pneumonia. He called this process **transformation**.

In the 1940s, researchers at Rockefeller Institute were working to develop a way to treat pneumonia; Oswald Avery, Colin MacLeod, and Maclyn McCarty removed and purified the main biomolecules (DNA, RNA, protein, lipids, and carbohydrates) from the heat-killed S cells. They tested these components one at a time to see which one could transfer genetic information to R cells that would allow them to make a capsule. After testing all the components, they discovered that only DNA was able to cause transformation. The result of their experiments clearly established that DNA is the carrier of genetic information and that it could be transferred from one bacterial cell to another.

The discovery that DNA carries genetic information was at first controversial because most scientists thought protein carried genetic information. In spite of this, the finding

Streptococcus pneumoniae bacterial species that causes pneumonia

transformation a heritable change caused by DNA that originates outside the cell

Figure BB2.2 **Griffith's Transformation Experiment** (a) Mice injected with live cells of the R strain do not get pneumonia and do not die. (b) Mice injected with live cells of the S strain get pneumonia and die. (c) Mice injected with heat-killed S cells do not get pneumonia and do not die. (d) Mice injected with live R cells *and* heat-killed S cells get pneumonia and die.

a. Mice injected with live cells of harmless strain R.

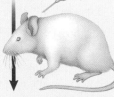

Mice do not die. No live R cells in their blood.

b. Mice injected with live cells of killer strain S.

Mice die. Live S cells in their blood.

c. Mice injected with heat-killed S cells.

Mice do not die. No live S cells in their blood.

d. Mice injected with live R cells *plus* heat-killed S cells.

Mice die. Live S cells in their blood.

Figure BB2.3 **James Watson and Francis Crick**
(a) Watson (on left) and Crick examine a model of DNA in 1953.
(b) Watson and Crick in 1993.

a.

A Barrington Brown/Photo Researchers

b.

Pierre Perrin/CORBIS

Figure BB2.4 **DNA Model** DNA is composed of two
nucleotide chains running in opposite directions and wound into a
helix. Inside the molecule, an A on one strand always pairs with a
T on the other strand, and a C on one strand always pairs with a G
on the other strand.

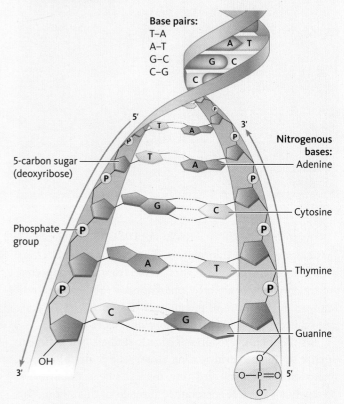

Base pairs:
T–A
A–T
G–C
C–G

5-carbon sugar
(deoxyribose)

Phosphate
group

Nitrogenous
bases:
Adenine

Cytosine

Thymine

Guanine

that DNA was responsible for transformation helped fuel efforts to understand the structure of this nucleic acid. From the mid-1940s through 1953, several laboratories made significant strides in unraveling the structure of DNA. These efforts resulted in the model of DNA structure proposed by James Watson and Francis Crick (Figure BB2.3) in 1953. The structure of DNA derived from their model is shown in Figure BB2.4.

BB2.2 What Are the Chemical Subunits of DNA?

When Watson and Crick began their work, it was already known that organisms contain two types of nucleic acids: **deoxyribonucleic acid (DNA)** and **ribonucleic acid (RNA)**. DNA is found in the nucleus, and RNA is found in both the nucleus and the cytoplasm.

Both are made up of subunits known as **nucleotides** (Figure BB2.5).

A single nucleotide has three components:

1. A nitrogen-containing base (there are two types of bases: **purines** and **pyrimidines**). The purine bases **adenine** (A) and **guanine** (G) are found in both RNA and DNA. The pyrimidine bases are **thymine** (T), found only in DNA; **uracil** (U), found only in RNA; and **cytosine** (C), found in both RNA and DNA. This means that RNA has four bases (A, G, U, C), and DNA has four bases (A, G, T, C).

2. A sugar (either **ribose**, found in RNA, or **deoxyribose**, found in DNA). The difference between the two sugars is a single oxygen atom. Ribose has an oxygen present as an –OH group, but the oxygen is missing in deoxyribose (*deoxy*, without oxygen), which has only an –H group.

3. A **phosphate group**. Phosphate groups are strongly acidic, which is why DNA and RNA are called acids.

The components of a nucleotide are assembled by chemically linking a base to a sugar, which in turn is linked to a

Figure BB2.5 Nucleotides The nucleotides of RNA (left) and DNA (right). All nucleotides contain a base (1), a sugar (2), and a phosphate group (3). DNA and RNA differ in the type of sugar, and in one base. Uracil is found only in RNA, and thymine is found only in DNA.

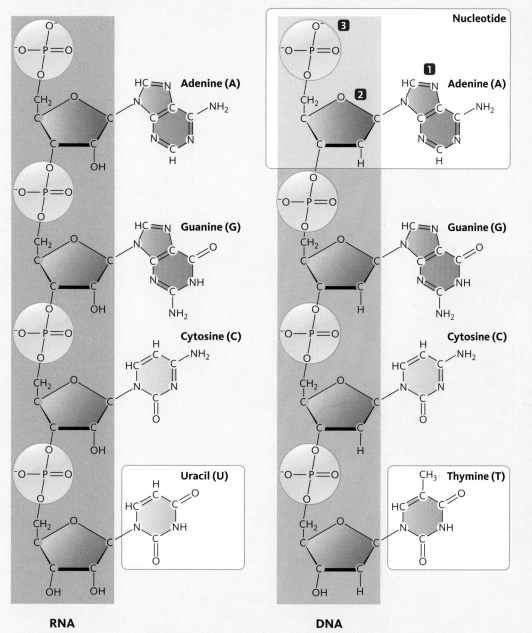

RNA

DNA

deoxyribonucleic acid (DNA) molecule consisting of anti-parallel strands of polynucleotides that is the primary carrier of genetic information

ribonucleic acid (RNA) a nucleic acid molecule that contains the pyrimidine uracil and the sugar ribose; the several forms of RNA function in gene expression

nucleotide basic building block of DNA and RNA; each nucleotide consists of a base, a phosphate, and a sugar

purine a class of double-ringed organic bases found in nucleic acids

pyrimidine a class of single-ringed organic bases found in nucleic acids

adenine a nitrogen-containing purine base found in nucleic acids

guanine a nitrogen-containing purine base found in nucleic acids

thymine a nitrogen-containing pyrimidine base found in nucleic acids

uracil a nitrogen-containing pyrimidine base found only in RNA

cytosine a nitrogen-containing pyrimidine base found in nucleic acids

ribose a pentose sugar found in RNA

deoxyribose a pentose sugar found in DNA

phosphate group a compound containing phosphorus chemically bonded to four oxygen molecules

x-ray diffraction method of studying the structure of a crystal using x-rays

phosphate group. As shown in Figure BB2.5, once made, nucleotides can be strung together in chains by linking the phosphate group of one nucleotide to the sugar of another nucleotide to form chains called *polynucleotides*.

BB2.3 How Are DNA Molecules Organized?

To create their model, Watson and Crick used information supplied from an image of DNA created by Rosalind Franklin (Figure BB2.6, page 98) using **x-ray diffraction**. (Figure BB2.7, page 98, shows the image created by Franklin.)

Figure BB2.6

Rosalind Franklin's x-ray diffraction work played a critical role in the discovery of the structure of DNA.

Science Source/Science Photo Library/Photo Researchers

Figure BB2.7 **One of Franklin's X-ray Diffraction Photos of a Crystal of DNA**

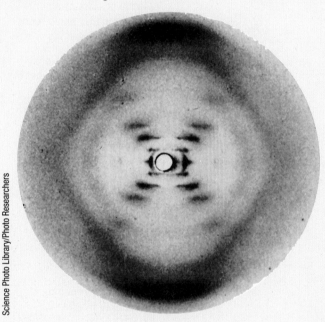

Science Photo Library/Photo Researchers

Figure BB2.8 **Nucleotide Chains in a DNA Molecule** The two chains run in opposite directions; a G in one chain pairs with a C in the opposite chain, and A pairs with T. The dotted lines between the bases represent the weak chemical bonds that hold the chains together.

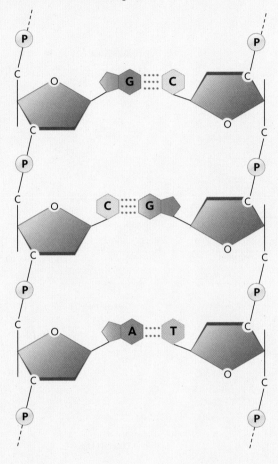

From observing Franklin's DNA image and from information about the base composition of DNA from a variety of organisms, Watson and Crick developed a model for the structure of DNA. Their model has the following features:

- DNA consists of two polynucleotide chains and is a ladder-like molecule, with the sugars and phosphates forming the side rails, and the bases as the rungs. The chains run in opposite directions, meaning they are anti-parallel (Figure BB2.8).

- The two polynucleotide chains are coiled to form a double helix (Figure BB2.9).

- In the double helix, the sugar and phosphate groups are on the outside of the molecule and form the backbone of the chain. The bases are on the inside.

- Within the molecule, an A in one chain always pairs with T in the opposite chain, and C always pairs with G; this combination is known as a base pair.

- This predictable base pairing makes the two chains of DNA *complementary* in base composition. This means that if one strand of nucleotides has the sequence A C G T C, you know the other strand must be T G C A G.

Figure BB2.9 **The Watson-Crick Model of DNA** (a) Two nucleotide chains are wound around a central axis forming a helix. Weak chemical bonds (dotted lines) between the bases hold the two chains together. (b) In DNA, A always pairs with T on the opposite strand, and C always pairs with G.

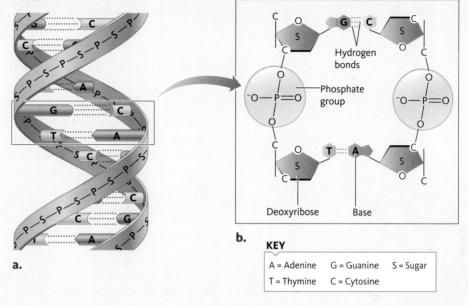

KEY

A = Adenine	G = Guanine	S = Sugar
T = Thymine	C = Cytosine	

The structural features of the Watson-Crick DNA model offer an explanation for several properties of genes:

1. *Genetic information is stored in the sequence of bases in the DNA.*

2. *The model offers a molecular explanation for mutation.* Because genetic information can be stored as a chain of bases in DNA, any change in the order or number of bases in a gene can result in a mutation that may produce a genetic disorder.

3. *The complementary sequence of bases in the two strands of DNA explains how DNA copies itself before each cell division.* In copying the DNA, each strand can be used as a template to reconstruct the base sequence in the opposite strand.

BB2.4 How Is DNA Organized in a Chromosome?

One long DNA molecule is coiled and combined with proteins to form a chromosome. The total length of the DNA in all 46 human chromosomes is about 6 ft. (1.8 m). To fit all this DNA inside the nucleus of a cell, which is only about 6 microns in diameter (there are 1 million microns in a meter), the DNA has to be folded over and over to compact the DNA by about 10,000 times. The diagram in Figure BB2.10 (page 100) shows how this is accomplished. Follow along with the description.

First, about 200 base pairs of DNA are wound around the outside of a protein cluster (histone core) to form structures called **nucleosomes** (Figure BB2.10a). Short threads of DNA connect the nucleosomes (Figure BB2.10b). At the next level, nucleosomes are coiled around each other to form a thin cylinder (Figure BB2.10c). Then the cylinders are coiled upon themselves to form thicker and thicker fibers, which again are coiled upon themselves, eventually forming the loops and fuzzy-looking fibers we can see as part of chromosomes in the electron microscope (Figure BB2.10d). Photomicrographs of a single replicated chromosome and several replicated human chromosomes are shown in Figure BB2.11a and b (page 100).

nucleosomes structures formed when DNA is coiled around histones

Nucleosome

Protein cluster
(histone core)

a.

Threads of DNA that
connect nucleosomes

b.

c.

d.

Figure BB2.10 **Folding of DNA in Chromosomes** The DNA is wound around clusters of **histone** proteins (a), which are connected by short stretches of DNA (b). The nucleosomes are coiled to form cylinders (c), which are then coiled further several times to form the fibers (d) that are visible in chromosomes.

BB2.5 How Are RNA Molecules Organized?

The second type of nucleic acid, RNA, is found in the nucleus *and* the cytoplasm. In human cells, RNA transfers genetic information from the nucleus to the cytoplasm, participates in the formation of proteins, and is a component of ribosomes (see "Biology Basics: Cells and Cell Structure").

RNA (Figure BB2.12) differs from DNA in two respects: the sugar in RNA is ribose (instead of deoxyribose in DNA), and the base uracil takes

Figure BB2.11 (a) In a replicated chromosome, each chromatid contains a double-stranded DNA molecule. The chromatids are joined at the centromere. (b) A colorized scanning micrograph showing several replicated human chromosomes.

a.

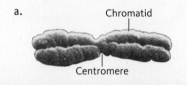

Chromatid

Centromere

b.

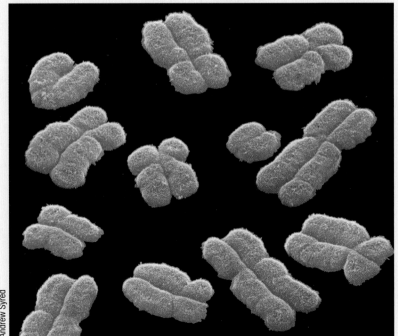

Andrew Syred

the place of the base thymine. Table BB2.1 shows these differences between RNA and DNA.

As shown in Figure BB2.12, RNA is usually single-stranded, and there is no complementary strand.

Table BB2.1 **Differences between DNA and RNA**

	DNA	RNA
Sugar	Deoxyribose	Ribose
Bases	Adenine	Adenine
	Cytosine	Cytosine
	Guanine	Guanine
	Thymine	Uracil

Figure BB2.12 RNA is a single-stranded molecule composed of nucleotides that contain the sugar ribose and the base uracil instead of thymine (which is found in DNA).

BB2.6 How Is Genetic Information Carried in a DNA Molecule?

DNA is found in the nucleus. It functions as a storehouse of genetic information. Based on their model for the structure of DNA, Watson and Crick proposed that genetic information is encoded in the sequence of base pairs in a DNA molecule.

The amount of information contained in DNA ranges from a few thousand nucleotides and 5 to 10 genes in some viruses to more than 3 billion nucleotides and about 20,000 genes in humans, and even more in some plants and amphibians.

A gene typically consists of hundreds or thousands of nucleotides. Each gene has a beginning and end, marked by specific nucleotide sequences. There is one molecule of DNA in each chromosome, which can contain thousands of genes. As we will see in Chapter 5, a change in any base pair in a gene can be considered a mutation.

BB2.7 What Are the Two Main Functions of DNA?

DNA has two major functions in the cell: (1) to replicate itself and (2) to carry the information for proteins and RNA molecules necessary for life. Here, we will explore how DNA replicates itself. In Chapter 5, we will discuss how DNA encodes information for proteins and the steps in forming these proteins. Later chapters will examine how an understanding of the structure and properties of DNA led to the development of recombinant DNA technology, its applications in medicine and forensics, and finally the Human Genome Project.

BB2.8 How Is DNA Replicated?

Before cell division begins, all cells must copy their DNA so that each daughter cell will receive an exact copy of the genetic information in the chromosomes. When a DNA molecule unwinds, each half serves as a template or pattern for making a new, complementary strand.

DNA replication occurs in all living things from bacteria to humans. The process is shown in Figure BB2.13 (page 102).

First, the double-stranded DNA molecule unwinds, opening the molecule to the action of an enzyme, **DNA polymerase**, which links nucleotides (already present in the nucleus) together to form a complementary new strand that binds to the template strand. When replication is completed, each DNA molecule will contain one old strand (the template strand that was copied) and one new strand (that is complementary to the old strand).

Thinking about this in terms of chromosomes, recall that each chromosome contains one double-stranded DNA molecule running from end to end. When DNA replication is complete, each chromosome consists of two DNA molecules (Figure BB2.14, page 102). These two molecules are visible in

histone protein that DNA coils around to form nucleosomes

DNA polymerase enzyme that catalyzes the synthesis of DNA using a template DNA strand and nucleotides

Figure BB2.13 **DNA Replication**

During replication, the polynucleotide strands of a DNA molecule separate, and each serves as a template for the synthesis of new strands (gold). After replication, each DNA molecule will contains one old polynucleotide strand (blue) and one new polynucleotide strand (gold).

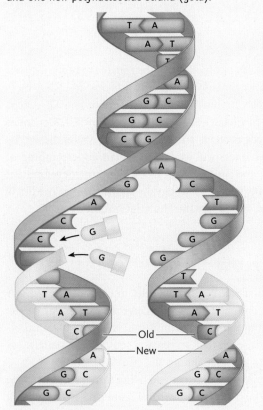

During DNA replication, the DNA strands unwind and each strand serves as a template for replication. The sequence of the new strands (gold) is complementary to the template strands (blue).

xeroderma pigmentosum a genetic disorder that results from defects in DNA repair

Figure BB2.14 An artist's depiction of the double-stranded DNA molecules contained in each chromatid of a replicated chromosome.

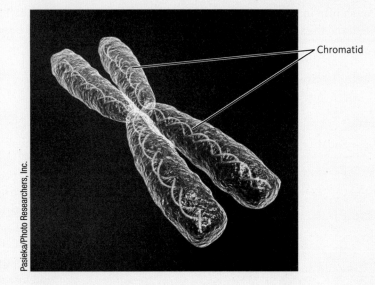

Chromatid

Pasieka/Photo Researchers, Inc.

the microscope as structures called *chromatids* that are joined to each other at a common centromere. Refer to the photomicrographs in Figure BB2.11 to see replicated chromosomes with the chromatids joined at the centromere. Each chromatid contains a DNA molecule composed of one old strand and one new strand. When the centromeres divide during mitosis, each chromatid becomes a separate chromosome.

Can DNA replication cause mutations? DNA replication is
not a perfect process. The enzyme DNA polymerase moves along the DNA at a speed of about 20 bases per second in humans, and sometimes the wrong nucleotide is inserted into the newly made strand. In most cases, DNA polymerase detects the error and repairs the mistake. Sometimes, however, the error is undetected and, therefore, not repaired. This creates a mutation.

In addition, DNA can be damaged when it is exposed to physical or chemical agents that cause mutations. For example, as shown in Figure BB2.15, when DNA is exposed to ultraviolet light (UV), changes can occur in the base pairing. If these errors are not detected or repaired properly, a mutation results.

What can happen if errors are not repaired? Cells have
a number of DNA repair systems, but even with all these checkpoints, errors in replication or DNA damage can still produce mutations. This can happen when the genes for the repair systems are themselves damaged. One example of a condition that results from mutations in DNA repair systems is **xeroderma pigmentosum (XP)**, a recessively inherited disorder (see Table 4.1). People with XP are extremely sensitive to UV light and develop skin cancer on sun-exposed areas of the body (see Figure BB2.16). About half of all XP

Figure BB2.15 **The Process of DNA Repair** When DNA is damaged by ultraviolet light (1), a section of one nucleotide chain is removed by a repair enzyme (2), and the remaining chain (3) is used as a template by DNA polymerase to repair the DNA (4).

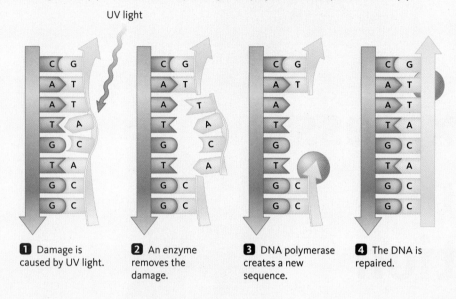

UV light

1 Damage is caused by UV light.

2 An enzyme removes the damage.

3 DNA polymerase creates a new sequence.

4 The DNA is repaired.

Figure BB2.16 **A Child with Xeroderma Pigmentosum**
This genetic disorder is associated with extreme sensitivity to ultraviolet radiation The boy's face and chest are covered with lesions, many of which may develop into cancer.

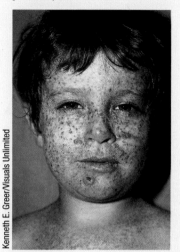

Kenneth E. Greer/Visuals Unlimited

individuals develop skin cancer by 8 years of age, and they also have a 20-fold higher risk of developing other cancers, including brain tumors.

Preview

Looking ahead, Chapter 5 focuses on genes and how gene action is related to genetic traits, including genetic diseases. Chapter 6 describes mutation and epigenetic changes in DNA. Chapter 7 considers recombinant DNA technology and how it is used in biotechnology. Chapter 8 tells how molecular biology has revolutionized genetic testing and prenatal diagnosis. Chapter 9 discusses the methods and applications of DNA forensics in several fields. Chapter 10 summarizes the history and findings of the Human Genome Project and its present and future impact on medicine and society and examines the development of genomics.

5

Gene Expression and Gene Regulation

REVIEW

What is in the cytoplasm? (BB1.4)

How is DNA organized in a chromosome? (BB2.4)

What can we learn from examining human pedigrees? (BB4.2)

CENTRAL POINTS

- Genes are composed of DNA that encodes RNA and proteins.

- During transcription, genetic information in DNA is copied into messenger RNA.

- During translation, the information in mRNA is transferred to the amino acid sequence of proteins.

- Mutations are heritable changes in DNA.

- Changes in DNA can produce changes in proteins.

CASE A: Marcia Johnson's Surprising Test Results

Marcia Johnson thought she was going to her obstetrician for a routine checkup. She had just found out she was pregnant, so she expected to get her prenatal vitamins and go home. In the waiting room, a nurse approached her and asked if she would be willing to be part of a study and give a blood sample for a simple test. The test was for sickle-cell anemia. The nurse explained that researchers were trying to establish how many people in northern Illinois were carriers of the mutant gene that causes the disorder. Marcia thought, "Why not?" She signed the consent form, gave blood, and promptly forgot about it.

Two weeks later her obstetrician called to tell her that she was indeed a carrier for sickle-cell anemia. Marcia was surprised, but she had searched the internet after she gave the blood sample and had learned that carriers were common among those with West African ancestors. The doctor said not to worry, but that she should ask her husband, Mark, to be tested. Mark was tested and found out he was also a carrier.

At the visit when Mark received his results, they also had the first ultrasound of Marcia's pregnancy. She was carrying twins! Marcia and Mark were disturbed by what she had learned about sickle-cell anemia on the internet, and they made an appointment with her obstetrician to have a prenatal test on the fetuses. A week or so later, a genetic counselor recommended by her doctor called to tell them the results.

Some questions come to mind when reading this case. Before we address these questions, let's look at what genes are, how they work, and why a mutant gene may result in a genetic disease.

In "Biology Basics: Cells and Cell Structure," you learned that DNA has two major functions. In this chapter, we will look at one of these: how DNA codes for proteins.

A Researcher in a Molecular Biology Laboratory

5.1 How Do Genes Control Traits?

Each of the trillions of cells in the human body (except mature red blood cells) contains a nucleus. Inside the nucleus is a set of 46 chromosomes, carrying two copies of the 20,000 to 25,000 genes that make us human. Recall from Chapter 3 that we get one copy of our gene set from our mother and the other from our father. Both of these gene sets can be identical, or they can carry different forms of each gene (these different forms are called **alleles**).

Genes (our *genotype*) are composed of DNA. Most, but not all, genes contain the information to make a **protein**. By their functions, proteins contribute to the observable traits, or **phenotype**, that helps make us unique. One example of a phenotype is the symptoms associated with cystic fibrosis (Figure 5.1, page 106). The normal allele of this gene codes for a plasma membrane protein that controls the flow of chloride ions into and out of a cell. The most common mutant allele of this gene produces a protein (CFTR) that cannot function. Those who carry two copies of the mutant allele lack this protein in all cells of the body. They show the symptoms of cystic fibrosis, including the production of thick mucus that blocks the ducts of some glands and lung passages. Mucus accumulation causes a life-threatening disorder associated with severe nutritional deficiencies and lung damage.

alleles different forms of the same gene that produce different phenotypes

protein a cellular macromolecule composed of amino acid subunits

phenotype the observable properties of an organism that are produced by the genotype

Figure 5.1 **A Child with Cystic Fibrosis** This child is using a bronchodilator, similar to those used by asthma patients. The mist she is inhaling helps open her airways, which are clogged by the thick mucus associated with cystic fibrosis.

© Karen Kasmauski/Corbis

Figure 5.2 **Protein Structure** The three-dimensional structure of a protein with the amino acid chain represented as a ribbon. Shape is important in protein folding and phenotype production.

PDB ID: 3JJU Pechkova, E., Tripatthi, S.K., Nicolini, C.

Figure 5.3 **The Chemical Structure of Amino Acids**
The 20 amino acids differ from each other in the chemical nature of the R groups. Peptide bonds are formed between the amino group of one amino acid and the carboxyl group of another amino acid.

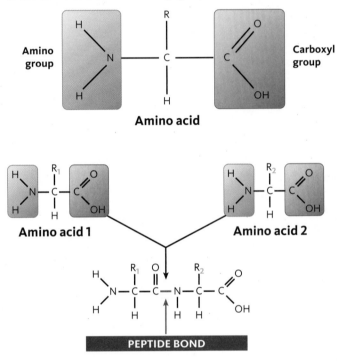

What is a protein?

A three-dimensional model of a protein is shown in Figure 5.2. Proteins take many forms and have many functions. They provide structure (hair, nails), serve as enzymes (proteins that control chemical reactions in the cell), can be chemical messengers (hormones such as human growth hormone and insulin), act as receptors (for sight, smell, hearing), and are carrier molecules (hemoglobin carries oxygen to cells in the body and is the molecule altered in sickle-cell anemia).

Proteins are composed of subunits called **amino acids** linked by chemical bonds (Figure 5.3) to form chains, called **polypeptides**. Once the polypeptide chain is complete, it folds into a three-dimensional shape and is called a protein. Twenty different types of amino acids are found in proteins. What makes each amino acid different is a chemical group (called an R group) attached to the backbone of the molecule. Two other chemical groups are found at the ends of all amino acids. One end carries an amino group (NH$_2$), and the other has a carboxyl group (COOH). These groups are chemically linked by the cell to form proteins (Figure 5.4).

Figure 5.4 **Linked Amino Acids**
Amino acids are linked together by peptide bonds to form proteins.

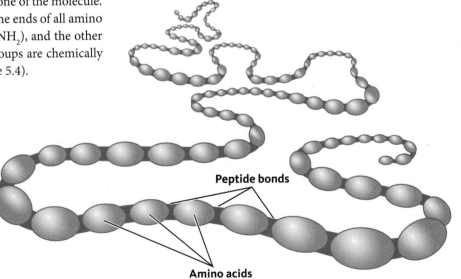

amino acid one of the 20 subunits found in proteins

polypeptides chains of amino acids

Table 5.1 **Essential Amino Acids**

At various stages in our lives, our bodies can produce 10 of the 20 amino acids. We must consume the other 10, which are listed here.

Amino Acid	Abbreviation
Arginine	Arg
Histidine	His
Isoleucine	Ile
Leucine	Leu
Lysine	Lys
Methionine	Met
Phenylalanine	Phe
Threonine	Thr
Tryptophan	Trp
Valine	Val

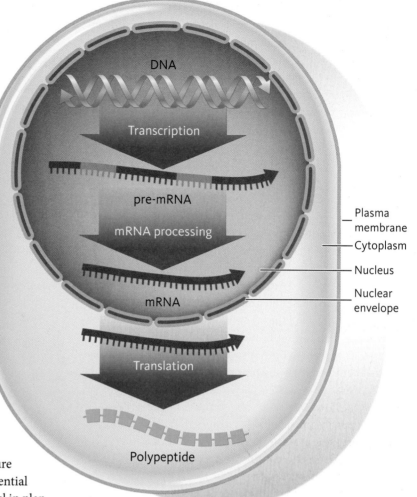

Figure 5.5 **The Flow of Genetic Information**
In the nucleus, one strand of DNA is transcribed into a strand of pre-mRNA. After processing, the finished mRNA moves from the nucleus to the cytoplasm, where the genetic information it carries is translated into the amino acid sequence of a protein.

To make all the proteins our bodies need, we must have a supply of all 20 amino acids. Some of these can be made by the cells of our body, but others must be included in our diet; these are called **essential amino acids** (Table 5.1). A balanced diet is necessary to ensure that we have an adequate supply of all the essential amino acids. Vegetarians must be especially careful in planning meals because some plants contain low levels of certain essential amino acids. Vegans, who consume no animal or dairy products, must select complementary protein sources to ensure that they eat a balanced diet that contains all the essential amino acids.

As we will see in a later discussion, after a polypeptide is made, it folds into a functional three-dimensional shape based on the order of amino acids, and is then called a protein. This protein then plays its role in producing a phenotype.

How are proteins made? In humans and other organisms whose cells have a nucleus, almost all the cell's DNA is found in the nucleus (some is in the mitochondria), and all proteins are manufactured by ribosomes, most of which are in the cytoplasm (see "Biology Basics: Cells and Cell Structure"). This means that the information encoded in DNA must somehow pass through the nuclear envelope and into the cytoplasm. Then it must move to the ribosomes. Chromosomes remain in the nucleus, but molecules of an intermediate, called **messenger RNA** or **mRNA**, carry genetic information from the nucleus to the cytoplasm.

The process of information transfer (Figure 5.5) from a gene (a DNA sequence) to a protein (an amino acid sequence) occurs in two steps, with RNA as an intermediate between the gene (DNA) and a polypeptide that folds to become a functional protein:

1. Transcription: DNA → mRNA
2. Translation: mRNA → polypeptide → protein

This process of information transfer is summarized in Figure 5.5.

essential amino acids those amino acids that cannot be synthesized and must be in the diet

messenger RNA (mRNA) a single-stranded RNA carrying an amino acid–coding nucleotide sequence of a gene

How is the information for proteins encoded in DNA? At first glance, it seems difficult to envision how the genetic information for so many different combinations of 20 amino acids can be carried in DNA. A DNA molecule contains only four different nucleotides (A, T, C, and G), and there are 20 different amino acids in proteins, so how can only four nucleotides encode the information for all 20 amino acids? The answer is that a sequence of three nucleotides in the DNA and in the mRNA copied from one strand of DNA carries the information that specifies each amino acid in a protein. This sequence of three nucleotides, called a **codon**, ensures that a specific amino acid is inserted at a specific location in a protein. Because the order of nucleotides in RNA is determined by the sequence of nucleotides in DNA, this means that the order of bases in DNA ultimately determines the order of amino acids in the protein a gene encodes. The type and location of amino acids in the protein determine its structure and function and its role in producing a phenotype. Not all of the bases in DNA contain information for proteins. In addition to the information that encodes an amino acid sequence, DNA also carries regulatory information that controls when and how much of a specific protein is made.

5.1 Essentials

- **Each gene encodes a gene product called a protein.**
- **Proteins give us the phenotype that makes us unique.**
- **Proteins are composed of one or more chains of amino acids.**
- **The genetic information in DNA is converted into the amino acid sequence of proteins in two steps: transcription and translation.**

codon a group of three nucleotides in mRNA that encode information for a specific amino acid

transcription the transfer of genetic information from DNA to mRNA

RNA polymerase an enzyme that synthesizes single-stranded RNA molecules using a DNA template

promoter a DNA sequence where RNA polymerase binds and starts transcription

termination sequence the nucleotide sequence at the end of a gene that signals the end of transcription

translation the transfer of genetic information in mRNA into the amino acid sequence of proteins

5.2 What Happens in Transcription, the First Step of Information Transfer?

In **transcription** (step 1), information encoded in the DNA sequence of a gene is copied into the sequence of bases in a messenger RNA molecule. Messenger RNA is similar to DNA (see "Biology Basics: DNA and RNA"), but its function is different.

Follow along with Figure 5.6 while reading this text:

Transcription begins when the chromosomal DNA at the location of a gene unwinds and one strand is used as a template to make an mRNA molecule. Once the DNA is unwound:

1. An enzyme, **RNA polymerase,** binds to a specific regulatory sequence (called a **promoter**) at the beginning of a gene.

2. Using the DNA sequence of the strand being copied as a template, RNA polymerase moves along the template, linking RNA nucleotides and forming an mRNA molecule.

The mRNA molecules produced by transcription are not exact copies of the gene's nucleotide sequence, but because of the rules of base pairing in nucleic acid molecules, they are called *complementary copies* (see inset). The rules of base pairing in DNA transcription are the same as in DNA replication, with one exception: an A on the DNA template ends up as a U in the mRNA. For example, if the nucleotide sequence in the DNA strand being copied is

C G G A T C A T

the mRNA will have the complementary sequence

G C C U A G U A

3. As the RNA polymerase moves along the DNA template making mRNA, it eventually reaches the end of the gene, marked by a specific nucleotide sequence (called a **termination sequence**). When the RNA polymerase reaches the end of the gene (termination), the polymerase detaches from the DNA strand.

4. The pre-mRNA molecule is released, and the separated DNA strands reform into a double helix.

After it is made, the pre-mRNA is processed to remove some sequences that do not code for amino acids in the protein product, and the ends of the mRNA are modified.

Once processed, the mRNA moves through pores in the nuclear envelope into the cytoplasm, where the protein will be assembled based on the information contained in the mRNA. This second step is called **translation**.

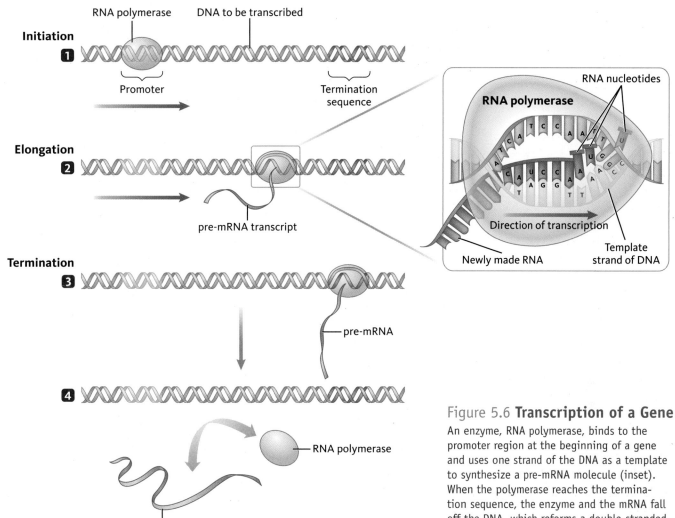

Initiation

RNA polymerase · DNA to be transcribed

Promoter · Termination sequence

Elongation

pre-mRNA transcript

RNA polymerase

RNA nucleotides

Direction of transcription

Newly made RNA · Template strand of DNA

Termination

pre-mRNA

RNA polymerase

Completed pre-mRNA

Figure 5.6 Transcription of a Gene

An enzyme, RNA polymerase, binds to the promoter region at the beginning of a gene and uses one strand of the DNA as a template to synthesize a pre-mRNA molecule (inset). When the polymerase reaches the termination sequence, the enzyme and the mRNA fall off the DNA, which reforms a double-stranded molecule.

5.2 Essentials

- **The DNA sequence of a gene is copied into the nucleotide sequence of messenger RNA.**

- **After the mRNA is processed, it moves from the nucleus to the cytoplasm, where protein synthesis takes place.**

5.3 What Happens in Translation, the Second Step of Information Transfer?

Translation converts the nucleotide sequence in mRNA into the amino acid sequence of a protein. This process requires the interaction of several different components, each of which has a separate, specialized job.

Let's begin with mRNA. In mRNA, each codon codes for a specific amino acid in the protein (Table 5.2), but as you can see from the table, many amino acids have more than one

codon. Some codons, the **stop codons** (UAA, UAG, and UGA), signal the end of protein synthesis. One codon, AUG, has two functions. It encodes the amino acid methionine, but also serves as the **start codon**, which marks the beginning of the coding sequence in mRNA. However, mRNA does not directly assemble amino acids into proteins. For this, interaction with two other components, tRNA and ribosomes, is required.

Transfer RNA (tRNA) molecules (Figure 5.7, page 110) perform two important tasks in translation: (1) they recognize and bind to a specific amino acid, and (2) they recognize the mRNA codon for that amino acid. At one end, each tRNA

stop codon codons (UAA, UAG, UGA) in mRNA that signal the end of translation

start codon the AUG codon that signals the location in mRNA where translation begins

transfer RNA (tRNA) a small RNA molecule that brings amino acids to the ribosome during protein synthesis

Figure 5.8 **Steps in the Process of Translation** Translation consists of three stages: (1) In initiation, a ribosome, an mRNA, and a tRNA carrying the amino acid methionine form a complex. (2) Next, a second tRNA with its attached amino acid enters the ribosome, and the second amino acid is linked to the first, forming an elongating chain as other amino acids are added. (3) When the ribosome reaches a termination codon, protein synthesis ends, and the components separate, releasing the amino acid polypeptide chain.

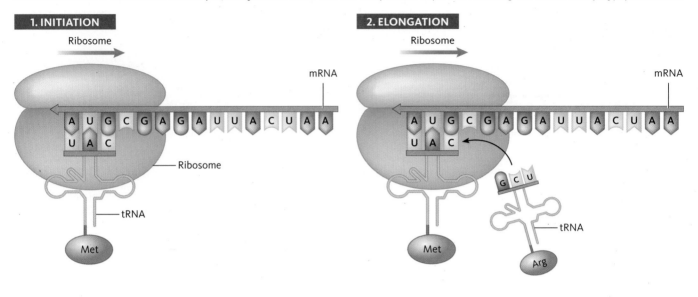

Table 5.2 **The Genetic Code for Amino Acids**
The messenger RNA codon for each amino acid.

Amino Acid	Abbreviation	mRNA Codons
Alanine	Ala	GCA GCC GCG GCU
Arginine	Arg	AGA AGG CGA CGC CGG CGU
Asparagine	Asn	AAC AAU
Aspartic acid	Asp	GAC GAU
Cysteine	Cys	UGC UGU
Glutamic acid	Glu	GAA GAG
Glutamine	Gln	CAA CAG
Glycine	Gly	GGA GGC GGG GGU
Histidine	His	CAC CAU
Isoleucine	Ile	AUA AUC AUU
Leucine	Leu	CUA CUC CUG CUU UUA UUG
Lysine	Lys	AAA AAG
Methionine	Met	AUG*
Phenylalanine	Phe	UUC UUU
Proline	Pro	CCA CCC CCG CCU
Serine	Ser	AGC AGU UCA UCC UCG UCU
Threonine	Thr	ACA ACC ACG ACU
Tryptophan	Trp	UGG
Tyrosine	Tyr	UAC UAU
Valine	Val	GUA GUC GUG GUU
Stop codons		UAA UAG UGA

Codon letters: A = adenine, C = cytosine, G = guanine, U = uracil.
*AUG also signals "start of translation" when it occurs at the beginning of a gene.

Figure 5.7 **The Structure of tRNA**
tRNA is a small molecule that contains a region called the anticodon that recognizes and binds to its complementary codon in mRNA. The tRNA also contains a region at the other end of the molecule that binds to the amino acid specified by the anticodon/codon combination.

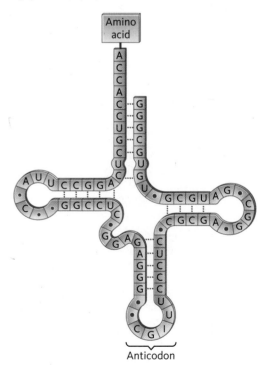

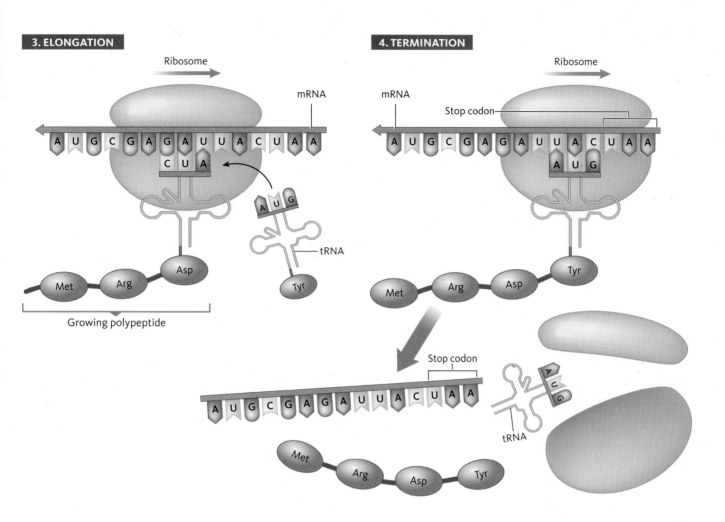

3. ELONGATION

Ribosome

mRNA

tRNA

Asp

Tyr

Met Arg

Growing polypeptide

4. TERMINATION

Ribosome

mRNA

Stop codon

Met Arg Asp Tyr

Stop codon

tRNA

Met

Arg Asp Tyr

molecule binds a specific amino acid and, at the other end, has a set of three nucleotides (called an **anticodon**) that can pair with the mRNA codon for that amino acid. As we will see in a later section, tRNAs deliver their attached amino acids to the ribosome in the order specified by mRNA codons. Once at the ribosome, the amino acids are linked by the formation of chemical bonds called **peptide bonds** into an amino acid chain. Once formed, the amino acid chain (called a *polypeptide*) folds into a three-dimensional shape to become a protein.

Ribosomes are organelles composed of two subunits and are the site of protein synthesis. Ribosomes can float free in the cytoplasm or be attached to the outer membrane of the rough endoplasmic reticulum (RER). Follow along with Figure 5.8 while reading about the process of translation.

1. As translation is initiated, an mRNA molecule is loaded onto ribosomes. Translation begins at a specific end of the mRNA molecule. The first codon translated, the start codon, is always AUG. It encodes the amino acid *methionine*, which is the first amino acid in all human proteins.

2. When the second amino acid specified by the mRNA is positioned in the ribosome, an enzyme within the ribosome forms a peptide bond between the two amino acids. Afterwards, the tRNA from the first amino acid is released and moves out of the ribosome.

anticodon a group of three nucleotides in a tRNA molecule that recognize a specific codon in mRNA

peptide bond the chemical bond that links amino acids together

ribosome an organelle that carries out protein synthesis

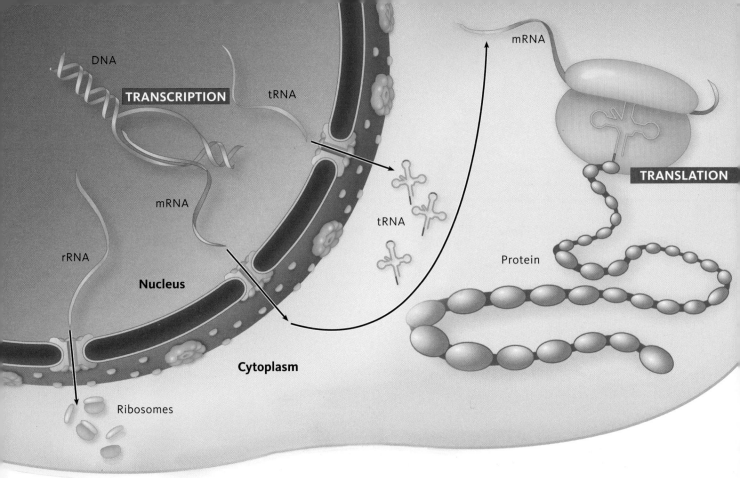

Figure 5.10 **Summary of Gene Expression** In the first step, which takes place in the nucleus, the information encoded in a gene is transcribed into the nucleotide sequence of an mRNA molecule. tRNA molecules are also made in the nucleus. Both the mRNA and tRNA are exported to the cytoplasm, where they bind to ribosomes and participate in protein synthesis, a process called translation.

Next, the ribosome moves down the mRNA to the next codon, and the process repeats itself, adding amino acids to the growing protein, a process called elongation.

3. Protein synthesis continues until the ribosome reaches a stop codon. Stop codons (UAA, UAG, and UGA) do not code for amino acids. When the stop codon is reached, protein synthesis ends, the ribosome detaches from the mRNA, and the completed polypeptide chain is freed and folds to form a functional protein. A photo of translation in the cell cytoplasm is shown in Figure 5.9; a summary of transcription and translation is shown in Figure 5.10.

5.3 Essentials

- **In the cytoplasm, each amino acid specified by an mRNA codon is linked to other amino acids to make a polypeptide chain.**

- **When complete, the polypeptide folds into a functional, three-dimensional molecule called a protein.**

- **The action of proteins helps produce phenotypes.**

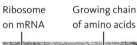
Ribosome on mRNA   Growing chain of amino acids

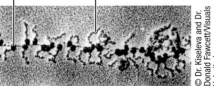

Figure 5.9 **Visualization of Translation** In this photo taken with an electron microscope, the ribosomes, seen as dense dots, are attached to the strand of mRNA. As the ribosomes move from left to right, the chain of amino acids increases in length.

5.4 Does a Cell Have Ways of Turning Genes On and Off?

Each of our chromosomes carries hundreds of the 20,000 to 25,000 genes that make us what we are, but not all those genes are active in every cell. In fact, almost all genes are switched off most of the time. Liver cells, for example, do not express the genes for eye color, and cells of the brain do not make proteins that digest food. In any given cell, only about 5–10% of the genes are active. The process of switching genes on and off is called **gene regulation**.

Gene regulation can occur at many places during transcription and translation, and even after translation is finished (Figure 5.11). One of the most important ways of regulating gene expression is by controlling access to the promoter region at the beginning of a gene. Recall that in a chromosome, DNA is wound around histone molecules (see "Biology Basics: DNA and RNA"). If the DNA is wound tightly around the histones, any genes in that region cannot be expressed because the promoter is not available to bind to RNA polymerase. If the histones are chemically modified, the DNA partly unwinds, and the gene promoter becomes accessible to RNA polymerase (Figure 5.12, page 114) and can be expressed. The modification of histones is reversible, allowing genes to be selectively turned on or off.

Instead of chemically modifying the histones, gene activity can also be regulated by adding chemical groups to the bases in the promoter region, effectively silencing the genes. So even if the histones are modified, and the promoter is accessible, RNA polymerase cannot bind to these chemically altered promoter regions. This form of regulation, which is reversible, is called epigenetic regulation and will be discussed in Chapter 6.

One other recently discovered way genes are regulated comes from the discovery that the cells contain small regulatory RNA molecules that bind to mRNA molecules. There are two main types of regulatory RNA molecules: **micro RNA (miRNA)** and **small interfering RNA (siRNA)**. In the cytoplasm, miRNA prevents translation by binding to mRNA and blocking ribosomes from attaching to the mRNA. This does not

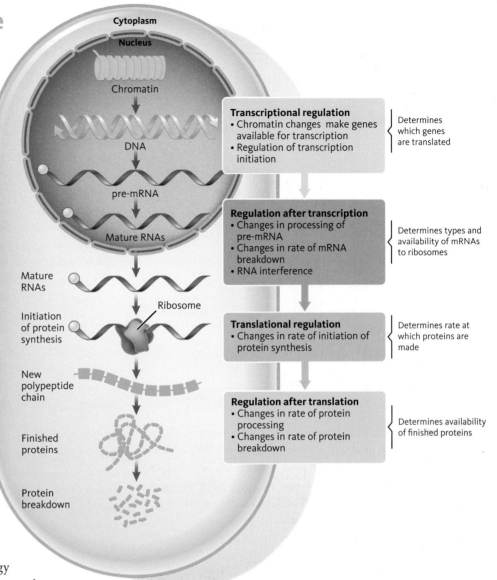

Figure 5.11 **Gene Regulation** The regulation of gene expression can occur at several levels in the cell. In the nucleus, control of transcription and the processing and transport of mRNA are important steps in regulating genes. In the cytoplasm, inactivation and degradation of mRNA as well as the control of translation play roles in regulation.

gene regulation the mechanisms that control the activity of genes

micro RNA (miRNA) a short RNA molecule that regulates gene expression by binding to mRNA

small interfering RNA (siRNA) a short RNA molecule that regulates gene expression by destroying mRNA

Figure 5.12 Histone modifications expose the promoter region of a gene and make transcription possible.

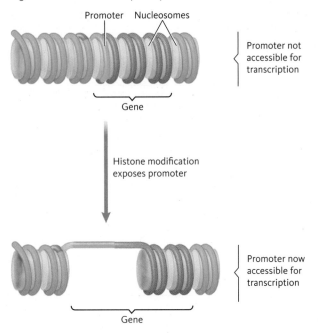

Promoter   Nucleosomes

Promoter not accessible for transcription

Gene

Histone modification exposes promoter

Promoter now accessible for transcription

Gene

destroy the mRNA but only prevents it from being translated. The other type of RNA, siRNA, binds to mRNA, which is then cut into pieces and destroyed before it can be translated. The regulation of gene expression by small RNA molecules is called **RNA interference (RNAi)**. In the future, RNAi may be used to regulate the expression of genes associated with specific genetic diseases.

In addition to events inside the cell, signals from outside the cell including hormones, nutrients, and other chemicals can cause changes in the number of active genes and the type and number of proteins produced. We will explore some of these forms of regulation in Chapter 12 when we discuss the causes of cancer.

5.4 Essentials

- **In the nucleus, most genes are inactivated.**
- **Access to promoter sequences by RNA polymerase determines whether or not a gene is expressed.**
- **Access to gene promoters can occur by chemical modification of histones and unwinding of DNA from the histones.**

RNA interference (RNAi) partial or complete gene silencing by small RNA molecules

mutation a heritable change in DNA

5.5 How Does an Altered Protein Result in a Genetic Disorder?

Remember from looking at Table 5.2 how similar the sequences of many codons are? A change in only one base pair in a gene can change a single amino acid in a protein. In other cases, a deletion of three base pairs can remove a single amino acid from a protein. In either case, the result can be a change in the structure and/or the function of the encoded protein. These and other changes in DNA are **mutations** and produce changes in the genotype. We will discuss mutation in more detail in Chapter 6. Here, our emphasis is on proteins and how changes in proteins produce an altered phenotype.

As we will see, even small changes in a protein (as small as one amino acid) can cause a change in the phenotype, often producing a genetic disorder. We will look at three examples where changes in the nucleotide sequence of a gene and a change in the encoded protein of an amino acid lead to a genetic disorder. In the first example, the change causes the protein to misfold; in the second, the size of the protein is reduced; and in the third, protein function is impaired.

How do changes in protein folding affect the phenotype? During and after synthesis, a protein becomes functional by folding into a three-dimensional shape that is determined by its amino acid sequence. A change in the amino acid sequence can cause a change in the three-dimensional shape, which ultimately depends on changes in the bases in DNA.

Earlier in this chapter, we described the phenotype of cystic fibrosis. In CF, the altered phenotype affects glands that produce mucus (described in Chapter 4). As a result, obstructive lung disease and lung infections result in premature death. The normal allele produces a 1480 amino acid protein located in the cell's plasma membrane. Here, it regulates the flow of chloride ions in and out of the cell. In most cases of CF, a mutation deletes one amino acid from the protein (at position 508). As a result, the protein does not fold properly; the misfolded protein is broken down and never reaches the plasma membrane. This results in faulty chloride transport in epithelial cells of lungs and other organs, which causes CF.

Protein misfolding is at the heart of many genetic disorders, including some forms of Alzheimer disease, Parkinson disease, Marfan syndrome, Tay-Sachs disease, and many others. In these disorders, defective folding causes the destruction of the protein or prevents the protein from functioning normally. In other disorders, such as mad cow disease and its human equivalent, Creutzfeldt-Jakob disease (CJD), a defectively folded protein can cause other proteins in the body to refold (Figure 5.13), producing disease-causing prions.

What causes shortened proteins to be produced? A look at the codon sequences in Table 5.2 shows that in some cases, changing only one base pair in a gene can create a stop codon, the signal to end protein synthesis. As a result, even though the mRNA has a normal length, translation ends prematurely because a mutation has created a stop codon before the end of the mRNA. The shortened protein often fails to function or has only partial function. The result is a change in phenotype we recognize as a genetic disorder.

Familial hypercholesterolemia (FH) is a dominantly inherited disorder associated with elevated blood levels of cholesterol and an increased risk of heart disease. The normal FH allele encodes a cell surface receptor of 860 amino acids that binds to LDL, the major cholesterol transport protein. The receptor moves LDL into the cell where the cholesterol is degraded. One mutation in the FH gene creates a stop codon at amino acid 167. As a result, translation produces a short, nonfunctional receptor that cannot bind to LDL. People heterozygous for this mutation produce only half the normal number of receptors, causing LDL levels in the blood to be twice as high as they should be. The high levels cause cardiovascular disease.

Which protein is altered in sickle-cell anemia? Remember from Chapter 4 that sickle-cell anemia is inherited as an autosomal recessive trait. Affected individuals can have a wide range of symptoms, including weakness, abdominal pain, kidney failure, and heart failure (Figure 5.14, page 116) that can lead to early death if left untreated. This painful and disabling condition is caused by a mutation in the gene that encodes **hemoglobin**, the oxygen-carrying protein in the blood.

Figure 5.13 **Changes in Protein Shape Can Produce Disease** At left is the normal shape of a prion protein. Most of the protein is folded into helices. If the protein refolds into another shape, as shown at right, with fewer helical regions and more regions folded as sheets, it becomes a disease-causing agent responsible for a progressive and fatal neurological disorder.

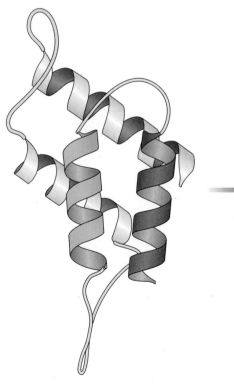

Normal folding pattern

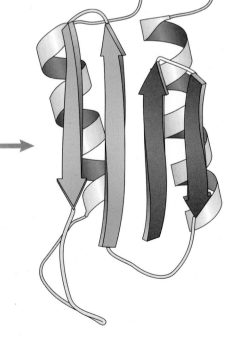

Refolded disease-causing form

BY THE NUMBERS

4 million
Number of base pairs in a bacterial cell

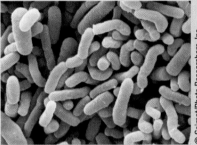

© Scimat/Photo Researchers, Inc.

3.2 billion
Number of base pairs of DNA in a human cell

146
Number of amino acids in the beta globin molecule

1
Number of mistakes in a gene that can cause sickle-cell anemia

hemoglobin the oxygen-carrying protein in red blood cells

Figure 5.14 **Symptoms of Sickle-Cell Anemia** Sickle-cell anemia begins with a mutation in DNA that results in the production of abnormal hemoglobin. The effects of the mutation at the molecular, cellular, and organ levels appear in many different ways that can cause illness.

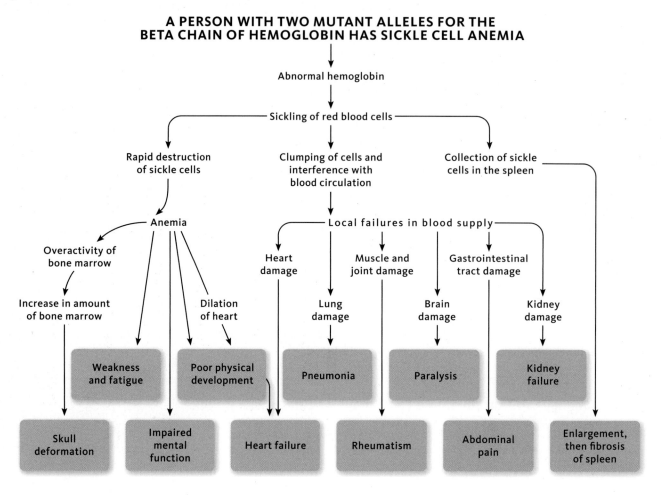

A PERSON WITH TWO MUTANT ALLELES FOR THE BETA CHAIN OF HEMOGLOBIN HAS SICKLE CELL ANEMIA

Each hemoglobin molecule in adults (called HbA) is composed of two different proteins: **alpha globin** and **beta globin** (Figure 5.15). The mutation in sickle-cell anemia affects the beta globin protein of hemoglobin. The difference between normal beta globin and the beta globin in sickle-cell anemia is a change in one of the 146 amino acids in the protein. After oxygen is transported from the lungs and unloaded into cells, hemoglobin molecules containing the altered beta globin stick together, forming long fibers inside the cell (Figure 5.16).

Figure 5.15 **Hemoglobin Molecule** The normal hemoglobin molecule is composed of two alpha globin subunits and two beta globin subunits. In sickle-cell anemia, the alpha globin subunits are normal, but the beta globin subunits carry a mutation that causes the clinical symptoms of the disease.

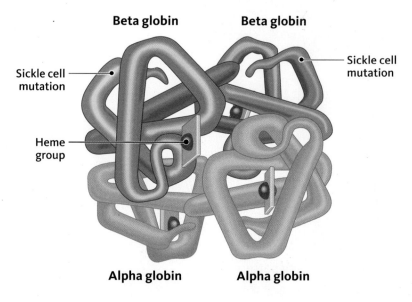

alpha globin one of the proteins in hemoglobin

beta globin one of the proteins in hemoglobin

Figure 5.16 **Polymerized Hemoglobin Molecules**

In sickle-cell anemia, the abnormal beta globin subunits cause the hemoglobin molecules in red blood cells to stick together, forming long fibers of different diameters that distort the shape of the cell into a sickle shape.

Polymerized hemoglobin molecule

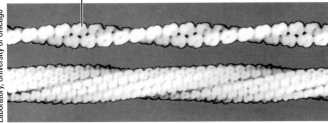

Figure 5.17 **Red Blood Cells** A colorized photo taken with an electron microscope showing normal, saucer-shaped red blood cells and sickle-shaped cells distorted by the formation of long fibers of abnormal hemoglobin. Compare the shapes of the normal cells and the sickled ones.

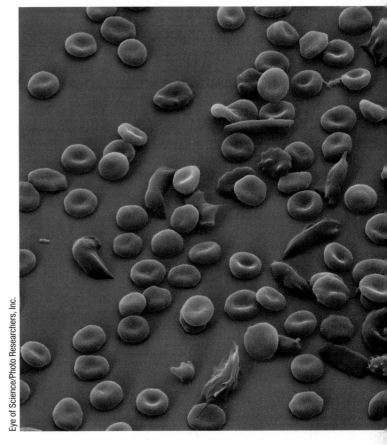

The fibers (shown above) distort and harden the membrane of the red blood cell, twisting the cell into a sickle shape (Figure 5.17). The deformed blood cells easily break down and no longer function. The loss of red blood cells reduces the oxygen-carrying capacity of the blood and results in anemia. The sickled cells also clog capillaries and small blood vessels, producing pain and tissue damage. Blockage of blood vessels in the brain can cause strokes and paralysis.

What kind of mutation is responsible for sickle-cell anemia?
The mRNA for the normal form of beta globin contains the codon GAG (for the amino acid glutamine) at amino acid number 6 (Figure 5.18). A single nucleotide change in DNA alters the mRNA codon for amino acid 6 from GAG to GUG. Instead of glutamic acid, the mutant form of beta globin has valine as amino acid 6. This single amino acid substitution changes the way beta globin interacts with other hemoglobin molecules, causing them to clump together to form fibers.

To put it simply, all the symptoms of sickle-cell anemia come from a change in *one* amino acid out of the 146 found in the beta globin gene. This is caused by a change in *one* base pair in the DNA of the beta globin gene.

5.5 Essentials

- **A mutation that affects only one nucleotide out of thousands in a gene can cause a genetic disorder by changing the amino acid sequence of a protein.**

- **A protein with an incorrect amino acid sequence can produce disease symptoms, as in sickle-cell anemia.**

Figure 5.18 **The Mutation in Sickle-Cell Anemia**
The DNA code (top line), the mRNA codons (middle line), and the corresponding amino acid sequence for the first eight amino acids of the normal (top) and sickle-cell (bottom) beta globin protein. The change in a single base (T → A) in the DNA coding for amino acid number 6 changes the amino acid at this position in the beta globin protein from Glu → Val and causes the symptoms of sickle-cell anemia.

Normal hemoglobin A	1	2	3	4	5	6	7	8
DNA	CAC	GTG	GAC	TGA	GGA	CTC	CTC	TTC
mRNA	GUG	CAC	CUG	ACU	CCU	GAG	GAG	AAG
Amino acid	Val	His	Leu	Thr	Pro	Glu	Glu	Lys

Hemoglobin in sickle cell anemia	1	2	3	4	5	6	7	8
DNA	CAC	GTG	GAC	TGA	GGA	CAC	CTC	TTC
mRNA	GUG	CAC	CUG	ACU	CCU	GUG	GAG	AAG
Amino acid	Val	His	Leu	Thr	Pro	Val	Glu	Lys

What can be done to treat sickle-cell anemia?

The genetic counselor should tell the Johnsons that treatment for sickle-cell anemia is available by transfusion (Figure 5.19) and by medications. First, she will show them a drawing of the hemoglobin molecule similar to the one in Figure 5.15. Then, she will explain that everyone carries more than one copy of the gene that controls the formation of beta globin, the protein that is mutated in sickle-cell anemia. Lastly, she will outline the treatment options that are available.

Figure 5.19 A Treatment for Sickle-Cell Anemia

This child is being treated for sickle-cell anemia using a transfusion of blood containing normal hemoglobin.

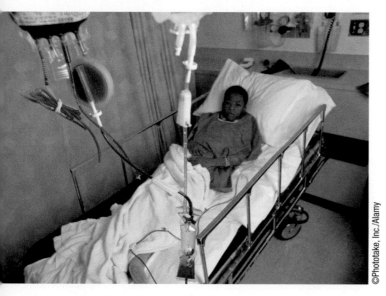

Everyone carries three slightly different copies of the gene for beta globin throughout his or her life: the embryonic version, the fetal version, and the adult version. Early in development, the embryonic version of the beta globin gene is active. This beta globin combines with alpha globin to form embryonic hemoglobin (HbE). When fetal development begins, the embryonic beta globin gene is switched off, and a fetal version of the beta globin gene is switched on, forming fetal hemoglobin (HbF). Just before birth, the fetal beta globin gene is switched off, and the adult beta globin gene becomes active, producing adult beta globin, which unites with alpha globin to form adult hemoglobin (HbA). These changes in gene activity in the beta globin genes are called gene switching. The mutation causing sickle-cell anemia is found only in the adult version of the beta globin gene and is not present in the embryonic or fetal beta globin genes, even in those with sickle-cell anemia.

Recently, in India and Saudi Arabia, it has been discovered that some people with sickle-cell anemia also have a separate genetic condition that keeps the fetal beta globin (HbF) gene active after birth. These individuals have both adult hemoglobin (HbA) and HbF molecules in their bloodstream. Because the HbF functions normally, these people have a very mild form of sickle-cell anemia. This finding sparked the idea that if the fetal version of the beta globin gene could be switched on in sickle-cell patients, it might be used as a treatment to decrease the symptoms of the disease.

The first breakthrough came from using a drug called *hydroxyurea* (Figure 5.20) to treat leukemia and other blood disorders. In patients treated with hydroxyurea, the fetal beta globin gene is active and produces HbF. This finding led to clinical studies attempting to treat sickle-cell anemia with hydroxyurea. These studies were successful, leading the U.S. Federal Drug Administration (FDA) to approve hydroxyurea as the first drug for use in treating sickle-cell anemia.

Unfortunately, about one-third of those with sickle-cell anemia do not respond to treatment with hydroxyurea, and in cases where the fetal beta globin gene is active, the drug only reduces, but does not eliminate, the painful and life-threatening symptoms. Evidence shows that the higher the levels of HbF in a sickle-cell patient, the more the symptoms are

CASE A: QUESTIONS

Now that we understand how proteins are made and how mutations can result in genetic disorders (specifically sickle-cell anemia), let's consider some of the questions raised in Marcia and Mark Johnson's case.

1. What should Marcia do before she calls her physician?

2. As we discussed in the case involving Martha Lawrence and Dr. Gould (Chapter 3), we know that problems can arise with the doctor's role in cases involving genetic testing. Does the doctor have to tell Marcia she is a carrier (heterozygous) for sickle-cell anemia? Why or why not?

3. Does the doctor have to tell Mark that he is a carrier?

4. Sickle-cell anemia does not affect an entire chromosome. How will this impact Marcia Johnson's decision?

The Johnsons met with the genetic counselor, who revealed the results of the tests on the twin fetuses. Both twins, being identical girls, have sickle-cell anemia. Now there are even more questions.

5. What are the Johnsons' options?

6. Do some research on how sickle-cell anemia affects individuals and their families. Should this information affect the Johnsons' decision? Why or why not?

7. Should all pregnant women have their fetuses tested for sickle-cell anemia? Why or why not?

8. Should everyone be tested to determine whether he or she is a carrier for sickle-cell anemia before marrying or having children? Why or why not?

Figure 5.20 Hydroxyurea can be used to treat sickle-cell anemia.

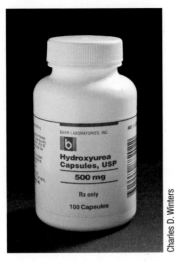

Charles D. Winters

reduced. Researchers are now working to find drugs that work in all affected individuals and that raise the levels of HbF high enough to eliminate the clinical symptoms of this disorder.

5.5 Essentials

- **In sickle-cell anemia and other genetic disorders, a change in *one* base pair in the DNA can cause an abnormal protein and an abnormal phenotype.**
- **Other genetic disorders can be caused by deleting nucleotides from a gene (cystic fibrosis).**

5.6 What Are the Legal and Ethical Issues Associated with Genetic Discrimination?

In the Law and Ethics section of Chapter 4 (Section 4.4), we discussed the confidentiality of medical records and how the law addresses it. More than 10 years ago, a federal law was passed (Health Insurance Portability and Accountability Act of 1996, or HIPAA) that required not only that a patient's records be kept confidential, but also that each patient be informed of this in writing.

The HIPAA law applies to Marcia Johnson's case as well. It is clear that she signed an informed consent form that allowed for the testing, but what might happen to these test results? The nurse who collected Marcia's blood was obviously collecting samples from many people, each of whom had to give his or her consent. The form that everyone signed was also required by the HIPAA law to disclose how the results of the test were to be used.

The next step in our case was to test Marcia's husband, Mark. What were his privacy rights? A consent form probably wasn't necessary here because his doctor would do the testing, not the group doing the study. Private tests fall under the

---

CASE B:
The Johnsons' Next Step

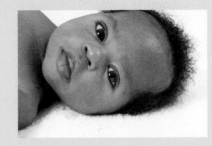

© Banana Stock/Jupiter Images

Now the twins are 15 months old. Having known before the birth that the twins would have sickle-cell anemia, the Johnsons and their doctors were able to plan specialized medical care from the moment they were born. The twins are healthy, but the Johnsons continue to monitor their condition carefully. They have made quite a few visits to the doctor and to the hospital for tests, treatment, and blood transfusions in the first 15 months. The Johnsons' insurance covered most of the cost of the medical care, but it was still expensive.

Now the Johnsons have a new dilemma: Marcia just confirmed through a home pregnancy test that she is pregnant again. Marcia and Mark are worried. What if this baby also has sickle-cell anemia?

1. **What is the first thing Marcia should do?**
2. **What would you advise her to do if you were her doctor?**
3. **What laws will protect Marcia's decision on what to do with her unborn fetus?**
4. **Go back to Chapter 3 and list what tests Marcia should have done.**
5. **What might happen if her doctor says, "Don't worry—everything will be all right"?**

HIPAA rules for confidentiality of medical records, including test results.

But, an interesting problem might arise here. If Marcia didn't want to share her results with anyone, then the lab, the study, or the doctor could not release them. This would apply to Mark or anyone else. Privacy is an important part of our medical treatment because most doctors believe that in order for patients to trust them, they must keep their confidential information out of the hands of anyone, even the patient's relatives, children, or spouse.

Employers might also want health and genetic information. If Marcia's and Mark's carrier status were known by insurance companies and employers, they might be discriminated against. As we discussed in Chapter 1, this can be a serious problem and is against the law.

Table 5.3 addresses some of the ethical and legal questions regarding discrimination.

Table 5.3 **Some Legal and Ethical Issues Associated with Genetic Discrimination**

Question	How Are These Questions Decided?	Related Case or Legal Issue
If someone collects your blood for a genetic test, can he or she use the results in a study?	Information on how one's blood tests are used should be spelled out in any consent form one signs.	We can volunteer our information, but even then HIPAA requires the results be kept anonymous.
Can the information used in a study be made available to anyone who wants it?	The FDA and drug companies are required to keep the results of large-scale studies private. This includes individual test results.	Most lawsuits involving clinical trials are about side effects and deaths. The rules of such studies require that patients' information be kept private.
What if Marcia didn't want Mark to know her results?	Marcia's result is her private business. Doctors cannot give this information to anyone unless the patient is in danger.	In Chapter 4 we discussed the case of *Tarasoff v. the Regents of the University of California*. In this case, it was held that doctors should share information only if the third party is in imminent danger.

Spotlight on Law

Norman-Bloodsaw v. Lawrence Berkeley Laboratory
135 F.3d 1260, 1269 (9th Cir. 1998)

AND

EEOC v. Burlington N. Santa Fe Railway Co.
No. 02-C-0456 (E.D. Wis. 2002)

In 2000, President Clinton signed Executive Order 13145 to prohibit discrimination in federal employment based on genetic information, and in 2005, the Genetic Information Nondiscrimination Act (GINA) was introduced that expanded this prohibition to all employers. It was signed into law on May 21, 2008, by President George W. Bush.

This action was a result of several legal cases in which genetic testing was used to discriminate against individuals. Two of these cases are *Norman-Bloodsaw v. Lawrence Berkeley Laboratory* (135 F.3d 1260, 1269 [9th Cir. 1998]) and *EEOC v. Burlington N. Santa Fe Railway Co.*, No. 02-C-0456 (E.D. Wis. 2002).

Lawrence Berkeley National Laboratory is the oldest national research laboratory in the United States. Operated by the University of California, it conducts research on human genetics, breast cancer, astrophysics, and nuclear science, among other subjects. For years, hundreds of employees gave blood samples as part of the hiring and employment process.

One day, Vertis Ellis, an administrative assistant, looked at her medical records, which were delivered to her office by mistake. Inside she found the results of syphilis tests taken without her knowledge in medical exams during her 29 years of employment. When she asked about these tests, laboratory officials acknowledged testing the blood and urine samples of its employees. They said tests for syphilis, sickle-cell trait, and pregnancy were done under the guidance and approval of the U.S. Department of Energy. They claimed the testing had stopped. Neither Ellis nor any other employee had either given permission for such tests or received results.

In September 1995, Marya S. Norman-Bloodsaw, Ellis, and six other employees filed a class action suit naming the laboratory, the secretary of the U.S. Department of Energy, and the regents of the University of California as defendants. The employees claimed that the testing occurred without their knowledge or consent and without any later notification that the tests had been conducted. The defense argued that the employees had consented to a complete medical exam, and the blood tests were part of that.

RESULTS

The district court in San Francisco dismissed the case, stating, "The three tests in question were administered as part of a comprehensive medical examination to which plaintiffs had consented."

The employees appealed to the U.S. Court of Appeals in San Francisco, which took into consideration three facts in the case:

1. Only black employees were tested for sickle-cell anemia.
2. Only female employees were tested for pregnancy.
3. No safeguards were taken to keep the results of these tests private.

The Court of Appeals asked: what was the government's interest in taking these tests? It found that the government

...d no interest in the results of the tests because they had ...effect on the job performance of the employees. Then the ...urt found that because only women and black employees ...re given some of the tests (pregnancy and sickle-cell ane-...a), this was discrimination based on sex and race.

...r second case is *EEOC v. Burlington N. Santa Fe Railway Co.*, ...02-C-0456 (E.D. Wis. 2002).

...y Avery, an employee of the Burlington Northern and ...ta Fe Railway, became unable to work because of a severe ...e of carpal tunnel syndrome. In this condition, a wrist ...ve is compressed, causing numbness, tingling, and muscle ...akness in the hand. After surgery he returned to work but ...s asked to supply seven vials of blood for tests. He was ...ious what these tests were and asked his wife, a nurse, ...at she thought. She wondered why they needed so many ...od samples. In discussions with his friends at work, he ...covered that other employees with carpal tunnel syndrome ...d also been asked for large numbers of blood samples, but ...ers, without this condition, were not.

...en asked to come back for another test, Gary, worrying, ...not go. The next day, he received notice that he would ...fired if he failed to submit a blood sample. Gary feared ...his job and filed a complaint with the Equal Employment ...portunity Commission (EEOC) and his union. The EEOC ...ulates the Americans with Disabilities Act (ADA), which ...hibits employers from discriminating against people with ...abilities.

...en the EEOC investigated, it found that the company ...s testing for a hereditary form of carpal tunnel syndrome ...led hereditary neuropathy with liability to pressure palsies ...NLPP). This inherited condition (the gene is on chromo-...ne 17) is an autosomal dominant disorder that affects ...proximately 16 out of 100,000 people, predisposing them ...get carpal tunnel syndrome. Two weeks after finish-...its investigation, the EEOC filed a lawsuit on behalf of ...ir employees, including Gary, to end genetic testing on ...ployees.

...e EEOC argued that the company based employment deci-...ns on the results of genetic testing, and therefore violated ...e ADA. Employers may require employees to submit to a ...dical examination only if the examination is job related ...d consistent with business necessity. Any test that may ...dict future disabilities is unlikely to be relevant to the ...ployee's present ability to perform his or her job.

Burlington argued that the incidence of carpal tunnel syndrome among its employees had recently increased. The company was testing employees to find out whether this increase was related to a genetic condition. It claimed that the results of genetic testing were used not to screen potential employees but to decide whether to pay disabled employees worker's compensation. If a worker had the genetic condition, it was preexisting and the company would not have to pay worker's compensation because the carpal tunnel syndrome was not work related.

RESULTS

The case was eventually settled without a trial. In the settlement, Burlington agreed to do the following:

1. immediately stop the genetic testing
2. pay the employees' legal costs and offer them an apology
3. turn over all test results to the employees
4. promise to keep all information obtained strictly confidential

QUESTIONS

1. List three ways in which these two cases are similar.
2. In *Burlington*, the company seemed to be doing something that might help its employees. Do you think this is the case? Why or why not?
3. In *Norman-Bloodsaw*, the test results were done at the time of employment and after hiring. Does this make a difference? Why or why not?
4. If the law required everyone to have a genetic test before he or she could apply for a marriage license, would that be discrimination? If so, which of these two cases could attorneys use to argue their point? Why?
5. Compare the two cases in the following table:

Case	State or Federal	Genetic Condition	Main Issue
Burlington			
Norman-Bloodsaw			

6. Do you agree with both case results? Why or why not?
7. How would you change these cases if you could?

The Essential 10

1. Each gene encodes a gene product called a protein. (Section 5.1)

   There are thousands of types of proteins.

2. Proteins produce the phenotype that makes us unique. (Section 5.1)

   Each protein has a specific function within the body.

3. Proteins are composed of one or more chains of amino acids. (Section 5.1)

   There can be hundreds of amino acids in a protein.

4. The DNA sequence of a gene is copied into the nucleotide sequence of messenger RNA. (Section 5.2)

   Each set of three nucleotides that specify an amino acid in an mRNA molecule is called a codon. (Section 5.2)

5. In the cytoplasm, each amino acid specified by an mRNA codon is linked to other amino acids to make a protein. (Section 5.3)

   The amino acids are assembled in the correct order so the protein can fold into the proper shape and function correctly. (Section 5.3)

6. Proteins bind to promoter sequences to turn genes on or off. (Section 5.4)

   Proteins can also bind to a cell's membranes and cause a chain reaction of events that turn genes on and off. (Section 5.4)

7. Genes can be changed by mutations, which change the presence and/or the function of the proteins they produce. (Section 5.5)

   Changes in these proteins can cause changes in phenotype that we recognize as a genetic disorder. (Section 5.5)

8. A gene mutation can cause one amino acid of a protein to be substituted for another amino acid. (Section 5.5)

   If this substitution occurs, the protein may not work correctly.

9. The incorrect protein can produce disease symptoms, as in sickle-cell anemia. (Section 5.5)

   Many genetic conditions are caused by something as small as a change in one amino acid in a protein. (Section 5.5)

10. In sickle-cell anemia and other genetic disorders, a mistake in *one* base pair in the DNA can cause an abnormal protein and an abnormal phenotype. (Section 5.5)

    This single change can cause many different effects in the body of a person with these conditions.

Review Questions

1. Using the codon/amino acid list, make up a protein by listing 10 codons and their respective amino acids.

2. How many different proteins can be made if the protein contains only three amino acids?

3. What is the difference between transcription and translation?

4. What is the role of tRNA in translation?

5. Why does the major part of protein synthesis occur in the endoplasmic reticulum?

6. What is the role of protein folding in protein function?

7. Explain why promoter regions are important in gene regulation.

8. Describe how RNA interference (RNAi) works.

9. Why are mutations important?

10. In many single-gene defects, a change in *one* base pair can change the outcome. Explain how this happens. List the steps.

11. In sickle-cell anemia, where are the mutations in the hemoglobin gene?

12. What are repeats, and how do they cause disease?

13. What can cause mutations?

14. If a mutation happens in your grandmother, does it automatically pass down to you? Explain your answer.

15. Why couldn't we live on the same diet as a cow?

Application Questions

1. We have discussed sickle-cell anemia in this chapter. Do you think that Marcia and Mark Johnson's decision would have been more difficult if their children had a different condition? Which ones would have made the decision more difficult?

2. How many amino acids could be encoded by DNA if one nucleotide specified one amino acid? How many could be encoded if a sequence of two nucleotides specified one amino acid?

3. Using the following mRNA sequence, derive the DNA template strand, the tRNA strand, and the amino acid sequence encoded by the mRNA:

   DNA:

   mRNA: UAC UCU CGA GGC

   tRNA:

   amino acid sequence:

4. List and describe the stages in the flow of genetic information.

5. If an essential amino acid is not present in a cell, synthesis of proteins requiring that amino acid would stop. Aside from dietary intake, what options does the cell have as a way of providing a supply of this amino acid?

6. Research the role of R groups in forming the three-dimensional structure of proteins. How can a change in an R group affect protein structure?

7. How might a change in protein folding bring about a change in phenotype?

8. Research how the misfolded prion protein causes mad cow disease, and why Creutzfeld-Jakob disease is considered similar to it.

9. When offered a choice of having a test to find out about the genetic status of their fetus, do you think that most people would accept? Why or why not?

10. Familial hypercholesterolemia is one of the most common genetic disorders with a frequency of 1 in 500. Most individuals with FH are heterozygotes. What are the chances that two heterozygotes would mate and produce a child carrying two normal FH alleles?

11. Knowing that heterozygotes for a mutation in the FH gene have a risk for cardiovascular disease, what blood levels of cholesterol can you predict for someone who is homozygous for two FH mutations? What about his or her risk of cardiovascular disease? What would you predict about the age at which cardiovascular disease might develop in homozygotes?

12. Does the genetic code offer some protection against the effects of nucleotide sequence causing changes in phenotype? How so?

13. Has the treatment for sickle-cell anemia changed over the last ten years?

14. A person who is heterozygous for sickle-cell anemia is said to have *sickle-cell trait*. Summarize and categorize it.

15. Find three other conditions where the gene and its products have been identified. Fill in the chart below:

Name of Condition	What Chromosome Is the Gene Found On?	What Is the Protein?	When Was It Discovered?

16. Look back at Table 5.2. You can see that some of the mRNA codons are used in many different proteins. Estimate how many possible combinations of codons you would need to make a protein of amino acids that do not duplicate.

17. In Huntington disease, the more repeats a person has, the earlier the onset of symptoms. Based on what you know about the science of repeats, why do you think this happens?

18. If hydroxyurea is used as a treatment for sickle-cell anemia, would this stop babies from being born with the condition? Why or why not?

Online Resources

Preparing for an exam? Log on at www.cengagebrain.com for study tools to help you assess your understanding. If assigned by your instructor, the Case A and Spotlight on Law activities for this chapter, "Marcia Johnson's Surprising Test Results" and "*EEOC v Burlington N. Santa Fe Railway Co., No.02-C-0456 (E.D. Wis. 2002,)*" will also be available.

In the Media

Toronto Star News, May 28, 2007

Toronto Playwright Uses Her Own Experience to Enlighten and Entertain

Nicholas Davis

Christine Nicole Harris was an active child. She did track and field, piano, theatre arts, and ballet. But, as she grew, things changed.

When she got older, she had to stop most of her physical activity. As a teen she cut back a lot. This was because of a disease she was diagnosed with as a baby. Her parents found out she had sickle-cell anemia shortly after she was born.

When they came here in the late 1970s from Jamaica, they had never heard of sickle-cell anemia. And after they had her, they decided to not have any more children.

To access this article online, go to www.cengagebrain.com.

QUESTIONS

1. Why do you think Christine's parents decided not to have any more children?

2. Christine is considered a role model for other people with sickle-cell anemia. Do you think she has an obligation to share her story?

San Francisco Examiner, November 5, 2007

Hospitals Will Begin Banking Blood from Newborns

John Upton

California hospitals will start banking blood from umbilical cords and placentas, with permission, to help cure blood diseases and other illnesses and as a source of stem cells.

This will create a California government-managed bank of donated cord blood, which will be used to find donor cord blood to match with patients who have any of roughly 70 diseases, including sickle-cell anemia and immune deficiencies.

To access this article online, go to www.cengagebrain.com.

QUESTIONS

1. Do you think that some parents might not give permission to take cord blood from their babies? Why?

2. If a law were passed that said that all babies' cord blood had to be taken at birth, do you think people would be for it or against it? Why?

6

Changes in DNA: Mutation and Epigenetics

REVIEW

How is DNA organized in a chromosome? (BB2.4)

How is DNA replicated? (BB2.8)

How are dominant traits inherited? (4.2)

CENTRAL POINTS

- Mutations are heritable changes in DNA.

- Some mutations cause changes in the phenotype while others do not.

- Often it is difficult to find mutations in the human genome.

- Radiation or chemicals may cause mutations in the genome.

- Epigenetic changes alter gene expression but do not alter DNA sequence.

- Epigenetics helps us understand why human behavior is so varied.

CASE A: A Grandmother's Question

In Chapter 5, Mark and Marcia Johnson used prenatal diagnosis and found out their twins have sickle-cell anemia. This information was difficult to hear and demanded a huge decision on the part of the couple. But another question arose at the same time. Sharon Tate, Marcia's mother, was wondering where the mutant allele for sickle-cell anemia came from. She needed to understand how recessive genetic traits are inherited.

She was confused, however, because no one in her family had ever suffered from sickle-cell anemia. She wondered, "Was I the one who gave Marcia the gene for sickle cell?" Without telling Marcia, she made an appointment with a genetic counselor and thought about Marcia's father. Frank died when Marcia was only 9, and Sharon had no idea what genes were being passed down in his family.

Some questions come to mind when reading about Sharon. Before we can address those questions, let's look at the events that alter DNA sequence and those that alter gene expression.

A grandmother meets with her daughter's genetic counselor.

Watson and Crick Examining a Model of DNA Structure

6.1 What Is a Mutation?

The set of genetic information carried by an individual is called a **genome**. Within the genome, a process called **mutation** can change genes. Mutations are changes in the nucleotide sequence of DNA that can be passed on to future generations. They can occur in any cell of the body (liver, breast, kidney, etc.) or in the cells that form sperm and eggs. A mutation in a body cell will be passed on by cell division to other body cells, but it cannot be transmitted to future generations. As we will see in Chapter 12, mutations in body cells are the underlying cause of many types of **cancer**, a form of uncontrolled cell division.

Mutations that take place in germ cells that form sperm and eggs *are* passed on to all the cells in members of future generations. In this chapter, we will focus on this type of mutation. First, we will discuss mutations that change the nucleotide sequence or number of nucleotides in a single gene. Next, we will consider other types of mutations and the cellular and environmental factors that cause mutations.

Once a gene is mutated, it can produce a partially functional protein, a nonfunctional protein, or, in some cases, no protein at all. Some mutations do not alter a gene or its protein product, but instead affect the timing and level of gene expression (see Chapter 5). Still other mutations change the DNA nucleotide sequence but make no change in the resulting protein. These mutations are invisible in a way because they don't show up in the phenotype.

genome sum of the genetic information carried by an organism

mutation a heritable change in the nucleotide sequence of DNA

cancer malignant cell growth and metastatic tumor formation at remote sites

6.1 Essentials

- **Mutations are heritable changes in DNA sequence.**
- **Mutations in body cells are not passed on to offspring, whereas mutations in cells that form sperm and eggs are passed on to offspring.**

6.2 What Causes Mutation?

Mutations can be caused by several different processes. Some of these include errors during DNA replication, as well as the actions of environmental agents (called **mutagens**), including radiation and chemicals.

During DNA replication (see "Biology Basics: DNA and RNA"), an enzyme, DNA polymerase, uses one strand of the DNA as a template and synthesizes a complementary strand. This process forms a new DNA molecule composed of one old strand (the template) and one new strand.

Rarely, an incorrect base pair is inserted in the new strand. In Figure 6.1 (part 1), the template strand has an A at the point where a new nucleotide will be added to the new strand. During replication, DNA polymerase should insert a T at that location, but instead, a mistake is made and a C is added (Figure 6.1, part 2). If this error is not corrected or repaired when the new strand is used as a template, the C will direct the insertion of a G in the new strand, and the result will be the replacement of an A/T base pair with a C/G base pair, creating a mutation. Fortunately, as we will see in a later section, cells have several systems to detect and repair such errors.

In addition to mutations produced during replication, several environmental agents can damage DNA and cause mutations. These include radiation and chemicals in the environment. We are constantly exposed to radiation from natural **background radiation**, and from ultraviolet (UV) light, which is a component of sunlight (also the light in tanning beds). Certain medical procedures such as x-rays and CT scans also expose us to radiation.

In addition to radiation, reactive chemicals in the environment can also cause mutations. Some of these include certain molecules found in plants, cigarette smoke, and vehicle exhaust, as well as chemicals used in food processing, chemotherapy, and industrial processes.

mutagens physical or chemical agents that cause mutations

background radiation radiation in the environment that we are exposed to in our everyday lives

free radicals reactive molecules that have unpaired electrons

rems unit of radiation effect equal to a standard dose of x-rays

millirems unit of radiation effect that is 1/1000 of a rem

Figure 6.1 Mutations Caused by Errors in DNA Replication Mutations occur when DNA polymerase incorrectly reads the nucleotide on the template strand and inserts the wrong nucleotide into the new strand of DNA.

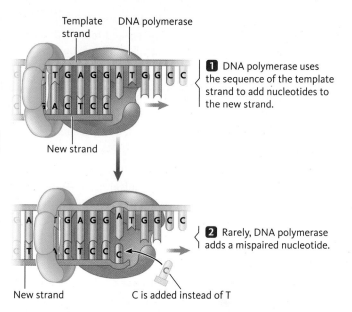

1 DNA polymerase uses the sequence of the template strand to add nucleotides to the new strand.

2 Rarely, DNA polymerase adds a mispaired nucleotide.

C is added instead of T

How much radiation are we exposed to?

Exposure to radiation is unavoidable. Everything in the physical world contains sources of radiation: the air we breathe, the food we eat, even the bricks in our houses. In addition, we are constantly bombarded with radiation from space. These sources are background radiation.

We are also exposed to radiation that results from human activity, including medical procedures, nuclear testing, nuclear power, and consumer goods. Radiation can cause damage to cells in several ways. As radiation strikes molecules in the cell, they are changed and form **free radicals** (Figure 6.2), highly reactive molecules that can produce mutations if they interact with DNA.

Radiation doses are measured in units called **rems**. Rems are defined as the amount of radiation that causes as much biological damage as a standard dose of x-rays. Cells begin to die from radiation damage at doses of about 100,000 rems. Because most people are exposed to very low doses of radiation, doses are usually measured as **millirems** (mrem), where 1000 mrem equals 1 rem.

In the United States, the average radiation dose is about 360 mrem per year, 81% of which is from background radiation (Figure 6.3). Even though this dose is low (Table 6.1), should it be a concern? Research shows that at doses below 5000 mrem, the major risk is mutations in body cells that increase the risk of cancer. Overall, however, this effect is very small and very difficult to detect.

Figure 6.2 **Creation of a Free Radical** (a) A stable oxygen atom with no unpaired electrons; (b) loss of an electron generates a situation with unpaired electrons, creating a free radical; (c) a free radical created from an oxygen atom.

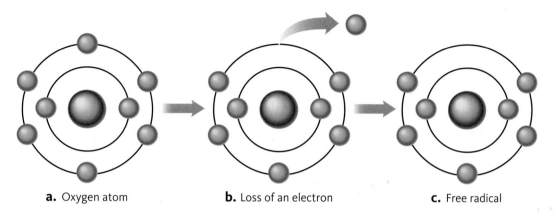

a. Oxygen atom   **b.** Loss of an electron   **c.** Free radical

How do chemicals cause mutations?

In our industrialized society, scientists have identified over 6 million chemical compounds. About 500,000 of them are used in manufacturing processes, packaging, food production, building materials, and so forth. Unfortunately, we know little or nothing about whether most of these chemicals can cause mutations. Based on experience, we know that some *classes* of chemicals do cause mutations, including the following:

1. **Base analogs** are chemicals that structurally resemble the bases (A, T, C, G) normally found in DNA. If they are present in a cell during replication, these analogs may be incorporated into DNA in place of the normal bases. When DNA replicates, analogs present in a replicating DNA strand can cause an incorrect base to be inserted into the newly synthesized strand, causing a mutation.
2. Chemicals from food preservatives, created by the breakdown of nitrates (such as nitrous acid) may modify the structure of bases in DNA and lead to the insertion of incorrect bases when DNA replicates.

Table 6.1 **Radiation Doses from Everyday Activities**

Radiation Source	Dose in Millirems (mrem)
Annual dose from background radiation, Boston, MA	102
Dental x-rays	9–10
Full-body CAT scan	12,000

Figure 6.3 **Radiation Exposure** In this pie chart, the amount of radiation we are exposed to is separated into specific percentages. Background radiation includes cosmic radiation, radon, and other forms of terrestrial radiation.

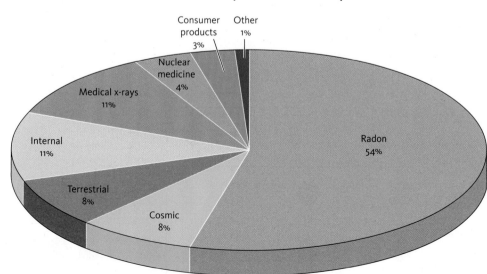

Sources of Radiation Exposure for the U.S. Population

Consumer products 3%
Other 1%
Nuclear medicine 4%
Medical x-rays 11%
Internal 11%
Terrestrial 8%
Cosmic 8%
Radon 54%

base analogs chemicals that are structurally similar to nucleotide bases (A, G, T, C)

Figure 6.4
Frameshift Mutation The insertion of a base (in red) causes a shift in the reading frame of mRNA codons, resulting in the addition of different amino acids to the growing polypeptide chain.

mRNA transcribed from the DNA

DNA TEMPLATE STRAND

Resulting amino acid sequence

C G U	G G U	U A U	U G G	A A U
G C A	C C A	A T A	A C C	T T A

Arginine   Glycine   Tyrosine   Tryptophan   Asparagine

Altered message in mRNA

A BASE INSERTION (RED) IN DNA

The altered amino acid sequence

C G U	G G U	U U A	U U G	G A A	U
G C A	C C A	A A T	A A C	C T T	A

Arginine   Glycine   Leucine   Leucine   Glutamic acid

3. Pesticides contain chemicals that are large, flat, ring-containing molecules that can insert themselves into the DNA helix and distort it. This changes the spacing between nucleotide pairs. During replication, the extra distance between nucleotides sometimes causes DNA polymerase to insert an extra nucleotide in the new strand, causing a different type of mutation called a **frameshift mutation** (Figure 6.4).

6.2 Essentials

- **Mutations can be caused by mistakes during DNA replication.**
- **Environmental agents such as chemicals and radiation can also cause mutations.**

6.3 How Do We Know When a Mutation Has Occurred?

Because mutations occur at the level of DNA, unless we sequence the DNA, we only know that a mutation occurs if there is a change in phenotype. If someone with a genetic disorder is born into a family with no history of that disorder,

frameshift mutation a mutation caused by the insertion or deletion of nucleotides

parents want to know how this happened. Is the cause genetic or nongenetic? For example, a woman who drinks alcohol while pregnant may give birth to a child with certain facial defects, growth problems, and nervous system abnormalities. This condition could be fetal alcohol syndrome (FAS), a nongenetic disorder caused by environmental factors (alcohol consumed by the mother). However, the phenotype of fetal alcohol syndrome closely resembles those seen in several genetic disorders, including Williams syndrome and Noonan syndrome (Table 6.2). To determine whether a condition is caused by genetic or nongenetic factors, a family history should be taken, a pedigree constructed, and a genetic test done, if one is available.

If pedigree analysis shows that a genetic disorder has appeared in a family with no history of the condition, geneticists generally assume that a mutation has occurred. Look at the pedigree in Figure 6.5. In this family, a new trait, severe blistering of the feet, appeared in one child (II-5), although the parents were unaffected. This female child transmitted the trait to three of her five children in generation III. The trait was then passed to offspring in generation IV. This pattern is consistent with the inheritance of an autosomal dominant trait. Therefore, geneticists would conclude that female II-5 carries a new mutation because neither of her parents suffered from the condition. If DNA-based genetic testing were available for this condition (see Chapter 8 for a discussion of genetic testing), it could be used to confirm this diagnosis.

Figure 6.5 A New Trait
A dominant trait, foot blistering, appeared in a member (II-5) of a family with no previous history of this condition. The trait is passed to later generations in a way that is consistent with the inheritance of an autosomal dominant trait.

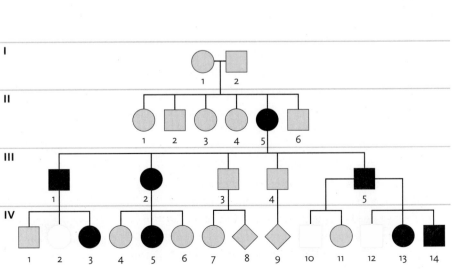

Table 6.2 Overlapping Phenotypes in Fetal Alcohol Syndrome and Two Genetic Disorders

Syndrome	Symptoms	Similarities to FAS
Fetal Alcohol Syndrome 	Flat nasal region between eyes Skin folds at corner of eyes No groove at base of nose on upper lip Small eye openings Small upper lip	
Noonan syndrome 	Flat nasal region between eyes Skin folds at corner of eyes Widely spaced eyes	Flat nasal region between eyes Skin folds at corner of eyes
Williams syndrome 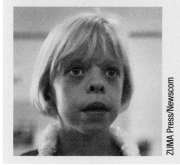	Flat nasal region between eyes Skin folds at corner of eyes Short distance between upper and lower eyelid Long groove from nose to upper lip	Flat nasal region between eyes Skin folds at corner of eyes Small eye openings

Recessive mutations on the X chromosome can often be detected by looking at the phenotypes of males in a pedigree. Recall from Chapter 4 that males carry only one copy of all genes on the X chromosome and mutant alleles on the X show up in the male offspring as genetic disorders. However, it can be difficult to determine whether a heterozygous mother (who does not have the condition) is the source of a mutation she has transmitted to her son or whether she inherited the mutant allele from her mother and is a carrier.

In one of the most famous examples of this situation, the X-linked disorder hemophilia (review Chapter 4) spread through the royal families of western Europe and Russia in the 19th century because of intermarriages. The origin of this disorder was traced to Queen Victoria of England (Figure 6.6, page 132). Because there was no previous history of the disease in her family, and one of her sons was affected, it was obvious that she was heterozygous for this mutation. Based on pedigree analysis of earlier generations in her family, it seems likely that the mutation occurred in a gamete from one of her parents, but we can only speculate as to which parent.

If, however, an autosomal recessive trait suddenly appears in a family, it is usually very difficult to identify when the mutation first occurred. Newly mutated recessive alleles can be carried through several generations in the heterozygous condition before becoming visible in a homozygous individual, as apparently happened in Marcia and Mark Johnson's case.

Mutations produce new alleles, some of which cause genetic disorders; others contribute to the genetic diversity we see in those around us. These include traits rather than disorders. While some mutations result in disorders such as sickle-cell anemia, the condition that Marcia Johnson (see Case A) was so worried about, mutations also produce alleles

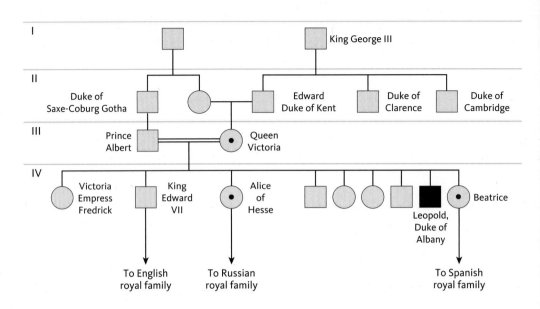

Figure 6.6 Pedigree of the Family of Queen Victoria of Great Britain The pedigree shows the queen's immediate ancestors and children. Because she transmitted the mutant allele for hemophilia to three of her children, she was probably a heterozygote rather than the source of the mutation.

I

II

Duke of Saxe-Coburg Gotha

Edward Duke of Kent

Duke of Clarence

Duke of Cambridge

King George III

III

Prince Albert

Queen Victoria

IV

Victoria Empress Fredrick

King Edward VII

Alice of Hesse

Leopold, Duke of Albany

Beatrice

To English royal family

To Russian royal family

To Spanish royal family

that confer advantages on those who carry them. We will discuss how natural selection acts on new alleles in Chapters 15 and 16.

6.3 Essentials

- **Pedigree analysis can often be used to help determine if a phenotype has a genetic basis.**
- **Mutated recessive alleles can be carried in a family for generations before appearing as a genetic disorder.**

6.4 How Often Do Mutations Occur in Human Genes?

The rate of mutation—that is, how often new mutations appear—is difficult to measure for a number of reasons. For example, if a mutation has no visible effect on the phenotype, we would not know a mutation has occurred. How then can we measure the overall rate of mutation in humans?

To answer that question, let's start by looking at how much DNA is carried in human cells. The nucleus of sperm and eggs contains about 3 billion nucleotides, and the nucleus of a diploid human cell contains about 6 billion nucleotides of DNA. Estimates of the rate of mutation, based on comparing the genomic DNA sequences of individuals, show that the mutation rate in humans is about 1 mutation for every 100 million base pairs. This means that each child is born with about 65 to 70 new mutations that are not present in either parent.

Should we be worried about 70 new mutations in each generation? Probably not. As we will see in Chapter 10, genes make up only about 1.5% of human DNA, so most mutations will not show up in our phenotypes.

Do different genes have different rates of mutation? Although we can estimate the overall rate of

CASE A: QUESTIONS

Now that we know more about mutation, let's answer some questions about Sharon's case:

A grandmother meets with her daughter's genetic counselor.

1. What questions should the genetic counselor ask Sharon? List three.

2. What test can be done to determine if Sharon is a carrier of sickle-cell anemia?

3. If tests show that Sharon is not a carrier, does this mean her husband must have been a heterozygous carrier? Why or why not? What other scenarios might explain her father's *not* being the carrier?

4. What steps would have to be taken to see if Frank was the carrier?

5. If Sharon found out she was the carrier, how do you think she might feel?

mutation, several factors affect the mutation rate in specific genes. Some of these factors are:

1. **Gene Size.** Larger genes are bigger targets for mutation. The longest human gene, encoding a muscle protein associated with muscular dystrophy, is 2,220,223 nucleotides long; the smallest human gene, encoding a growth factor, is 252 nucleotides long. It stands to reason that the longest gene will have a higher mutation rate than the smallest. Large genes can have mutation rates 1000 times higher than average, and small genes can have mutation rates 10 times lower than average.

Figure 6.7 **A Series of Trinucleotide Repeats (red arrows)** Expansion of these repeats in subsequent generations causes earlier onset of the disease and more severe symptoms.

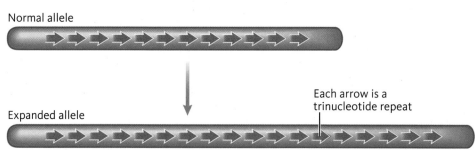

Normal allele

Expanded allele

Each arrow is a trinucleotide repeat

2. **Nucleotide repeats**. Some genes contain nucleotide repeats that can increase in number with each generation (Figure 6.7). For example, the mutant allele for myotonic dystrophy, a form of muscular dystrophy, contains more triplet nucleotide repeats than the normal allele. The normal gene contains 5 to 37 repeats of a CTG sequence. Mutant alleles that result in myotonic dystrophy contain 50 to 72,000 copies of this sequence, and the symptoms become more severe as the number of repeats increases. This is also true of Huntington disease (see Chapter 4), where additional repeats of the CAG sequence cause increased severity and earlier onset.

3. **Nucleotide composition**. As we will see later in this chapter, genes that have many adjacent C/G base pairs have a higher rate of mutation.

In the next section, we will see what changes in DNA are produced by mutation, and how these result in genetic disease.

6.4 Essentials

- **There are a number of ways to determine mutation rate in human genes.**
- **Certain types of genes have increased mutation rates.**

6.5 How Does a Mutation Show Up in DNA?

As we already know, when a mutation occurs, there is a change in the nucleotide sequence of DNA. At the molecular level, these changes can be as simple as the substitution of one nucleotide for another, or they can involve hundreds or thousands of nucleotides. There are several basic types of mutations, including:

1. single nucleotide substitutions
2. nucleotide insertions and deletions
3. trinucleotide repeats

Table 6.3 shows an example of each of these mutation types.

An increase in trinucleotide repeats as the cause of a genetic disorder was first reported for the *FMR-1* gene on the X chromosome (Figure 6.8, page 134). In the normal *FMR-1* allele, a CGG repeat occurs from 6 (CGGCGGCGGCGGCGGCGG) to 52 times. These repeats do not cause any abnormal phenotype and are located in a noncoding region of the gene. If a mutation expands the number of CGG repeats to 80 to 100 copies, there will still be no change in the phenotype. However, these alleles are unstable and are likely to undergo further expansion as they are passed to new generations. Mutant alleles with more than 230 copies of the CGG repeat inactivate the *FMR-1* gene and cause mental retardation.

Table 6.3 **Types of Mutations**

Type of Mutation	Definition	Example
Substitution in nucleotide sequence	One nucleotide is changed, which changes a single amino acid in a protein.	In sickle-cell anemia, a change from CTC to CAC in the beta globin gene causes valine to replace glutamic acid in beta globin (see Chapter 5).
Insertions and deletions	Mutation causes base pairs to be added or deleted, causing large-scale changes in an amino acid sequence, or shortened, nonfunctional proteins.	Some mutations in *CFTR*, the gene associated with cystic fibrosis, produce a shortened, nonfunctional protein.
Nucleotide repeats	An increase in the number of trinucleotide repeats can cause a genetic disorder.	The mutant allele for Huntington disease has many CAG repeats, and the number of repeats determines the severity and onset age of the symptoms.

Figure 6.8 **A Photomicrograph of an X Chromosome** The arrow shows the location of the *FMR-1* gene.

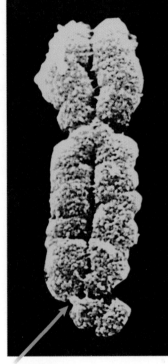

C. J. Harrison

Caused by a mutation at FMR-1 gene

6.5 Essentials

- **There are three general types of mutations.**
- **Mutations are very common and don't always show up as an observable phenotype.**

senescence a resting stage in which a cell does not divide

apoptosis programmed cell death

6.6 How Are Mutations Repaired?

Not all mutations will become a permanent change in the genome. All cells have systems that repair mutations and other forms of damage to DNA. Without repair systems, accumulated mutations and DNA damage would destroy much of the DNA in the cell. Cells that accumulate a lot of mutations and damage can have several fates:

1. Cells can stop dividing and become dormant, a condition called **senescence**.

2. Control systems in the cell can induce cell suicide, a process called **apoptosis.** If your skin has ever peeled after a sunburn, you have seen what happens to cells with DNA so damaged by ultraviolet light from the sun that suicide is the only option.

3. The accumulated mutations can cause the cell to escape the normal controls over cell division and become cancerous.

What type of repair systems exist? Human cells carry over 150 genes devoted to DNA repair. These genes encode enzymes that monitor and repair DNA damage and mutations.

During replication, DNA polymerase adds about 20 nucleotides per second to a newly synthesized strand. Once in a while, an incorrect nucleotide is added to the newly synthesized strand, producing a potential mutation. However, DNA polymerase not only synthesizes a new strand, but it can also proofread its work. If an incorrect nucleotide is added to the new strand, the enzyme can detect the error, reverse direction, and remove the incorrect nucleotide (Figure 6.9). Then the enzyme inserts the correct nucleotide and resumes replication. The few mistakes that elude proofreading remain as mutations (Figure 6.10).

Figure 6.9 Self-correcting DNA Polymerase The enzyme DNA polymerase can proofread the base pairing between the template strand and the new strand. If the enzyme inserted an incorrect base, it is removed and replaced with the correct base.

Figure 6.10 Replication Error Outcomes Base pair substitutions that occur during DNA replication have two possible outcomes. The mutation can be detected by DNA polymerase proofreading and corrected (*bottom left*) or remain undetected and become a mutation.

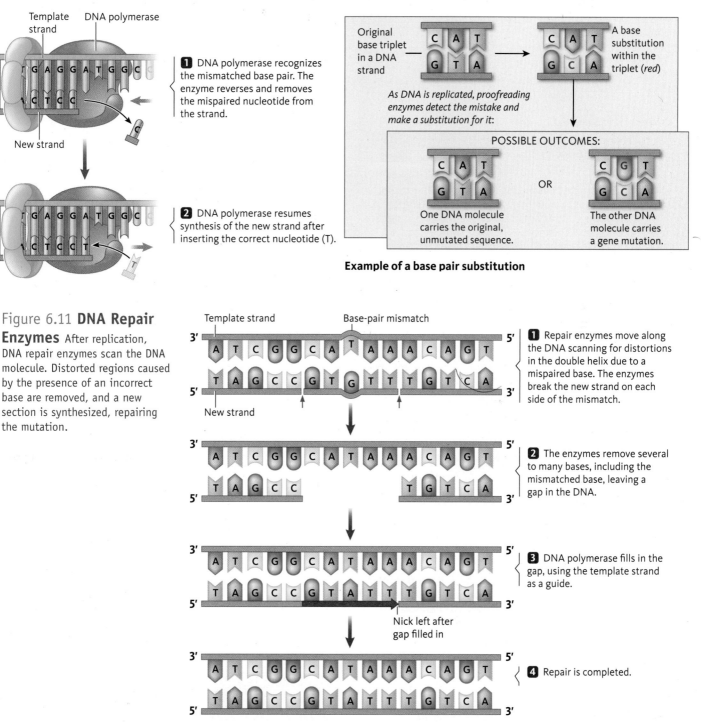

Template strand DNA polymerase

1 DNA polymerase recognizes the mismatched base pair. The enzyme reverses and removes the mispaired nucleotide from the strand.

New strand

2 DNA polymerase resumes synthesis of the new strand after inserting the correct nucleotide (T).

Original base triplet in a DNA strand

A base substitution within the triplet (*red*)

As DNA is replicated, proofreading enzymes detect the mistake and make a substitution for it:

POSSIBLE OUTCOMES:

OR

One DNA molecule carries the original, unmutated sequence.

The other DNA molecule carries a gene mutation.

Example of a base pair substitution

Figure 6.11 DNA Repair Enzymes After replication, DNA repair enzymes scan the DNA molecule. Distorted regions caused by the presence of an incorrect base are removed, and a new section is synthesized, repairing the mutation.

Template strand Base-pair mismatch

New strand

1 Repair enzymes move along the DNA scanning for distortions in the double helix due to a mispaired base. The enzymes break the new strand on each side of the mismatch.

2 The enzymes remove several to many bases, including the mismatched base, leaving a gap in the DNA.

3 DNA polymerase fills in the gap, using the template strand as a guide.

Nick left after gap filled in

4 Repair is completed.

Although very few replication errors escape the proofreading system, cells contain another system for detecting and repairing these mutations. Improper base pairing distorts the structure of DNA, and DNA repair enzymes recognize these distortions. Once the distortion is recognized, these repair enzymes cut the DNA strand on either side of the mismatched nucleotide, removing a portion of the DNA strand (Figure 6.11). DNA polymerase then fills in the gap, and another enzyme closes up the sugar-phosphate backbone. Still other systems repair damage caused by ultraviolet light. Because DNA repair is under genetic control, mutations in repair genes themselves can cause genetic disorders such as xeroderma pigmentosum, as discussed in "Biology Basics: DNA and RNA."

CASE B:
Why Do I Feel This Way?

Ruth had always been sensitive to smells. When she walked down the street in a crowd, she could smell people's colognes and perfumes. She hated it.

rodneyforte.com/Alamy

In stores and at movies she had to move away from some people. It was always hard to tell people not to wear deodorants and perfumes around her. Often they didn't understand.

As she got older, this situation got much worse, and it was hard for her to go out in public. When she was watching television one day she was surprised to see a show on public television about how chemicals affect us. Two people on the panel suffered from a condition called Multiple Chemical Sensitivities (MCS).

People who have this condition are extremely sensitive to chemicals in the environment. They also have a heightened sense of smell. Ruth thought this description fit her perfectly. The expert interviewed on the show was explaining how chemicals can actually enter cells and make changes in the DNA sequence. Ruth wondered, "Has this happened to me?" The expert had called this process epigenetics, and he outlined recent research showing a link between exposure to certain chemicals and epigenetic changes in DNA.

1. Ruth's physician did not think that epigenetic changes had anything to do with her condition, but he thought she might be the "canary in the mine." Do some research on this phrase and discuss how it fits Ruth.

2. On the internet, Ruth found research results indicating that exposure to petrochemicals early in life might be a cause of MCS. She had worked as a lab assistant in college and had been exposed to lots of different chemicals. Was her sensitivity an epigenetic condition? Would any changes caused by the chemicals be passed on to her future children? Why or why not?

3. If epigenetic changes do not cause MCS, do you think someone with Ruth's history of chemical exposure would be more likely to develop sensitivity to chemicals?

4. Should Ruth take any special precautions if she were to get pregnant?

5. Today more people understand about MCS because more people are suffering from it even though they don't know the name of the condition. How should the public be made aware of the damaging effects of chemicals in the environment?

6. One day Ruth called a local restaurant run by a famous chef to make a reservation. The woman on the phone accepted the reservation and said, "Please do not wear any cologne when you come to dinner." That amazed her. She wondered if bans on the wearing perfumes in certain places could become part of the law, just as smoking is prohibited in most indoor public places. What is your opinion?

6.6 Essentials

- Many accidental changes in the DNA are fixed by repair mechanisms built into the cells themselves.
- Not all mutations are detected and fixed by repair systems.

6.7 What Is Epigenetics?

Although we know a lot about gene regulation and how gene sets are turned on and off, it is clear that some things are difficult to explain. For example, identical twins have identical genotypes, but they do not always have identical phenotypes. In other cases, even though we inherit one copy of each gene from our mother and one from our father, in some cases, only the maternal copy or the paternal copy is expressed, while the other copy is always silent.

Epigenetics is the study of the ways in which chemical modifications to DNA and its associated proteins alter gene expression without changing the nucleotide sequence of the DNA. An **epigenetic trait** is a stable phenotype that results from changes in gene expression. Unlike mutations, which involve a change in the DNA sequence, epigenetic modifications do not change the nucleotide sequence, but they do affect how genes behave. The **epigenome** refers to the epigenetic state of a cell. During its life span, an organism has one genome, but the epigenome and the pattern of gene expression can change many times during that lifespan, depending on environmental conditions.

How is DNA modified by epigenetic changes?   Before we discuss epigenetic changes, let's briefly review two concepts: (1) how DNA is organized into chromosomes (see "Biology Basics: DNA and RNA") and (2) how genes become activated and inactivated.

In the nucleus, DNA is wound around clusters of histone proteins to form nucleosomes (Figure 6.12). Nucleosomes are coiled over and over, eventually forming the fibers that we

epigenetics the study of heritable changes in the genome other than changes in DNA sequence

epigenetic trait phenotype that is produced by epigenetic changes to DNA

epigenome the epigenetic state of a cell

promoter regulatory region located at the beginning of a gene

methylation addition of a methyl group to a DNA base or a protein

genetic imprinting selective expression of either the maternal or paternal copy of a gene

see as parts of chromosomes. The control region of a gene is called the **promoter**. If the promoter region is tightly wound around a histone, the gene cannot be expressed because the promoter is not available to bind to RNA polymerase. However, if the histone molecules associated with the promoter are chemically modified, the association with the DNA is loosened, making the promoters available for transcription. These modifications are reversible, and by tightening the grip of histones on DNA, genes can be turned off.

In addition, other epigenetic modifications can add or remove chemical groups to DNA bases in and near a promoter region, and these modifications can also activate or silence genes (Figure 6.13). This modification involves the addition or removal of methyl groups to or from the DNA (Figure 6.14) in a process called **methylation**. Methylation of DNA adds a methyl group ($-CH_3$) to cytosine. DNA methylation occurs almost exclusively on cytosine bases adjacent to a guanine base (a CpG sequence). Both histone modification and DNA methylation are reversible and do not change the base sequence of the modified DNA (and, therefore, are not mutations).

Unlike the genome, which is largely identical in all cells of an organism, epigenetic modifications are specific to a cell type, and during embryonic development, these changes help define which genes will turn on and which will not. If you think about it, if all genes are the same in every cell, we wouldn't want them all turned on at the same time. For example, what might happen if the gene that encodes for a digestive enzyme was active in the eye? Epigenetics helps explain how our cells can selectively turn on or turn off different gene sets in response to factors in the environment. It also explains how identical twins with identical genotypes can have different phenotypes. The discovery that environmental factors lead to epigenetic modifications of DNA has transformed our understanding of how the genome interacts with the environment. Epigenetic research will likely lead to many more exciting breakthroughs in genetics.

What is imprinting?
Although our cells carry two copies of each gene, in some cases, only one copy is active, and the other is silenced. This selective use of either the maternal or paternal copy of a gene is called **genetic imprinting** and is caused by epigenetic changes to DNA. Imprinting is not a mutation or a permanent change in a gene or a chromosome region. What is affected is the expression of a gene, not the gene itself.

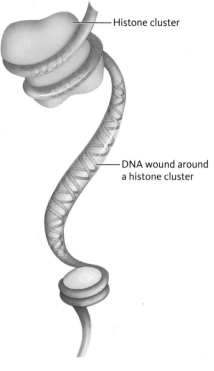

Figure 6.12 **DNA Wound Around a Histone Complex, Forming a Nucleosome**

Histone cluster

DNA wound around a histone cluster

Figure 6.14 **Methylation**
DNA methylation adds methyl groups to bases (mostly cytosine) and to histone proteins. The methyl group is connected to them at the red chemical bond. Methylation is a form of epigenetic modification, which is reversible. In other words, methyl groups can be removed from DNA and histone proteins.

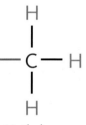

Methyl group

Figure 6.13 **Gene Silencing** (a) Normally the bases in and near the promoter region are not chemically modified, allowing the adjacent gene to be transcribed. (b) In gene silencing by epigenetic modification, the bases in and near the promoter region are chemically modified by methylation. This prevents the binding of RNA polymerase and other proteins, and the gene is not transcribed (it has been silenced).

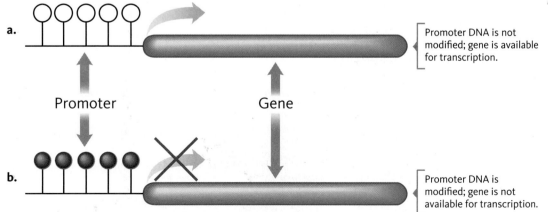

a.

Promoter   Gene

Promoter DNA is not modified; gene is available for transcription.

b.

Promoter DNA is modified; gene is not available for transcription.

Imprinting does not affect all genes. Only genes in certain regions of seven human chromosomes are known to be imprinted. Some of these are regions on the short arm of chromosome 4, the long arm of chromosome 15, and both arms of chromosome 18.

How is genetic imprinting involved in genetic disorders?

Abnormalities of genetic imprinting are responsible for several genetic disorders. The best studied of these is Beckwith-Wiedemann syndrome (BWS), inherited as an autosomal dominant trait (Figure 6.15).

BWS is a cell growth disorder that causes enlarged organs, high birth weight, and predisposition to cancer. BWS is not caused by mutation nor is it associated with any chromosomal aberration (see Chapter 3). Instead, it is a disorder caused by abnormal patterns of imprinting that in turn are caused by improper epigenetic modifications of certain genes. Genes that cause BWS are located in two small clusters on chromosome 11 (Figure 6.16).

Figure 6.15 **Beckwith-Wiedemann Syndrome (BWS)** BWS is associated with overgrowth of tissues and organs. (a) Overgrowth of the tongue makes it too big for the mouth; (b) the edge of the ear is overgrown and enlarged; (c) overgrowth of the intestines causes them to protrude through the navel.

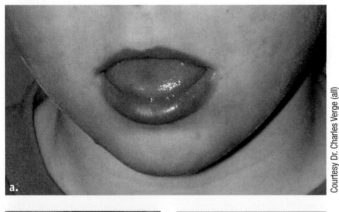

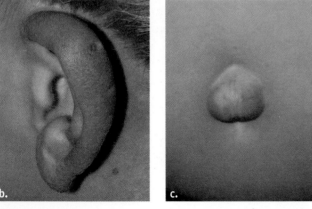

Courtesy Dr. Charles Verge (all)

Figure 6.16 **A Map of Chromosome 11** The arrow shows the location of the BWS gene.

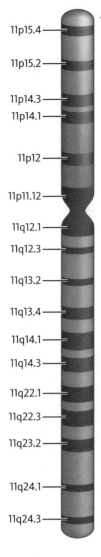

These clusters contain more than a dozen imprinted genes. Some of these genes have the paternal copy active and the maternal copy silenced by epigenetic modification. Other genes in these clusters have active maternal copies, and the paternal copy is silenced by epigenetic modification. All of these genes regulate growth during prenatal development.

In many affected individuals, there is no epigenetic modification (imprinting) of one maternal gene in one of these clusters. As a result, instead of having only the paternal copy active, *both* copies of the gene are activated, causing overgrowth of certain tissues affected in this disease.

What other conditions are associated with epigenetic changes?

Epigenetic modifications of DNA are a response to certain environmental signals and change gene expression in ways that may persist through several generations (see "Spotlight on Science and Law" in this chapter). This can lead to a number of human genetic disorders, including Prader-Willi syndrome (Figure 6.17) and Angelman syndrome (Figure 6.18), as well as Beckwith-Wiedemann syndrome, discussed earlier. Because epigenetic changes turn genes on and off, they have also been suspected in some forms of cancer. In Chapter 12, we will discuss how an understanding of the way epigenetic changes cause cancer is leading to new methods of treatment.

6.7 Essentials

- Epigenetic modifications alter gene expression without permanent changes to DNA.
- Environmental factors can cause epigenetic changes in gene expression.
- Abnormal epigenetic modifications cause genetic disorders.
- Epigenetic changes are not mutations.

Figure 6.17
A Person with Prader-Willi Syndrome

Figure 6.18
A Child with Angelman Syndrome

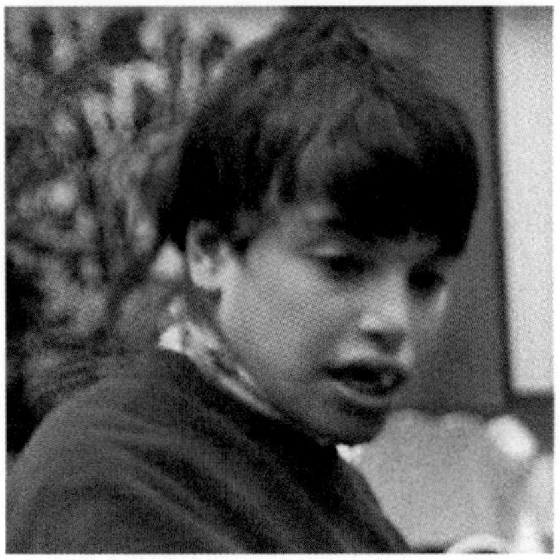

study the impact of environmental factors on the genome at the molecular level. The results of these projects may shed much-needed light on how the environment we are exposed to early in life can affect the diseases we are predisposed to in adulthood.

Are there drugs that use epigenetics to treat diseases? At the present time, there are four epigenetic drugs on the market, all of which are used to treat cancers. In addition, about 30 new epigenetic drugs are being developed. While many of these will also be used in cancer chemotherapy, others will target cardiovascular diseases and neurodegenerative disorders, including Huntington disease and Alzheimer disease. Development of these drugs is slow, because blocking or reversing epigenetic alterations in the genome to treat disease may activate or silence other, non-target genes that might create unanticipated and undesirable side effects. However, the success of epigenetic cancer drugs already in use suggests that epigenetic drug development will produce a new generation of therapies for human disorders.

6.8 Essentials
- The study of epigenetics will throw light on a large number of questions about diseases and changes in the genome.

6.8 What Part Will Epigenetic Research Play in the Future?

Several projects to map the human epigenome are underway. One of these is called the NIH Roadmap Epigenomics Project, based on the idea that many aspects of overall health and susceptibility to disease are the result of epigenetic regulation or misregulation of gene activity. The Human Epigenome Atlas will be one of the results of this project. The project is now collecting and cataloging information on the human epigenome, locating which genes on which chromosomes can be epigenetically modified. This information will be used as a reference standard to study epigenetic changes caused by environmental factors as well as changes observed in various genetic disorders.

Another project, called the Human Epigenome Project, is a multinational, public/private consortium that aims to identify, map, and establish the functions of all DNA methylation patterns in the human genome. It is hoped that analysis of these methylation patterns will establish that genetic responses to environmental cues via epigenetic mechanisms are a pathway to disease.

Although these projects are in their early stages, information already available strongly suggests that we are on the threshold of a new era in genetics, where we will be able to

6.9 What Are the Legal and Ethical Aspects of Mutations and Epigenetics?

When chemicals and other pollutants are shown to cause epigenetic changes in the genome, will lawsuits be the next step?

Class-action lawsuits, which involve large numbers of plaintiffs, have been going on for a long time. Exposure to asbestos (Figure 6.19, page 140) has been proven to cause **mesothelioma**. Mesothelioma is a cancer of the lungs that takes years to develop but then grows and spreads very quickly and is almost always fatal. Even though companies that used asbestos in their manufacturing knew that it was dangerous, they did not tell their employees. When the employees became sick years later, they sued the employers. These lawsuits, called **liability** suits, are still going on.

class-action lawsuits lawsuits that involve many plaintiffs
mesothelioma form of lung cancer caused by exposure to asbestos
liability legal term meaning legally responsible

Figure 6.19 **Asbestos Warning Sign** Occupational exposure to asbestos is a risk factor for mesothelioma.

DANGER
ASBESTOS
CANCER AND LUNG DISEASE HAZARD
KEEP OUT
AUTHORIZED PERSONNEL ONLY
RESPIRATORS AND PROTECTIVE CLOTHING ARE REQUIRED IN THIS AREA.

Bronwyn Photo/Shutterstock.com

causation a legal term meaning that one thing causes another

settlement amount of money paid out to avoid taking a case to court

Liability means that the defendant (in these cases, the companies) is being held responsible for the illnesses of the plaintiffs. In such cases, the most important point plaintiffs must prove is **causation**. Showing that the employers knew that working with asbestos poses a high risk of mesothelioma is the best argument. In Chapter 12, we will see that this point was very important in the lawsuits against tobacco companies brought by cancer patients.

There have been hundreds of thousands of liability claims against asbestos companies and manufacturers of asbestos products over the past 30 years. There have also been thousands of **settlements**. In a settlement, the companies and employees agree on an amount of money that will satisfy both parties, and so the lawsuits do not go to court for trial. Because many cases have been settled, some companies set up trust funds to pay off claims without a court fight.

This type of lawsuit is widely known by the public for three reasons. First, the press covers these lawsuits because of the sympathetic nature of the plaintiffs. Second, the suits usually ask for large amounts of money because it has to be divided among many (sometimes hundreds) of plaintiffs. And, third, the attorneys who specialize in these cases are often very flamboyant.

One such attorney, Jan Schlichtmann, and his partner, Jerome Facher, were portrayed in a book, *A Civil Action* by Jonathan Harr (see Section 12.7), which was later made into a movie. This book detailed one class-action lawsuit against a company that buried pesticides on its property, and these pesticides eventually poisoned the water of an entire community in Massachusetts. The case is known as *Anderson v. Cryovac*. Table 6.4 presents some of the issues that arise from class-action lawsuits.

Table 6.4 **Legal and Ethical Questions in Class-Action Suits**

Question	How Are These Questions Decided?	Legal Issue
What is the major argument of the plaintiffs in these cases?	By the judge and the attorneys for the plaintiffs	The company's actions injured me.
What is the major argument of the defendants in these cases?	By the judge and the attorneys for the defendants	We had no idea that our product was dangerous.
What if an injured person does not want to join a class-action suit?	The individual decides what he or she wants to do. The injured person might sue alone, drop the case, or negotiate a settlement with the defendant.	This issue is sometimes decided when an attorney or other injured parties convince the person to join the lawsuit.
What are the negatives of class-action suits?		Too many plaintiffs, not enough money, the suits take a very long time, and people sometimes die before settlements are reached.
What are the positives of class-action suits?		Attorneys take no fees, the plaintiffs do not have to get involved, there is quite a bit of money to be divided.

Spotlight on Science and Law: The Case of DES

Our spotlight for this chapter falls on a case that began over 60 years ago and is still continuing today. It involves **diethylstilbestrol (DES)**, a synthetic estrogen, and the companies that manufactured, marketed, and advertised it to physicians (Figure 6.20) for use with pregnant patients and those who were trying to get pregnant.

From the 1950s through the 1970s, DES was prescribed in increasing doses to women in the belief that it would prevent miscarriages. Scientifically, it seemed plausible; if a woman had more estrogen in her body, the lining of the uterus was less likely to become detached, saving a pregnancy. In the beginning this proved to be true, and many women gave birth to seemingly healthy babies. However, in the long term, there were many problems.

Research has identified three generations affected by treatment with this drug. They include the woman given the drug (generation 1), their daughters and sons (generation 2, called DES daughters and DES sons) and the third generation (called DES granddaughters and DES grandsons). A fourth generation has not been yet been studied.

When DES daughters showed up in physicians' offices with a rare form of vaginal cancer, studies on the effects of DES treatment began. It was not until years later that the reproductive problems of DES sons were studied. Table 6.5 shows some of the results of these studies.

diethylstilbestrol (DES) a synthetic estrogen used as a drug and as a food additive

Figure 6.20 **Ad Promoting Use of Diethylstilbestrol (DES)**

Publicité américaine de 1957 © DES France

Table 6.5 **Generational Effects of DES**

	First Generation	Second-Generation Females	Second-Generation Males	Third-Generation Females	Third-Generation Males
Physical	None	Uterine shape abnormalities	Increased risk of and urogenital abnormalities at birth	More likely to have irregular periods	Hypospadia
Behavioral	None	Increased likelihood of a woman being lesbian or bisexual[1]	Increased likelihood of a man being homosexual, transgendered, or feminized.[2] Increased risk of undescended testicles, infertility, and epididymal cysts	Not known	Not known
Medical	Increased risk of breast cancer	Vaginal cancer (40x risk), increased risk of breast cancer, infertility, ectopic pregnancy, and fibroid tumors	Not known	Increase in risk of infertility and ovarian cancer	Not known
Suspected symptoms		Autoimmune conditions	Increase in testicular cancer	Not known	Not known

1. Ehrhardt, A. A., H. F. L. Meyer-Bahlburg, L. R. Rosen, J. F. Feldman, N. P. Veridiano, I. Zimmerman, and B. S. McEwen (1985). "Sexual orientation after prenatal exposure to exogenous estrogen," *Archives of Sexual Behavior* 14 (1): 57–77. doi:10.1007/BF0154135.

2. Kerlin, S. P. (2006–08). "Prenatal exposure to diethylstilbestrol (DES) in males and gender-related disorders: Results from a 5-year study," International Behavioral Development Symposium 2005.

In the 1970s, the negative publicity surrounding the discovery of the long-term effects of DES resulted in a huge wave of lawsuits in the United States against its manufacturers. These culminated in a landmark 1980 decision of the Supreme Court of California, *Sindell v. Abbott Laboratories,* in which the court said a **market share liability** existed against all DES manufacturers even if the mother could not identify who made the actual pill they took.

market share liability a legal term referring to the division of large settlements in class-action lawsuits among the manufacturers of the product

QUESTIONS

1. The effects of DES on subsequent generations have be identified as epigenetic changes. Do you agree? Why o not?

2. Is it possible for grown children to know if their moth were given DES? How could they find out?

3. To what other types of lawsuits could market share lia be applied?

4. What do you think mothers who took DES felt when th children were diagnosed with these conditions?

10

The Essential

1 Mutations are heritable changes in the nucleotide sequence of DNA. (Section 6.1)
They are passed on if they occur in eggs or sperm.

2 Not all mutations are visible in the human phenotype. (Section 6.1)
Invisible mutations occur every day.

3 Some mutations are caused by mistakes during DNA replication. (Section 6.2)
This happens when the wrong nucleotide is inserted in a new DNA strand.

4 Radiation and chemical agents also cause mutations. (Section 6.2)
Background radiation is unavoidable.

5 It is often difficult to determine when and where a mutation occurs. (Section 6.3)
Pedigree construction is often used to discover when a mutation has occurred.

6 Mutations can also show up as increases in the number of trinucleotide repeat sequences. (Section 6.4)
Huntington disease is caused by an increase in CAG repeats.

7 Mutations can be repaired. (Section 6.6)
Many repair mechanisms check for mutations and fix them.

8 Epigenetics is a new and developing field of study. (Section 6.7)
Epigenetic changes in the genome may cause many effects in humans, both positive and negative.

9 Epigenetic changes in the genome are not permanent and can be reversed. (Section 6.7)
Turning genes on and off in different cells and at different times can change the phenotype.

10 In the future, mutations may be the reason for any number of class-action lawsuits. (Section 6.8)
Large numbers of people may have been adversely affected by chemicals in the environment.

Review Questions

1. What is a mutation?

2. What part do histones play in epigenetics?

3. How do mutations form?

4. Are most mutations repaired?

5. What is epigenetics?

6. Give an example of a trinucleotide repeat and the disease it is associated with.

7. How does DNA polymerase work to correct its own mistakes?

8. What is imprinting and how does it affect genetic conditions?

9. What are the three main types of mutations?

10. List two types of genetic mutations and tell how they occur.

Application Questions

1. As discussed, Huntington disease is caused by a significant increase in the number of trinucleotide repeats. The longer the repeat, the earlier the onset of symptoms. How do you think this happens?

2. In a real case, a man went for genetic testing and found he did not carry the cystic fibrosis gene, even though he and his wife had a son with CF. What does this mean about the paternity of his son? What might cause this to happen? If the gene were dominant, would it be more likely to be a mutation? Why or why not?

3. Research two mutations associated with expansion of trinucleotide repeats not listed in the chapter. Explain how they manifest themselves in the phenotype.

4. Make a chart of chemicals that can possibly harm developing embryos (these chemicals are called teratogens).

5. Do research on the NIH Epigenetics Roadmap Project and write a report on it.

6. Research the *Anderson v. Cryovac* lawsuit and write a short summary. Use the book *A Civil Action* as a source.

7. Read the book *A Civil Action* and then watch the movie. Discuss them within groups in your class. Answer the question: was the movie an accurate portrayal of the case and the attorneys?

8. In Chapter 5, we discussed *Norman-Bloodsaw v. Lawrence Berkeley Laboratory* (135 F.3d 1260, 1269 [9th Cir. 1998]) and *EEOC v. Burlington N. Santa Fe Railway Co.*, No. 02-C-0456 (E.D. Wis. 2002). How are these cases different from the asbestos cases mentioned in Section 6.9?

9. Create a chart about DES. In the chart, list how many DES daughters and sons have sued under the market share liability legal doctrine, the name of these cases, and the results.

. . . a serious mutagen, such as radon, were found in your home?

. . . you discovered *before* you became pregnant that you and your partner each carried a recessive gene for a genetic disorder?

. . . you discovered *after* you became pregnant that you and your partner each carried a recessive gene for a genetic disorder?

Online Resources

Preparing for an exam? Log on at www.cengagebrain.com for study tools to help you assess your understanding. If assigned by your instructor, the Case B and Spotlight on Science and Law activities for this chapter will also be available.

In the Media

NewsHerald.com, April 25, 2011

Research at the DDC Clinic May Have ID'd Gene Mutation Linked to Form of Dwarfism

Angela Gardner

A small clinic in Middlefield Township hopes to make a big difference for Amish families who suffer from genetic disorders. The latest research coming out of the clinic, called the Center for Special Needs Children, might help identify gene mutation responsible for a rare form of dwarfism, which has been found in many of Ohio's Amish communities.

Microcephalic osteodysplastic primordial dwarfism type 1 is a rare disorder that greatly slows an embryo's growth in the uterus and causes severe brain and organ abnormalities and deformities of the arms and/or legs, and could lead to death in infancy or early childhood.

Finding such a mutation will no doubt lead to a prenatal test for the condition.

QUESTION

If a prenatal test was developed, would this alter the number of babies born with this form of dwarfism? Why?

To access this article online, go to www.cengagebrain.com.

NHS Choices, April 22, 2011

Can Your Pregnancy Diet "Make Your Child Fat"?

"A mother's diet during pregnancy can alter the DNA of her child and increase the risk of obesity," reported BBC News.

The news story is based on a study that looked at maternal diet and how it might be associated with epigenetic changes in the offspring. Epigenetics is the study of how genes can be influenced by the environment, without their DNA sequence being directly changed. Researchers measured DNA methylation, a form of epigenetics. They wanted to see if the measurement showed anything about the weight of a child at 6 and 9 months.

The researchers found an association between lower consumption of carbohydrates during early pregnancy and increased methylation of the RXRA gene, a gene associated with childhood fat retention. It was unclear whether the diet of the mothers during pregnancy caused the methylation of the gene.

QUESTION

Most people would not want their child to be overweight. What might be asked of pregnant women to stop this from happening?

To access this article online, go to www.cengagebrain.com.

7

Biotechnology

REVIEW

How are DNA molecules organized? (BB-2.3)

How do genes control traits? (Section 5.1)

CENTRAL POINTS

- Recombinant DNA technology can link together DNA from different organisms.
- Biotechnology uses recombinant DNA technology to make products.
- Bacteria, plants, and animals can be modified using biotechnology.
- The safety of transgenic organisms continues to be debated.
- Human proteins can be produced by biotechnology for disease treatment.
- Many biotechnology inventions have been patented.

CASE A: A Taller Son for Chris

Chris Crowley is concerned about his 10-year-old son, Mike, who is short—not just short for his age, but very short. Just today Mike came home from school crying. Some kids had made fun of him and pushed him around. This was not the first time.

Chris had gone through this himself. His adult height is only 5 feet 6 inches (1.6 m) tall, which is fairly short for a man. During his teen years, his friends had made fun of him, and he didn't want Mike to go through that.

Chris had recently read an article about human growth hormone (hGH), a drug that might help Mike. hGH is safe and available because now it is made using **recombinant DNA technology**. Children given injections of the drug before puberty sometimes grew 5 to 6 inches (12.7–15.2 cm). Athletes also used it to increase their strength for certain sports. Chris thought that if Mike could grow a few inches and be stronger, it would give him confidence and possibly a better life.

When Chris brought this up with Mike's pediatrician, Dr. Sanchez, he was surprised

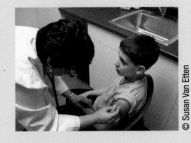

when she said that Mike wasn't a candidate for hGH treatment. The reason was that Mike was still within the "normal" range of height for his age, and the hormone is used to treat only the shortest 1–2% of all children. She explained that even though Mike was shorter than his friends, he might still grow to be within the normal range of adult height. Moreover, the side effects of this drug are still unknown. Chris left the doctor's office feeling somewhat confused and let down.

Some questions come to mind when reading this case. Before we can address these questions, let's look at what biotechnology and recombinant DNA are, how they are used to create drugs, and their other uses.

A High Resolution Photo of DNA

7.1 What Is Biotechnology?

Biotechnology couples biological organisms with genetic technology to create products, processes, and services. Many fields contribute to biotechnology, including microbiology, biochemistry, genetics, medicine, environmental sciences, and computer science. The relationships among some of these areas in biotechnology are shown in Figure 7.1, page 148. Underlying biotechnology are several processes, collectively called genetic engineering. These processes include recombinant DNA technology, gene modification, and gene technology. Genetic engineering allows scientists to cut genes out of DNA and to splice or recombine these genes into other DNA molecules. These recombinant DNA molecules can be transferred to organisms, giving them the ability to produce a specifically selected new product (for example, a bacterial cell making a human protein). The ability of a genetically engineered organism to make a product is useful in medicine, agriculture, forensics, industry, and other areas.

Biotechnology has many uses, one of which is making human proteins for medical purposes, including human growth hormone (hGH). As we will see, this application uses several aspects of biotechnology, including recombinant DNA technology, microbiology, and medicine.

Before the development of biotechnology, proteins used to treat disease—such as insulin, hGH, and blood-clotting factors—were collected from many sources, including animals in slaughterhouses, human cadavers, urine, and donated human blood. In some

recombinant DNA technology the process by which recombinant DNA is created

biotechnology scientific process that uses recombinant DNA to make products

Figure 7.1 **The Range of Methods and Uses of Biotechnology**

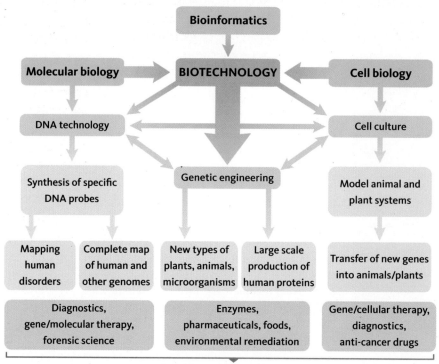

Resulting products and services

Table 7.1 **FDA-Approved Uses for Human Growth Hormone**

Children
hGH insufficiency or deficiency
Turner syndrome
Prader-Willi syndrome
Chronic renal insufficiency
Infants born small for gestational age
Adults
hGH deficiency
AIDS-associated wasting syndrome

transgenic organism animals, plants, or microbes that have been genetically altered using recombinant DNA techniques

cases, proteins from these sources were contaminated with disease-causing agents that exposed people to serious and potentially fatal risks. For example, before biotechnology was used, some batches of clotting factor made from donated blood were contaminated with HIV (human immunodeficiency virus), the virus associated with AIDS (acquired immunodeficiency syndrome). Many people treated with these contaminated batches of clotting factor became infected with HIV and developed AIDS.

Until 1985, children with growth problems were treated with growth hormone isolated from pituitary glands removed from human cadavers. The available supply was limited, and only a small number of children could be treated. More than two dozen of the 7000 children treated contracted a deadly brain condition because the hGH was contaminated with prions, a disease-causing agent similar to the one that causes mad cow disease.

In 1985, the FDA approved the use of hGH produced using biotechnology and removed the cadaver-derived hGH from the market. Combining technology with genetics allowed human growth hormone to be produced in bacteria, with two immediate benefits: there is no possibility of prion contamination, and unlimited amounts of the hormone can potentially be produced. In fact, partly because of these advantages, the U.S. Food and Drug Administration has expanded the number of conditions that can be treated with this hormone (Table 7.1). hGH can now be used to treat children with very short stature (children shorter than Chris Crowley's son) and not just those with serious growth disorders.

How is biotechnology used to make human growth hormone?

As mentioned above, recombinant DNA technology is one of the methods used in biotechnology. To make human growth hormone, scientists isolated the human gene for hGH from a human cell and transferred it to a bacterial cell, creating a **transgenic organism** (an organism that carries a gene from another species). The transgenic bacterial cell and its descendants synthesize hGH, which is recovered and purified for medical use.

More recently, researchers successfully transferred the gene for hGH into cows, which secrete the hormone into their milk. The milk from these cows is collected, and the hormone is extracted and purified. It is estimated that a herd of 15 transgenic cows would be able to meet the world-wide demand for hGH. Biotechnology is being used to transfer genes for other medically important human proteins into animals and plants, where these proteins are synthesized in large quantities before extraction and purification for use in treating diseases.

7.1 Essentials

- **Recombinant DNA techniques can be used to transfer genes among species.**
- **Transgenic organisms can produce human proteins that are used in medical treatment.**

7.2 What Is Recombinant DNA Technology?

Recombinant DNA can be defined as the combination of DNA from two or more different organisms. Several steps are required to create human proteins such as hGH in transgenic organisms; these steps are part of recombinant DNA technology. To see how this process works, follow along with Figure 7.2.

- Identify the gene for growth hormone in humans.
- Extract DNA from human cells that were grown in the laboratory.
- Treat the DNA with a **restriction enzyme**, a protein that cuts the DNA into fragments at specific sites (step 1).

Often these cuts leave an overhanging single-stranded region called a "sticky end" (Table 7.2).

- Cut a circular DNA molecule (called a **plasmid**) with the same restriction enzyme (step 2). Many species of bacteria carry these small, circular DNA molecules. Researchers have genetically modified plasmids to create carrier molecules called **vectors** that are used in recombinant DNA technology. This plasmid DNA will be used to carry the inserted human gene into bacterial cells.
- The fragments of human DNA and fragments of plasmid DNA (called vectors) created by restriction enzyme digestion are mixed and spliced together to form **recombinant DNA molecules** (steps 3 and 4).
- The recombinant plasmids carrying human DNA are transferred into bacterial cells (step 5).
- The bacteria that carry the human gene for growth hormone are identified. The process has been engineered so that bacterial cells carrying the growth hormone gene will synthesize the human protein. Once isolated, these bacteria are used to synthesize human growth hormone, which is recovered, purified, and used for medical treatments.

restriction enzyme a protein that, when mixed with DNA, cuts the DNA in a specific place

plasmid circular DNA found in bacteria

vector molecules used for carrying a DNA segment into a host cell, where it is copied

recombinant DNA molecule a DNA molecule that has had a new section added

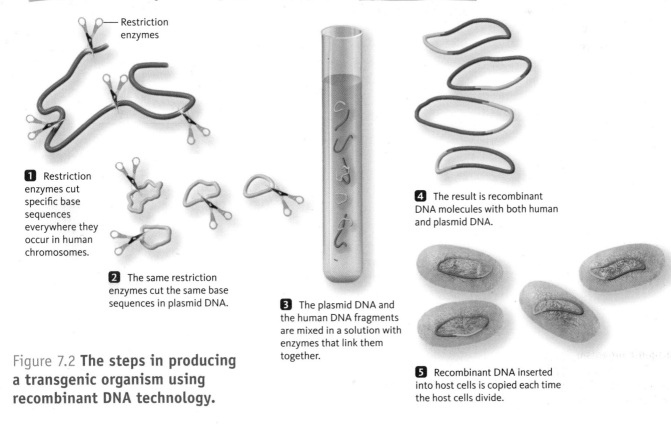

1 Restriction enzymes cut specific base sequences everywhere they occur in human chromosomes.

2 The same restriction enzymes cut the same base sequences in plasmid DNA.

3 The plasmid DNA and the human DNA fragments are mixed in a solution with enzymes that link them together.

4 The result is recombinant DNA molecules with both human and plasmid DNA.

5 Recombinant DNA inserted into host cells is copied each time the host cells divide.

Figure 7.2 **The steps in producing a transgenic organism using recombinant DNA technology.**

Table 7.2 Restriction Enzymes

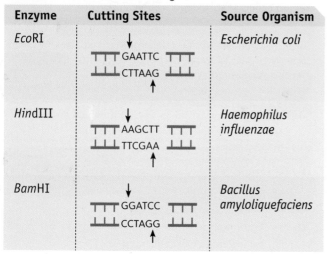

Enzyme	Cutting Sites	Source Organism
EcoRI	GAATTC / CTTAAG	Escherichia coli
HindIII	AAGCTT / TTCGAA	Haemophilus influenzae
BamHI	GGATCC / CCTAGG	Bacillus amyloliquefaciens

Table 7.3 Some Drugs Produced by Biotechnology

Drug	Uses
Abatacept	Rheumatoid arthritis
Alefacept	Psoriasis
ATryn	Anti-thrombin deficiency
Erythropoietin	Anemia
Etanercept	Immune disorders
Factor VIII, X	Blood-clotting disorders
Follicle-stimulating hormone	Fertility treatment
Infliximab	Immune disorders
Interferons	Leukemia
Hepatitis B vaccine	Hepatitis B immunity
Human growth hormone	Growth deficiency
Insulin	Diabetes
Trastuzumab (Herceptin)	Breast cancer

Although we have used the production of human growth hormone as an example, the same process is used to transfer genes from many different organisms into bacteria, yeast, plants, and animals. These transgenic organisms are used to produce a variety of products such as vaccines or proteins for medical use. In agriculture, transgenic plants are used directly as crops with nutritional enhancements or herbicide resistance. Many products we eat, use in our homes, take as medicines, and even wear are or will be produced by biotechnology (Table 7.3).

7.2 Essentials

- **Restriction enzymes bind to DNA at specific recognition sequences and cut the DNA at these sequences.**
- **Recombinant DNA is created when DNA from two or more different organisms is joined.**
- **Biotechnology is the commercial use of recombinant DNA techniques to produce drugs, enhanced crop plants and farm animals, industrial chemicals, waste treatment, bioremediation, and many other uses.**

7.3 How Is Biotechnology Used to Make Transgenic Plants?

Transgenic crops have been grown in the United States since 1996. The graph in Figure 7.3 shows the dramatic increase in the production of transgenic crops between 1996 and 2006 in the United States. Products made from corn and soybeans, as well as cottonseed and canola oils, currently account for the majority of foods that contain transgenic ingredients.

Biotechnology is widely used in agriculture to make transgenic crop plants with new characteristics (Figure 7.4). Genes transferred to the crop plant can originate in another plant, an animal, or even fungi or bacteria. One or more new genes are used to give the transgenic plant a unique trait, such as resistance to **herbicides** (chemical weed killers), insects, or viral or fungal diseases.

herbicides poisons that kill plants

transgenic crops crops that have been altered by introduction of DNA from another organism

Figure 7.3 The Use of Transgenic Crops in the United States between 1996 and 2006

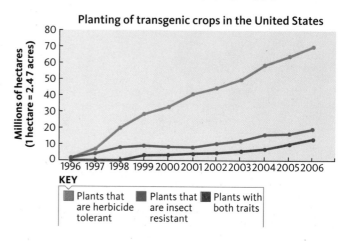

Planting of transgenic crops in the United States

KEY
- Plants that are herbicide tolerant
- Plants that are insect resistant
- Plants with both traits

Figure 7.4 The Production of Transgenic Crop Plants

A specific gene is incorporated into a bacterial cell and used to transfer a genetic trait into a plant cell (1). The plant cells are grown in the laboratory (2), and then transferred to a field to form a mature genetically modified organism (3). Traits transferred to crops in this way are herbicide resistance and resistance to insects.

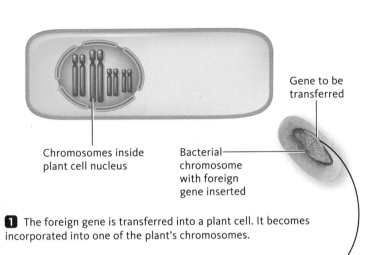

Gene to be transferred

Chromosomes inside plant cell nucleus

Bacterial chromosome with foreign gene inserted

1 The foreign gene is transferred into a plant cell. It becomes incorporated into one of the plant's chromosomes.

Gene being transferred

2 The plant cell divides to form an embryo that develops into a transgenic plant as shown below.

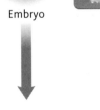

Embryo

3 This photo shows genetically modified plants that carry a new trait. These plants will be transferred to a field and grown in large numbers.

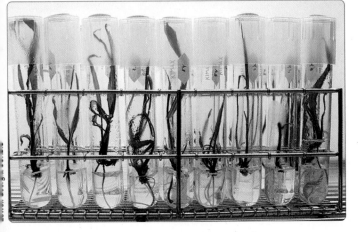

CASE A: QUESTIONS

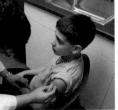

© Susan Van Etten

Now that we understand how some biotechnology processes work, let's look at some questions related to Mike's short stature.

1. After the doctor's visit, Chris decided he must deal with Mike's short stature without the use of hGH. Suggest four ways that Chris could deal with Mike's stature.

2. What do you think Chris should do?

3. Did Dr. Sanchez do the right thing in denying Mike hGH treatment?

4. Should parents be able to make all medical decisions for their children?

5. If Dr. Sanchez administered hGH to Mike, what might happen?

6. Should Mike be included in the decision whether or not to undergo hGH treatment?

7. Human growth hormone is legal in the United States for treating those with low hGH levels and children who are considered too short. If hGH has other potential uses, should it be used for those purposes? Why or why not?

8. Athletic associations including the International Olympic Committee (IOC) and the World Anti-Doping Agency (WADA) have concluded that use of hGH by athletes might give them an advantage over others; therefore, they have banned the use of this hormone. What do you think?

Some other examples of products created with biotechnology are shown in Figure 7.6, page 152.

Figure 7.5 The use of transgenic crops in 22 countries has increased every year following their introduction.

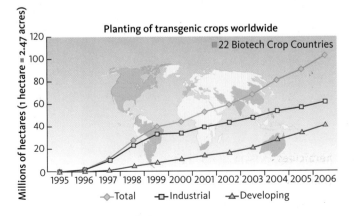

Planting of transgenic crops worldwide

22 Biotech Crop Countries

Millions of hectares (1 hectare = 2.47 acres)

1995 1996 1997 1998 1999 2000 2001 2002 2003 2004 2005 2006

—◆— Total   —□— Industrial   —△— Developing

Figure 7.6 **Some Products of Biotechnology**

Transgenic Plants and Animals

Photo courtesy of Professor Qing-Hu Ma, Institute of Botany, Chinese Academy of Sciences

Transgenic tobacco plants are also being used to produce hGH. Most work with transgenic plants to make human proteins is still in early stages of development, but soon fields of transgenic plants or small herds of transgenic animals may replace laboratories for making human proteins used in medical treatments.

Eric Carr/Alamy

One of the most successful uses of recombinant DNA technology is the synthesis of human insulin. Many people who have diabetes must take the protein insulin because their bodies do not produce it. Before insulin was made using recombinant DNA, it was extracted from the pancreases of pigs and cows. Some people with diabetes had allergic reactions to pig and cow insulin. Now that recombinant human insulin is available, such reactions are rare.

© Golden Rice Humanitarian Board

Golden rice is a transgenic strain of rice that contains two genes from daffodils and one gene from bacteria. With these added genes, the rice plants synthesize beta-carotene, a compound that turns the rice a golden color. When the rice is eaten, beta-carotene is converted into vitamin A. In many Asian countries there is a dietary deficiency of vitamin A. Golden rice strains are now available for planting in Southeast Asia.

Wild Type

© Susan Van Etten

Factor VIII is a protein that helps form blood clots. People with one form of hemophilia need Factor VIII because it is not present in their blood. Recombinant DNA techniques have been used to make human Factor VIII, so people with hemophilia do not have to use blood donors to get this needed protein. In the 1980s, many people with hemophilia contracted HIV from tainted blood products.

© Workbook Stock/Jupiter Images

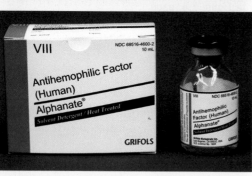

The human tissue-typing gene complex called HLA has been transferred to pig embryos. The adult pigs that result have organs compatible with humans. Transplanting these organs (called *xenotransplantation*) to needy recipients can increase the number of organs available for transplantation.

Transgenic plants resistant to herbicides are now available worldwide and are being used in almost two dozen countries (Figure 7.5, page 151). Transferred genes can also increase the nutritional value of crops, helping eliminate dietary deficiencies that are widespread in many parts of the world.

7.3 Essentials
- **Transgenic bacteria, plants, and animals are created with recombinant DNA techniques.**

7.4 Are Transgenic Organisms Safe?

The development and use of transgenic crops have generated controversy in many regions, including the United States, Europe, and developing African nations. Genetic engineering has been used to make crop plants resistant to herbicides, pests (as shown in the cotton plant photo in Figure 7.7), and disease, and to change nutritional value, chemical content, growing season, and crop yields. Accompanying these advances, many questions have been raised about the safety and environmental impact of transgenic crops, often called *genetically modified (GM) crops* or *genetically modified organisms (GMOs)*, such as the Flavr Savr tomato (Figure 7.8). Several organizations and movements actively oppose the introduction and use of transgenic crops. Controversies about transgenic crops focus on safety and labeling, and the rights of farmers, loss of traditional crops, and biodiversity, among others. As transgenic crops become more common and new ones are developed, we clearly need to identify and monitor any health and environmental risks and resolve the related economic and social issues.

What do we know so far about these risks of transgenic crops?
Most food produced from transgenic crops contains one or more proteins encoded by a transferred gene. Standards for evaluating the safety of foods from transgenic crops are in place in many European countries and the United States. In general, transgenic crops are considered safe if the proteins produced by the transplanted genes are not

Figure 7.7 (left) A cotton plant attacked and destroyed by insects. (right) A transgenic cotton plant engineered to be insect-resistant.

© Agricultural Research Service/USDA

Figure 7.8 **Transgenic Flavr Savr Tomatoes** These tomatoes were engineered to ripen on the vine before shipping and to withstand transport long distances without softening. Vine ripening was thought to improve taste and flavor. Consumer demand for this product was low; some thought it still tasted like the typical store-bought tomato. The product was eventually withdrawn from the market.

© Tom Myers/Photo Researchers, Inc.

poisonous and do not cause allergies any more than plants crossbred in the normal breeding method.

As an example, let's consider herbicide-resistant crops. The bacterial protein encoded by the transplanted genes is broken down in our digestive system when the plant is eaten. In studies with mice, the protein was not poisonous even in very high doses. This protein does not contain any amino acid sequences that are similar to any known toxins or allergens. After 15 years of widespread use, no human health risks associated with these crops have been identified.

However, several environmental risks related to transgenic crops have been identified, including the transfer of the transplanted genes to wild plants by crossbreeding. This may ultimately reduce biodiversity, with a possible negative impact on the environment.

This problem and others related to transgenic crops are the same as those we face using conventional crops. Even though transgenic crops have a different genetic history from that of conventional crops, their effects on the environment are very similar.

As gene transfer technology becomes more sophisticated, novel combinations of traits may be developed, requiring specific management plans. For example, if production of pharmaceutical products in plants becomes common, it will be necessary to develop safeguards to keep such plants out of the food supply and to ensure that transgenes for those proteins are not transferred to wild plants.

Humans have been genetically modifying agricultural plants and animals for more than 10,000 years using selective breeding and crossbreeding to produce the diversity of domesticated plants and animals we depend on for our food. Biotechnology has changed only the way and the rate at which these changes are made. It has increased the specificity and predictability of changes that can be made and also expanded the range of species that can donate genes. To continue improving plants and animals by conventional breeding and biotechnology, research, testing, and public education must address scientific questions, economic issues, health safeguards, and public perceptions.

7.4 Essentials

- **Safeguards are needed in developing and using transgenic organisms to produce food and drugs.**

7.5 Can Transgenic Animals Be Used to Study Human Diseases?

Information gained from the **Human Genome Project** (discussed in Chapter 10), which has sequenced all the human genes, combined with genome information from plants and other animals, shows that many genes found in the human genome are also present in other species, including the mouse and even the fruit fly, *Drosophila*. Because of these similarities, a growing set of transgenic animals (Figure 7.9) is being generated and used to explore mechanisms of human disease and to test drugs developed to treat genetic disorders. These **animal models** of human disease are important in research. Without an animal model, it is often difficult to study new drug treatments and the underlying causes of a disease. Models of specific human diseases can be created by transferring human disease genes into animals. These transgenic animal models can be used to:

- produce an animal with symptoms that mirror those in humans

- study the early stages of development and progress of a disease

- develop and test drugs that hopefully will cure or treat the animal model of the human disease

Eventually, this information and the drugs developed using animal models will be used to treat human diseases.

Because of the close genetic similarity between mice and humans, mice have been used extensively as animal models of human diseases (Figure 7.10).

Human Genome Project the project that is deciphering the organization, sequence, and function of all human genes

Drosophila a fruit fly used as a model genetic organism

animal models animals used in laboratory work to mimic human diseases

Figure 7.9 The mouse on the left was genetically engineered to be obese (normal mouse is on the right). The obese mouse is used as a model organism to study the genetic control of body mass and weight.

© Courtesy Salih Wakil, Ph.D., Baylor College of Medicine

Figure 7.10 **Some Genetically Engineered Mice Used as Model Organisms to Study Human Diseases**

© Catherine Chalmers

The *rhino mouse* is used to study immune deficiency conditions.

© Catherine Chalmers

The *curly tail mouse* is used to study neural tube defects such as spina bifida.

© Catherine Chalmers

The *obese mouse* is used to study products that can help with weight loss.

In one approach, a transgenic mouse model of Huntington disease (HD) has been developed and is being used to study this fatal genetic disorder. (Recall from Chapter 4 that members of Alan's family had this neurodegenerative disorder, which is inherited as an autosomal dominant trait.) HD transgenic mice (which have the mutant human allele for HD inserted in their DNA) are being used to study what happens to the brain in the earliest stages of the disease, something that is impossible to do in humans. In addition, HD mice are used to link changes in brain structure with changes in behavior.

Research on these mice has identified several molecular mechanisms that play important roles in the early stages of the disease. These model organisms are also being used to screen drugs to identify those that improve symptoms or reverse brain damage. Several candidate drugs have been identified, some of which are now being tested in human clinical trials as experimental treatments for HD. Similar methods have been used to construct animal models of other human genetic disorders using mice and other organisms, including the fruit fly, *Drosophila melanogaster*.

How is biotechnology used to treat genetic diseases?
It seems as though the use of recombinant DNA technology might be applied to cure genetic diseases. After all, if we can isolate a specific gene, why not just place it in a person who carries one or two mutant alleles of that gene and have the normal allele make the correct protein?

This simple idea is the basis for gene therapy trials being conducted around the world. However, in going from idea to application, two major problems exist. First, after a gene is isolated, it must be delivered to target cells and inserted into a chromosome. Then the gene must be switched on to direct the synthesis of enough of the normal protein to correct the genetic disorder. For example, someone with cystic fibrosis (CF) has problems with mucus secretion caused by ineffective chloride transport across the cell membrane. As you learned in Chapter 5, this can have life-threatening symptoms. After the gene for cystic fibrosis was identified, it was thought that transferring the normal allele into a child with CF might result in the uptake of the gene into cells and the production of a functional version of the chloride transport protein. However, in reality, this has proven to be far from the case. Some gene therapy clinical trials have been partially successful, while others have been tragic. In Chapter 10 gene therapy is explained in detail.

7.5 Essentials
- **Transgenic animals have been used to create animal models for human diseases.**
- **The same techniques are being tested to create cures for human genetic diseases.**

7.6 How Can Stem Cells Be Used in Biotechnology?

In Chapter 1, stem cells were identified as part of the developing embryo. These **embryonic stem cells** (ESCs) will form all the cells, tissues, and organs of the body. Because they are able to form so many different cell types, they are called

embryonic stem cells cells found in the inner cell mass of a developing embryo

CASE B:
Strawberries on Trial

© Index Stock Imagery/
Jupiter Images

In 1987, vandals slipped past guards, climbed over a high fence surrounding a field in California, spread rock salt, and sprayed an herbicide (a chemical that kills plants) on a large strawberry patch. Over 2000 strawberry plants were destroyed, and millions of dollars' worth of work was lost. This story may seem strange except for the fact that these strawberry plants had been sprayed with genetically engineered bacteria.

These bacteria, a strain of *Pseudomonas syringae* called "ice-minus," had been sprayed on the leaves of strawberry plants to prevent ice from forming on the leaves when the temperature dropped below freezing. Use of this transgenic bacterial strain extended the harvesting time of the strawberries and reduced plant loss caused by freezing. The strawberry plants and the strawberries themselves were not genetically modified.

After their arrest, the vandals said they would do it again if they had a chance.

1. **Why would these people want to destroy the strawberry fields?**

2. **This type of bacteria seems to be helpful to farmers; shouldn't we do all we can to help them improve crops and keep food prices under control? Why or why not?**

3. **Scientists have also produced transgenic corn plants by implanting a gene from fireflies that makes the plants glow when the firefly gene is expressed. What advantages do you see for the corn farmer from this change?**

4. **Erwin Chargaff wrote in *Science* on June 4, 1976, "Have we the right to counteract, irreversibly, the evolutionary wisdom of millions of years, in order to satisfy the ambition and the curiosity of a few scientists?" What are your thoughts on his statement? Are transgenic organisms changing the course of evolution?**

Figure 7.11 Embryonic Stem Cells (ESCs) (a) Photograph of a human blastocyst embryo showing the location of embryonic stem cells. (b) Embryonic stem cells can be recovered and grown in the laboratory to form ESC lines that are used in research to develop treatments for human diseases.

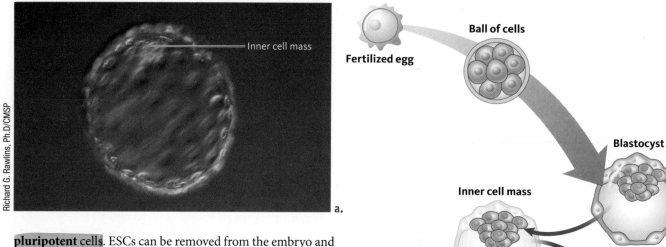

Inner cell mass

a.

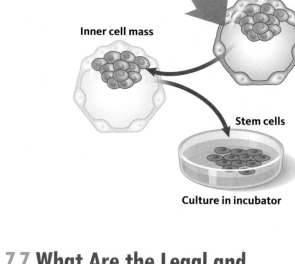

b.

Fertilized egg

Ball of cells

Blastocyst

Inner cell mass

Stem cells

Culture in incubator

pluripotent cells. ESCs can be removed from the embryo and grown into a cell line in a laboratory dish (Figure 7.11).

In the laboratory, embryonic stem cells can be reprogrammed to become many different cell types, including muscle, bone, skin, and nerve tissue. Embryonic stem cells are being studied to understand the early stages of the disease process and to develop therapies to treat diseases. A second type of stem cell, the **adult stem cell**, is found in many tissues of the adult body. These stem cells can also generate different cell types to replace those that are worn out or damaged. These cells are called **multipotent** cells because they can form only a limited number of cell types. Adult stem cells have been used for several decades to treat diseases. One type of adult stem found in bone marrow divides to form new blood cells to replace those that die. In Chapter 12, we will see how bone marrow cells are used to treat blood conditions such as leukemia.

Many biotech companies are developing therapies based on the use of embryonic and adult stem cells. Many of these therapies may lead to treatment for conditions including spinal cord injury, regrowth of burned tissue, and damaged heart tissue. Several clinical trials are now under way to determine whether embryonic stem cells can be used to treat diseases.

7.6 Essentials

- **Research on embryonic stem cells may result in new treatments for disease.**
- **Adult stem cells have several uses in treating diseases.**

pluripotent stem cells from the embryo that can grow into any type of cell

adult stem cells cells found in the adult human body that give rise to new but specific cell types

multipotent cells that can grow into only a limited number of cell types

7.7 What Are the Legal and Ethical Issues Associated with Biotechnology?

In addition to the scientific and societal problems with transgenic animals and plants discussed earlier in this chapter, there are legal and ethical issues. One involves patenting these organisms and the genes that are identified.

The U.S. Patent and Trademark Office (USPTO) decides whether new inventions can be patented and develops rules about what can be patented. Anything patentable must be novel, not obvious, and useful. These rules also state that anything that is naturally occurring cannot be patented. However, in 1972, after a landmark legal decision in a case called *Diamond v. Chakrabarty* (447 U.S. 303 [1980]), all that changed. Dr. Chakrabarty developed a bacterial strain that could break down oil. To do this, he crossed bacterial strains carrying different plasmids (gene-carrying DNA molecules found naturally in bacteria), eventually placing four different plasmids into a single strain of *Pseudomonas* bacteria. Each plasmid carried a gene that could degrade a different component of oil. Placed together in one strain, these bacteria could be used to clean up large oil spills.

At first, the USPTO refused to patent this bacterial strain on the grounds that the bacteria were naturally occurring, but the argument of Chakrabarty's attorneys won the case. Their argument, still used today, was that the strain containing all four of these plasmids in a single type of bacteria was unique and completely different from naturally occurring *Pseudomonas*. The patent was issued in 1972.

In a later case, Harvard University applied for and received a patent on a transgenic strain of mice called the **OncoMouse** (Figure 7.12). These mice carry a human cancer gene (**oncogene**) in their DNA. (In medical terminology, *onco-* means "cancer.") As a result of carrying this gene, the mice are more susceptible to cancer and are used to test chemicals for their ability to cause cancer. This case was the first in which a transgenic mammal was patented.

In the United States, the patent for the OncoMouse was issued in 1988. But in Canada, the case *Harvard College v. Canada (Commissioner of Patents)* was not decided until 2002. The Canadian courts ruled that a patent for a mammal could not be issued. Although recombinant bacterial strains have been patented in Canada, no patents for transgenic vertebrates have been issued.

Table 7.4 addresses some of the ethical and legal questions in the patenting of transgenic organisms and genes.

More recently, the USPTO has been trying to restrict some of the patents on genes and fragments of genes. It is now necessary for patent applications to identify the usefulness of the gene the applicant is seeking to patent.

Lawsuits sometimes arise when one company claims to have patented a specific allele and has created a test for that mutation. If another mutation in that same gene were discovered, could a new patent be issued for this mutant allele? If the companies were in different countries this might be possible, but until recently, if a U.S. company tried to patent the new mutant allele, it was unlikely that the patent would be issued.

In a recent case, *Re Kubin*, the USPTO turned down a patent on a gene discovered by Immunex, a private company. The gene encodes a protein found on white blood cells that plays a part in the immune response. The patent was turned down because another company had already discovered the protein, and finding the gene for that protein was a simple laboratory technique that could be done by any scientist with a lab manual. In other words, they said, the gene did not fulfill one of the main rules for patents—the substance must not be obvious.

Figure 7.12 **The OncoMouse was the first transgenic mammal to be patented.** This mouse strain has been genetically engineered to be susceptible to cancer. It is used to identify chemicals that cause cancer and to study the early events in the transformation of normal cells to cancerous cells.

SSPL/Getty Images

OncoMouse the first genetically modified mammal patented

oncogene a gene, when mutated, that can cause cancer within a cell

Table 7.4 Legal and Ethical Questions about Patenting Organisms

Question	How Are These Questions Decided?	Related Case or Legal Issue
Can we patent genes that are discovered in human DNA?	When researchers apply for patents with the USPTO, judges who specialize in patent law decide each case.	Biotech companies are discovering genes every day and patenting them, even though the genes are naturally occurring. Their argument, similar to the one used in *Chakrabarty*, is that when a gene is isolated and removed from an organism, the situation is different than that of a gene that occurs naturally in the DNA of a species.
Can we patent sections of genes that are found in human DNA?	Such patents are also allowed, even when the gene is not identified or its use is unknown.	Using the argument that when removed from the cell and its DNA, a section is different from that found in nature, biotech companies have gained ownership of DNA sequences and collect fees from other companies that want to use them. Some of these are only parts of genes that have not been identified.
Can the methods used in recombinant DNA techniques be patented?	Methods or processes can be patented if they are useful and unique.	When such methods are patented, no other company or lab can use the process without paying a fee or getting permission from the owner of the patent.
Should we be manipulating the genome of animals, plants, and people?	This ethical question is being debated every day. It is often said that this will create organisms that are very different from those that came about by evolution. Scientists usually have autonomy to work on their own, especially in privately owned biotech companies.	Laws that control what can be done in biotech labs may be written, and some already exist in countries other than the United States. One serious question here might be, who can control science?

In March, 2010, Judge Robert W. Sweet of the U.S. District Court for the Southern District of New York, in the case *Association of Molecular Pathology v. USPTO*, decided a case that may be the tipping point for patenting of genes. The judge reviewed the patent Myriad Genetics received on the *BRCA-1* and *BRCA-2* ovarian and breast cancer susceptibility genes (see Chapter 12). In his ruling, the judge stated that the argument that the gene is somehow changed and is, therefore, not a product of nature when it is removed from the genome is not adequate. Myriad Genetics is planning to appeal the ruling, and it will no doubt go to the U.S. Supreme Court. Such decisions might completely change biotech gene patenting in the near future.

7.7 Essentials

- **Many ethical questions about the use of recombinant DNA technology are being debated.**

Spotlight on Ethics

The Asilomar Conference, February 1975
Asilomar Conference Center, Monterey Peninsula, California

In the 1970s, recombinant DNA was in its infancy. After the discovery of the structure of DNA, the idea of recombinant DNA seemed to make sense. However, some experiments made scientists think that using these techniques might cause some problems.

The Asilomar Conference was called by Dr. Paul Berg from Stanford University and other prominent scientists of the day. Berg was one of the first individuals to develop recombinant DNA technology. In 1972 he used a restriction enzyme to cut the DNA of the monkey virus SV40 and used methods similar to those discussed in this chapter to link fragments of SV40 DNA to DNA from another virus, known as bacteriophage lambda. The final step would have involved placing the recombined genetic material into a laboratory strain of *E. coli* as a host cell to allow the recombinant DNA molecule to be copied. Berg, however, did not take this last step.

Berg halted his experiment because of pleas from several fellow investigators who feared that biohazards might be associated with a host cell carrying this recombinant molecule. SV40 causes cancerous tumors in mice, and *E. coli* is found in the human intestinal tract; investigators feared inserting SV40 DNA into host cells that might escape into the environment and infect laboratory workers, who could then develop cancer.

A group of leading researchers concerned about Berg's experiment sent a letter to the president of the National Academy of Sciences requesting that he appoint a committee to study the biosafety ramifications of this new technology. This committee met in 1974 and concluded that an international conference was necessary to resolve the issue, and that, until then, scientists should voluntarily stop all experiments involving recombinant DNA technology.

The full conference was called in 1975. Its main goal was to address the potential hazards presented by recombinant DNA technology. It was one of the first scientific conferences to invite testimony of the public. A group of about 140 professionals (primarily scientists, but also lawyers and physicians) participated in the conference to draw up voluntary guidelines to ensure the safety of recombinant DNA technology. Some of these guidelines were as follows:

1. Organisms containing recombinant DNA must be contained.

2. The level of containment should match the risk of the organism to humans, animals, and plants. These risks were ranked from low to high according to their potential for causing human disease.

3. Physical barriers should be used to minimize the escape of transgenic organisms.

4. The following types of experiments were prohibited:
   a. the use of any organisms that could live in the human body
   b. the transfer of any gene that produces poisons
   c. the synthesis of any products harmful to humans, animals, and plants

QUESTIONS

1. Many scientists who participated in the conference felt that by writing their own guidelines, they avoided legislation that would control what was done. How could legislation work for the benefit and detriment of science?

2. The conference was held around the time of the Watergate break-in and the ultimate resignation of President Nixon. Do you think the ethical concerns surrounding this scandal might have influenced the outcome of the conference?

3. Should scientists control their own work?

4. With the controversies about the use of stem cells being an issue with significant political overtones, do you think that guidelines written by scientists would eliminate or reduce the controversy?

5. Who should control science? Give three reasons for your answer.

6. The book *The Andromeda Strain* by Michael Crichton was published in 1969. Research its plot and then answer this question: Could the publishing of this *fiction* book have affected this conference? Why or why not?

1 Recombinant DNA techniques can be used to transfer genes among species. (Section 7.1)

In some cases, the transferred gene can be expressed and produce a protein product.

2 Recombinant DNA is created when two different types of DNA are joined. (Section 7.2)

These fragments can be small sections of DNA that carry one or more genes.

3 Restriction enzymes are used to cut DNA into fragments. (Section 7.2)

The discovery of restriction enzymes was one of the major scientific breakthroughs in the development of biotechnology.

4 Restriction enzymes recognize specific DNA sequences. (Section 7.2)

Because they are sequence-specific, these enzymes can be used to cut any DNA molecules carrying these sequences.

5 Biotechnology is the commercial use of recombinant DNA techniques to produce drugs and medical treatments. (Section 7.2)

Biotechnology has become a multibillion-dollar industry.

6 The first bacterial strain to be patented was one that degrades oil; this case is a landmark in patenting organisms. (Section 7.2)

Even though oil-eating bacteria are present in our natural surroundings, they had been changed to a different strain in the laboratory by matings to produce a unique combination of genes.

7 Transgenic bacteria, plants, and animals can be created with recombinant DNA techniques. (Section 7.3)

Transgenic organisms are used in agriculture, industry, and medicine.

8 Transgenic techniques are being used to produce food and drugs and to treat genetic diseases. (Section 7.6)

Scientists are studying the clinical and long-term effects of these products. (Section 7.7)

9 Transgenic animals have been created for use as animal models to study human diseases. (Section 7.5)

Animal models of human disease are used to evaluate the action of drugs to treat disease.

10 Many ethical questions about the use of recombinant DNA technology are still being debated, even after decades of widespread use. (Section 7.6)

When transgenic animals and plants are developed, they have been genetically altered, and, to some extent, might be considered "man-made."

Review Questions

1. Explain the meaning of the term *recombinant DNA*.

2. What are the major steps in creating recombinant DNA?

3. Humans have been genetically modifying plants using crossbreeding for a very long time. How is this different from creating plants using recombinant DNA?

4. Why is it possible to use an animal like *Drosophila* or a mouse to study a human genetic disease?

5. What is the importance of an animal model for studying human disease?

6. Name the bacteria used most commonly in recombinant DNA technology.

7. Explain why scientists held the Asilomar Conference.

8. Do you think scientists or someone else should be regulating the safety of recombinant DNA research? Why?

9. The discovery of restriction enzymes was the beginning of the recombinant DNA era. Why was this discovery necessary before DNA from different sources could be mixed to form recombinant DNA molecules?

10. Give an example of a vector. Why is it referred to as a vector?

11. Name one food that is being sold today that is derived from or contains products derived from the use of recombinant DNA technology.

12. Explain why the patenting of recombinant DNA organisms was so difficult in the beginning.

Application Questions

1. A number of individuals and organizations oppose the use of recombinant DNA technology in food production and related areas. List the pros and cons of this technology.

2. Research some of the groups that have been outspoken against recombinant DNA technology and discuss their reasoning.

3. In the United States genetically modified crop plants have been used on a large scale since 1996. Have any health-related problems or ecological problems been identified in that time? If so, what are they, and are they different from problems associated with the use of standard crops?

4. Before recombinant human insulin was developed, people with diabetes used the insulin extracted from cows and pigs. How is recombinant DNA–based insulin created?

5. Some athletes have admitted using hGH to enhance their performance. But recently, athletic competitions have banned its use. How can officials determine whether an athlete uses this product, considering that it is a human protein found in everyone's body?

6. Many people believe that if a product is developed that can help people feel better about themselves, it should be made available. Research the use of recombinant DNA–derived products such as epidermal growth factor (EGF) in cosmetics. Why should EGF-containing products be available, or why should they not? Give scientific information to support your answer. As a side note, EGF is also used as a treatment for a serious and potentially fatal gastrointestinal disorder in infants. Demand for EGF by the cosmetic industry has lowered the cost of producing EGF by 80% over conventional methods of extraction and purification. Does this information change your views in any way? Explain.

7. Research the history of the discovery of restriction enzymes. At the time, was there any thought that these enzymes could be used as the basis for recombinant DNA technology?

8. Think of three conditions that might be treated in the same way that hemophilia is treated today. List them and explain how recombinant DNA technology could be used to develop a treatment.

9. If biotechnology is used to treat a condition, does it matter how that condition is inherited?

10. In many areas of the world where rice is a major part of the diet, vitamin A deficiency is a leading cause of childhood blindness and death. A strain of rice called golden rice (because of its color) has been genetically modified to contain beta carotene, a precursor to vitamin A. Golden rice can supply the daily requirement for vitamin A intake in children. Should such nutritionally enhanced crop plants be allowed in the marketplace in spite of fears about genetically modified organisms and their use in humans? Why or why not?

11. Think of five ways we might use recombinant DNA technology in the future. Explore uses in areas such as medicine, agriculture, and the chemical industry. Be realistic.

12. It has been said that using recombinant DNA to produce clotting factor saved the lives of thousands of people with hemophilia. How did this happen?

13. Insulin was developed in the 1920s as a treatment for type I diabetes, which, at the time, was a fatal disease of childhood. Up until the 1980s, insulin was extracted from animal parts recovered from slaughterhouses. Research the social reaction to the introduction of insulin. Was there opposition to its introduction, or its source? Are there parallels between that time and the introduction and use of genetically modified organisms today?

WHAT WOULD YOU DO IF...?

. . . you were a legislator being asked to vote on a law to forbid the patenting of transgenic animals?

. . . you were a parent of a child with a serious genetic condition and were asked to have your future embryos examined in the lab to help identify the gene for this condition?

. . . you saw a GM tomato in the market and were curious about its taste?

. . . you were a legislator being asked to vote on labeling of transgenic food?

. . . your local grocery store hung up a sign that read, "We do not carry GM food"?

Online Resources

Preparing for an exam? Log on at www.cengagebrain.com for study tools to help you assess your understanding. If assigned by your instructor, the Case B and Spotlight on Ethics activities for this chapter, "Strawberries on Trial" and "The Asilomar Conference, February 1975" will also be available.

In the Media

The Body, March 30, 2011

Tennessee: Volunteers Needed for HIV Trials

Participants, who are HIV-negative men, will get a series of three immunizations with a recombinant DNA–based vaccine over eight weeks, followed by a single recombinant booster at week 24, or placebos.

Scientists have created an adenovirus that carries HIV gene segments. It will take two years to accumulate good data.

To access this article online, go to www.cengagebrain.com.

QUESTION

What three problems do you see for men who volunteer for this trial?

Wired, March 22, 2007

Better Teeth Through Biochemistry
Charles Graeber

A genetically engineered mouthwash may make it possible for you to have no cavities *ever* again.

Acid made by bacteria in your mouth causes cavities. So if you have no acid, you will have no cavities. The genetically engineered bacterial strain, a form of *Streptococcus*, has had the acid-producing gene deleted. This strain has been patented as OraGen. All the bacteria need to activate is some sugar.

If approved, it would work like this: A dental technician squirts a syringe of genetically modified mouthwash across your teeth. You sit for 5 minutes, chew on some sugary candy to activate the new bacteria, and then leave.

To access this article online, go to www.cengagebrain.com.

QUESTIONS

1. How do you think this would affect dentistry?

2. Would you use OraGen?

8

Genetic Testing and Prenatal Diagnosis

REVIEW

How is genetic information carried in DNA? (BB2.6)

How does an altered gene cause a genetic disease? (Section 5.6)

What can we learn from examining human pedigrees? (Section 4.2)

CENTRAL POINTS

- Testing is done to identify genetic disorders in fetuses, newborns, and adults.

- Phenylketonuria is diagnosed via blood samples taken from newborns.

- Adults can be tested for many genetic disorders.

- Tests are often done on large groups to obtain genetic information about populations.

- Results of genetic testing often create privacy issues.

CASE A: Hospital Tests Babies

Al and Victoria are in their early 30s and had been married five years when their first child was born. Victoria's pregnancy was perfectly normal, and she had continued working until one week before the baby's birth. She said she had never felt better. At birth, Al and Victoria's son appeared perfectly normal.

A few days after they took the baby home, the doctor called to say that the baby needed more tests. He asked that both parents come to the office visit because he wanted to talk to them. In his office the doctor said that a blood test done at the hospital showed that the child had a genetic disorder called phenylketonuria (PKU). If their child was not treated using a special diet, he would become mentally retarded. Al and Victoria had never heard of PKU, and this news upset them. They immediately called around to family members and found that only a few relatives on either side of the family had ever heard of PKU,

and none knew of any family member with this disorder. Al and Victoria wondered how a genetic disease that no one in their family had ever had could suddenly appear

A heel stick is the most common method of obtaining a blood sample for testing in newborns.

in their child. They made an appointment for another visit with the doctor and, in the meantime, began reading about PKU.

Some questions come to mind when reading this case. Before we can address these questions, let's look at what genetic testing is, how it is used to diagnose genetic diseases, and the future prospects for this technology.

Genetic Testing Usually Involves Taking a Blood Sample

8.1 What Is Genetic Testing?

This chapter will discuss several methods used to determine whether someone has or is at risk for a genetic disease. Before exploring the rationale, methods, and issues involved, let's define the differences between the two major types: genetic testing and genetic screening. **Genetic testing** determines whether someone has a certain genotype—in other words, what is his or her genetic makeup? Genetic testing identifies individuals in the following groups:

1. those who may have or may carry a genetic disease
2. those who are at risk of having a child with a genetic disorder
3. those who may have a genetic susceptibility to drugs and environmental agents

   Genetic screening is done on large populations rather than on individuals. The goals of genetic screening are more limited; it identifies people in the following groups:

1. those who may have or may carry a genetic disease
2. those who are at risk of having a child with a genetic disorder

   Genetic testing is most often a matter of individual choice, whereas genetic screening is often mandated by law. Genetic testing can have a serious impact on the lives of individuals and their families, as in the following situations:

1. Identification of a person who has or is at risk for a genetic disorder often leads to the discovery of other affected or at-risk family members.

genetic testing testing for genetic conditions

genetic screening large-scale genetic testing

2. Testing for some disorders can identify a person who is healthy now but will develop serious or fatal genetic disorders in later life. When the condition is fatal—for example, Huntington disease, as in Alan's family (see Chapter 4)—this information often has serious personal, family, and social consequences.

3. The results of genetic testing may have a direct impact on siblings and the children or grandchildren of the person being tested.

Genetic counselors work with people who are about to have or have had genetic testing to explain the tests and the results and to help them make decisions based on the results.

What types of genetic testing are available?

Several forms of genetic testing are currently being used:

1. **Prenatal diagnosis**—to determine the genotype of a fetus, such as testing for sickle-cell anemia.

2. **Carrier testing**—to test members of a family with a history of a genetic disorder such as cystic fibrosis to identify heterozygotes and to determine their chances of having an affected child.

prenatal diagnosis diagnosis of conditions in a developing fetus

carrier testing a genetic test for carrier status of a recessive disorder

presymptomatic testing genetic testing before symptoms manifest themselves

ultrasonography, or ultrasound a prenatal test that uses sound waves to visualize an unborn fetus

amniocentesis a prenatal test that involves removal of amniotic fluid

chorionic villus sampling (CVS) a prenatal test that removes a small sample of villi

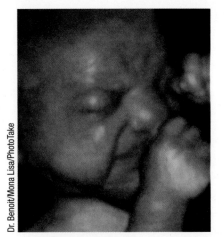

Dr. Benoit/Mona Lisa/PhotoTake

Figure 8.1
Three-dimensional ultrasounds show details of the fetus's facial features.

3. **Presymptomatic testing**—to identify individuals who will develop disorders in midlife, such as Huntington disease or adult polycystic kidney disease.

8.1 Essentials
- **Genetic testing determines what genotype a person has.**
- **Genetic testing can be done on fetuses, newborns, children, and adults.**

8.2 Why Is Prenatal Genetic Testing Done?

Prenatal genetic testing is used to detect genetic disorders and birth defects in a developing fetus. More than 200 single-gene disorders can be diagnosed by prenatal testing. Some of these are listed in Table 8.1.

In most cases, testing is done only when there is a family history or some other risk factor is identified (such as the age of the mother). Remember from Chapter 4 that families

Table 8.1 Some Genetic Disorders That Can Be Diagnosed by Genetic Testing

Disorder	Incidence	Inheritance Pattern
Cystic fibrosis	1 in 2000 Caucasians	Autosomal recessive
Congenital adrenal hyperplasia	1 in 10,000	Autosomal recessive
Duchenne muscular dystrophy	1 in 3500 male births	X-linked recessive
Hemophilia A	1 in 8500 male births	X-linked recessive
Alpha and beta thalassemia	Varies	Autosomal recessive
Huntington disease	1 in 10,000	Autosomal dominant
Polycystic kidney disease	1 in 1000	Autosomal dominant
Sickle-cell anemia	1 in 400 African Americans	Autosomal recessive
Tay-Sachs disease	1 in 3600 Ashkenazi Jews and French Canadians; 1 in 360,000 in the general population	Autosomal recessive

who carry conditions inherited as autosomal recessive disorders such as sickle-cell anemia may not know that the mutant allele runs in their family until a child is born with the condition. The parents may be tested to determine whether they are heterozygous carriers of the disease before they conceive. If tests reveal that both parents are carriers, the fetus has a 25% chance of being affected with the condition.

In Chapter 3 we discussed a technique known as **ultrasound** (see Figure 8.1), a method that can provide some information about the health of the fetus. But for conditions caused by chromosomal aberrations, such as Down syndrome (trisomy 21), chromosome analysis is the most direct way to identify an affected fetus. Because the risk of Down syndrome increases dramatically with the age of the mother (see Chapter 3), chromosomal analysis of fetal cells is recommended for all pregnancies in which the mother is age 35 or older. Prenatal testing for any genetic disease requires a sample of cells from the fetus, which can be obtained by using **amniocentesis** or **chorionic villus sampling (CVS)**, as discussed in Chapter 3.

a.

Transducer

Jose Luis Peaez, Inc./Blend Images LLC

Figure 8.2 Ultrasonography (a) Ultrasonography (ultrasound) is a noninvasive testing procedure based on sonar technology originally developed for the military to find submarines underwater. A transducer is placed on the abdomen over the uterus. The transducer emits pulses of high-frequency sound; the reflected sound waves are converted into images and displayed on a screen. Many women routinely have ultrasounds during pregnancy to monitor the development of the fetus, determine its sex, and check for problems. (b) Ultrasound can also be used to diagnose some genetic disorders, including Down syndrome. In the ultrasound below, extra skin at the fold of the neck is outlined in red. This condition is associated with a Down syndrome fetus.

b.

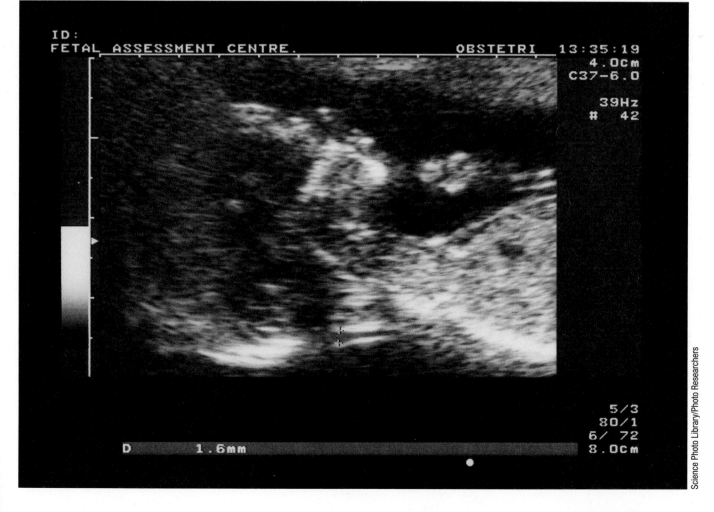

ID:
FETAL ASSESSMENT CENTRE.   OBSTETRI 13:35:19
                                                         4.0cm
                                                      C37-6.0

                                                         39Hz
                                                      #   42

                                                         5/3
                                                        80/1
                                                        6/ 72
D   1.6mm   8.0cm

Science Photo Library/Photo Researchers

How is ultrasound done?

The procedure for ultrasound is shown in Figure 8.2a, page 167. Ultrasound is one of the prenatal tests used to diagnose genetic disorders. It can be used to identify trisomy 13, trisomy 18, and, to a lesser extent, trisomy 21 (Down syndrome) because each of these disorders has unique physical traits that can be seen via ultrasound. For example, in Down syndrome, affected children often have a thick fold of skin in the neck region. This can be seen in ultrasound, and amniocentesis can confirm the diagnosis. An ultrasound image of a fetus with a neck fold (outlined in red) is shown in Figure 8.2b.

How is amniocentesis done?

The procedure for amniocentesis is shown in Figure 8.3 and was briefly described in Chapter 3. More than 100 disorders can now be diagnosed by amniocentesis. The cells retrieved by this method can be analyzed to detect biochemical disorders, chromosomal abnormalities, and the sex of the fetus.

When is amniocentesis used?

Amniocentesis carries a small risk of maternal infection and a slight increase

chorionic villi finger-like projections of the fetal chorion that work their way into the wall of the uterus

Figure 8.3 **Amniocentesis** As discussed in Chapter 3, amniocentesis is one way fetal cells can be collected for prenatal testing. First, the fetus and the placenta are located by ultrasound and a needle is inserted through the abdominal wall, avoiding the fetus and the placenta. Approximately 10–30 ml of amniotic fluid is removed. Amniotic fluid is mostly fetal urine containing cells shed from the fetus's skin. The cells are isolated from the fluid by centrifugation, grown in the laboratory, and used to construct a karyotype (see Chapter 3). Amniocentesis is usually not performed until the 16th week of pregnancy.

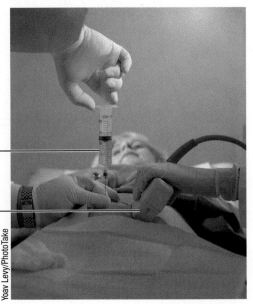

Amniotic fluid being removed from uterus

Ultrasound guiding the needle through the uterine wall

Yoav Levy/PhotoTake

in the probability of a spontaneous abortion. To offset these risks, amniocentesis is normally used only under certain conditions:

1. When the mother is age 35 or older. Because the risk of having a child with chromosomal abnormalities increases dramatically after age 35, amniocentesis is recommended for pregnant women who are age 35 or older. Most amniocentesis procedures are performed because of advanced maternal age.

2. When the family has a previous child with a chromosomal abnormality. The recurrence risk in such cases is 1 to 2%.

3. When a parent has a chromosomal abnormality. This can cause an abnormal karyotype in the child, and amniocentesis should be considered.

4. When the mother is a carrier of an X-linked disorder.

How is chorionic villus sampling (CVS) done?

Chorionic villus sampling (CVS) is another method of prenatal diagnosis. After implanation and formation of the chorion by the fetus, cords of cells from the chorion grow into the uterine wall, forming finger-like structures called **chorionic villi**. Eventually, the fetal chorionic villi and maternal tissue combine to form the placenta (Figure 8.4). Cells from the villi are removed by suction during CVS (Figure 8.5). Enough material is usually obtained to allow karyotype preparation, biochemical testing, or extraction of DNA for molecular analysis.

Because tissue is taken directly from the fetus itself, CVS has several advantages over amniocentesis:

1. CVS is performed earlier in the pregnancy (8 to 10 weeks) than amniocentesis (16 weeks). This allows the patient to find out test results earlier.

2. Because placental cells are already dividing, karyotypes are available within a few hours or a few days. With

Figure 8.4 **An Early Embryo Surrounded by Its Tissues** The chorionic villi are embryonic tissues that can be sampled for genetic testing.

Placenta fetal membranes | Chorionic villi | Yolk sac | Placental cord | Placental vessels | Placenta (developing)

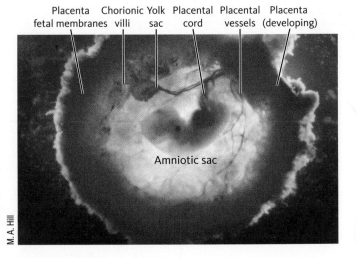

Amniotic sac

M. A. Hill

Figure 8.5 Chorionic Villus Sampling As discussed in Chapter 3, during this procedure, a flexible catheter is inserted through the vagina or abdomen into the uterus, guided by ultrasound images. A sample of the chorionic villi (fetal tissue that forms part of the placenta) is removed by suction. The cells of the villi are used for karyotype preparation, biochemical testing, or DNA extraction for genotype analysis.

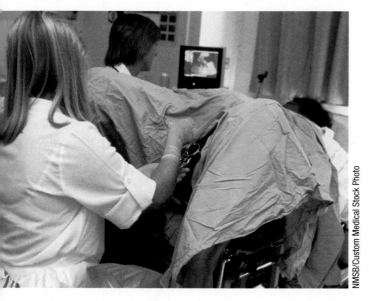

NMSB/Custom Medical Stock Photo

amniocentesis, fetal cells must be stimulated to divide and grown in the laboratory for several days before a karyotype can be prepared.

When is CVS used?

In the past, CVS was used less frequently than amniocentesis because of its increased risk of spontaneous abortion. However, if the procedure is performed at a major medical center, the risks are similar to those associated with amniocentesis. CVS offers early diagnosis of genetic diseases, and if termination of pregnancy is elected, maternal risks are lower at an earlier stage of pregnancy. The conditions under which CVS is used are similar to those for amniocentesis (Table 8.2), but because it can be used earlier in pregnancy, some couples choose CVS if they know they are at risk of having a child with a genetic condition.

Are there less invasive methods for prenatal genetic testing?

Realizing that amniocentesis and CVS carry a 0.2 to 0.3% risk of miscarriage, researchers are working to develop new, less-invasive ways of prenatal testing. As discussed in Chapter 3, one of these less-invasive methods involves recovering fetal cells or fetal DNA from the mother's blood, which minimizes risk to both the mother and the fetus because all that is necessary is a sample of blood from the pregnant woman.

Several types of fetal cells enter the maternal circulation, including:

1. placental cells
2. white blood cells
3. immature red blood cells with nuclei

These cells probably begin to enter the mother's bloodstream in detectable amounts between the 6th and 12th weeks of pregnancy.

A problem with using fetal cells from the mother's circulation is that only 1 in every 100,000 cells in the mother's blood is a fetal cell. Collecting enough fetal cells from a blood sample is one of the challenges facing those working to develop this technique. Fetal cells collected this way have been used to diagnose some genetic disorders, including sickle-cell anemia, and chromosomal abnormalities, but the method needs further development before being widely adopted.

The amniotic fluid surrounding the fetus also contains fetal DNA (called free fetal DNA or ffDNA) produced when fetal cells break down. The DNA is released as chromosomes become fragmented. The ffDNA can cross the placenta and enter the mother's bloodstream. Recent breakthroughs in the use of ffDNA to detect genetic disorders in the fetus are discussed in Chapter 17.

Table 8.2 Comparison of Prenatal Diagnosis Tests

	Ultrasound	Amniocentesis	CVS
When do we test?	Week 18	Week 16	Week 11
When do you get the results?	Immediately	Week 18	Week 12
What does it test for?	Structural problems such as limb growth, size of head, and heart abnormalities	Chromosomal abnormalities, biochemical disorders, fetal genotype	Chromosomal abnormalities, biochemical disorders, fetal genotype
With whom do physicians use this procedure?	Most pregnant women	Women age 35 and older	Women at high risk for having a child with a genetic disorder
Chance of miscarriage?	None	0.2–0.3%	0.2–0.3%

Figure 8.6 **An Early Embryo Composed of Several Blastomeres**

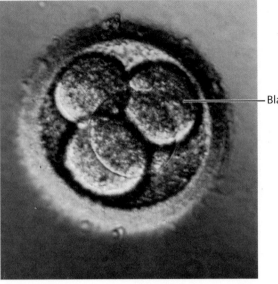

Blastomere

Dr. Yorgos Nikas/Photo Researchers

Figure 8.7 **A Blastomere Is Removed from an Embryo for Genetic Testing**

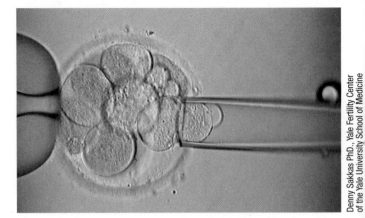

Denny Sakkas PhD., Yale Fertility Center of the Yale University School of Medicine

Can embryos be tested before they are implanted in the mother? In Chapter 1, we discussed the use of preimplantation genetic diagnosis (PGD) for sex selection. This test can also be used to diagnose genetic disorders in the earliest stages of embryonic development. Here we will focus on the use of PGD to detect genetic disorders.

In vitro fertilization (IVF) always precedes analysis by PGD. For IVF, eggs are collected, fertilized, and allowed to develop in a culture dish for several days. On about the third day after fertilization, the embryo consists of six to eight cells. Each of these cells is called a **blastomere** (Figure 8.6).

For PGD, one of the blastomeres is removed (Figure 8.7). DNA is extracted from this cell and analyzed to determine whether the embryo has a genetic disorder. Blastomere testing can be used for many common autosomal recessive disorders, such as **Tay-Sachs disease**, dominant disorders, and most X-linked disorders, including muscular dystrophy and hemophilia. Embryos that do not have a genetic disorder are implanted in the mother's uterus for development.

8.2 Essentials

- **Prenatal testing is usually done to detect genetic disorders in the fetus when there is a risk of such conditions.**
- **Preimplantation genetic diagnosis can detect genetic abnormalities *before* the embryo is implanted into the mother.**

blastomere one of the early embryonic cells

Tay-Sachs disease a recessive genetic trait that is almost always fatal

8.3 How Are Fetal Cells Analyzed?

The fluids and cells obtained by amniocentesis, CVS, and PGD can be analyzed (Figure 8.8) using several different methods, including karyotyping (see Chapter 3), biochemical analysis, and recombinant DNA techniques (see Chapter 7).

Direct DNA analysis is the most specific and sensitive method currently available. The accuracy, sensitivity, and ease with which recombinant DNA technology can be used to identify a genetic disease and susceptibilities carried by an individual have raised a number of legal and ethical issues that have yet to be resolved; these will be discussed later in this chapter.

8.3 Essentials

- **Prenatal testing can determine the genotype of a fetus by examining cells retrieved using several different procedures.**

Figure 8.8 **A Technician Prepares a Karyotype in a Cytogenetics Laboratory**

Chromosome spread

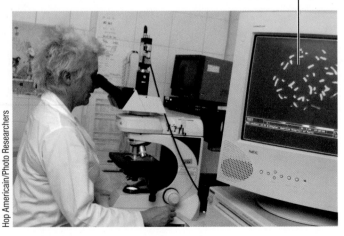

Hop Americain/Photo Researchers

8.4 How Can Prenatal Tests Diagnose Phenylketonuria?

Like all babies born in the United States, Al and Victoria's baby was tested for phenylketonuria at birth. However, if there is a family history of PKU, the test is often done prenatally. PKU is a metabolic disorder found in people across all ethnic and racial groups. It is inherited as an autosomal recessive trait; therefore, both parents must be heterozygous carriers for their child to inherit the disorder. The gene that causes PKU is called **PAH** and is located on chromosome 12 (Figure 8.9). People with this condition cannot convert an important amino acid, **phenylalanine**, into another amino acid, tyrosine.

The enzyme that converts phenylalanine into tyrosine is called phenylalanine hydroxylase (PAH). In PKU, a mutation in the *PAH* gene inactivates the enzyme. Therefore, phenylalanine cannot be broken down and builds up in the body's tissues, causing damage to the brain and other organs.

PKU is a genetic disorder, but it is also an environmental disease. If phenylalanine is removed from the diet, the child will develop normally, and there is no mental retardation. This is a successful treatment for the condition. Because the brain is fully mature by the early teen years, many people with PKU switch to a normal diet at this time.

However, if a woman with PKU gets pregnant, she must follow the PKU diet very carefully; otherwise, the child may be seriously affected because the excess amount of phenylalanine in the mother's blood would pass into the fetus, causing serious mental retardation, even if the fetus has a normal or a heterozygous genotype.

Can we test everyone for conditions such as PKU? Genetic testing on a large scale (genetic screening) is not always possible. For some disorders, such as the sickle-cell anemia that Martha Johnson's twins had (see Chapter 5), a single mutation is always responsible, so testing is efficient and uncovers all cases. However, several hundred different mutations have been identified in the *PAH* gene, and testing for all of these mutations is impractical. However, most states screen newborns for PKU using a simple blood test that measures phenylalanine levels. This test is simple and not based on DNA analysis; therefore, it is not testing for mutations in the *PAH* gene. Figure 8.10 shows a photo of a *Guthrie card* containing blood samples taken from a child soon after birth. The blood is transferred to the card and analyzed for PKU levels. The card is kept as part of the child's medical record.

8.4 Essentials
- **PKU is one of the few genetic conditions that has a treatment.**

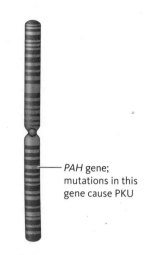

Figure 8.9 Chromosome 12, Showing the Location of the PAH Gene at 12q24.1
A mutant allele of this gene is responsible for phenylketonuria.

PAH gene; mutations in this gene cause PKU

Chromosome 12

Figure 8.10 Guthrie Card
Blood samples from newborns are placed on a card before being tested for PKU.

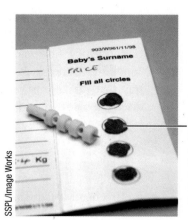

903/W961/11/98
Baby's Surname
PRICE
Fill all circles

SSPL/Image Works

This is how blood from heel sticks is stored on cards for analysis. The cards are kept as part of the infant's medical record.

PAH a gene associated with PKU

phenylalanine an essential amino acid

8.5 Can Adults Be Tested for Genetic Conditions?

Some conditions—such as Huntington disease, which runs in Alan's family (see Chapter 4)—show up much later in life but can be detected by testing long before symptoms appear; these are called presymptomatic tests. A genetic predisposition to breast cancer (discussed in Chapter 12) also falls into this category. This test can detect a predisposition to breast cancer before the disease develops. Adult-onset disorders, including Huntington disease, amyotrophic lateral sclerosis (ALS), adult polycystic kidney disease (ADPKD), and others, all begin later in life. As an example, the symptoms of ADPKD, which affects about 1 in 1000 individuals, usually appear between ages 35 and 50. This disease is characterized by the formation of cysts in one or both kidneys; these cysts grow and gradually destroy the kidney. Figure 8.11 shows a normal kidney (left) and one with ADPKD (right).

Treatment options for ADPKD include kidney dialysis or transplantation of a normal kidney, but many affected individuals die prematurely before a transplant becomes available. Because ADPKD is a dominant trait, anyone who is heterozygous for the mutant gene will be affected. Genetic testing can determine which family members carry a mutant allele and will develop this disorder. Testing for these and other disorders that appear in adults can be done at any age before (or after) the condition appears. The decision on whether to be tested for these traits is a personal choice.

8.5 Essentials

- **Adults can be tested for many genetic conditions. A person may want to find out if he or she is a carrier of a condition before marriage.**

Figure 8.11 (left) A normal kidney. (right) An enlarged kidney from someone with polycystic kidney disease.

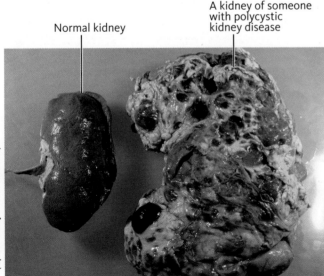

A kidney of someone with polycystic kidney disease

Normal kidney

Arthur Glauberman/Science Photo Library
—(represented by Photo Researchers)

8.6 What Is a Genetic Screening Program?

Some genetic screening programs, such as those for newborns, are mandated by law in the United States. All states and the District of Columbia require newborns to be tested for a range of genetic disorders, although the number and types of disorders screened for vary significantly from state to state. These programs began in the 1960s with screening for PKU and gradually expanded to all states and a wide range of genetic disorders. Al and Victoria's baby was tested under the screening program in their state.

Many states screen for only three to eight disorders (Figure 8.12). However, using new technology, labs can now screen for 30 to 50 metabolic disorders from a single blood sample. The number of tests currently mandated in each state is shown in the map. Parents should be aware of which disorders are included in their state's screening program.

Are there screening programs to detect adult carriers of genetic disorders?
As discussed in Chapter 4, a person can carry a gene for a recessive genetic condition but remain disease free. These individuals are heterozygous for the disorder and are called *carriers*. Carriers of certain recessive disorders, such as PKU and sickle-cell anemia, are found in higher frequencies in certain ethnic groups.

Population screening for carriers of genetic disorders is possible only under certain circumstances, including the following:

1. The disease must occur mainly in defined populations. For example, in the United States, sickle-cell carriers are most often found among African Americans. However, PKU, the condition present in Al and Victoria's baby, occurs across all racial and ethnic groups and would not qualify for carrier testing.

2. Tests for carriers must be available, fast, and fairly inexpensive.

3. Screening for these disorders must give at-risk couples several options for having only unaffected children.

At this point, there are no mandated programs for carrier screening; however, one disorder that meets these three conditions is Tay-Sachs disease, a fatal autosomal recessive trait that affects 1 in 360,000 individuals (Figure 8.13). This disease is associated with a disorder in the lysosomes in the body's cells (see "Biology Review: Cells and Cell Structure"). It can lead to mental retardation, blindness, and death by age 3 or 4.

In Jews of Eastern European ancestry (Ashkenazi Jews), the frequency of this disorder is almost 100 times higher than in the general population (Figure 8.14). In the 1970s,

Figure 8.12 **Genetic Screening Programs in the United States** The key shows how many tests are mandated by law in each state.

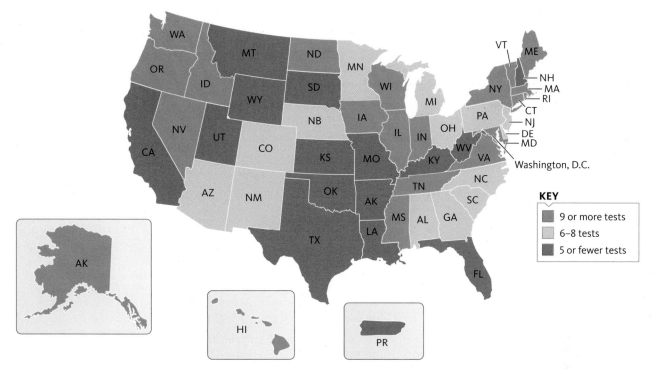

KEY
- 9 or more tests
- 6–8 tests
- 5 or fewer tests

Figure 8.13 **A Child with Tay-Sachs Disease** In this fatal genetic disorder, degeneration of cells in the nervous system leads to death by age 2 or 3.

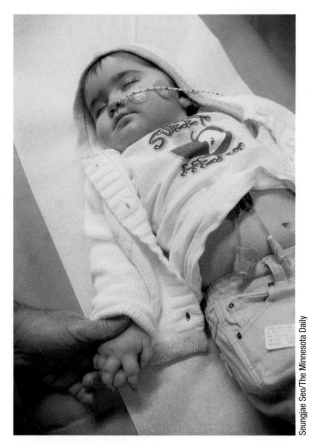

Seungjae Seo/The Minnesota Daily

Figure 8.14 **Members of Orthodox Jewish Communities Have Embraced Genetic Testing for Tay-Sachs Disease**

© blickwinkel/Alamy

1 in 4500

Frequency of births with PKU among people of Irish or Scotch descent

Warning for people with PKU

1 in 13,000 to 1 in 19,000

Frequency of births with PKU in the United States

1 in 300 to 1 in 500

Number of miscarriages after amniocentesis

1 in 300 to 1 in 500

Number of miscarriages after CVS

CASE A: QUESTIONS

Now that we understand how prenatal testing is done and why, let's look at some of the issues raised in Al and Victoria's case.

A heel stick is the most common method of obtaining a blood sample for testing in newborns.

1. Why was Al and Victoria's baby tested for PKU?

2. Neither Al's family nor Victoria's family has a history of PKU. Should Al and Victoria's baby have been tested at all? Explain.

3. Should Al and Victoria have the right to refuse this testing for their child? Why or why not?

4. Based on what you know about the law, do Al and Victoria have a right to sue the hospital? Why?

5. PKU is a treatable genetic disorder. Affected children must follow a strict diet that prevents many of the condition's symptoms. Does this information change your opinion about testing the baby?

6. Testing of newborns for PKU is done in every state; some states offer more extensive testing for newborns. Al and Victoria live in one of these states. Should they have been notified of the testing? Why or why not?

The sickle-cell anemia screening program generated unanticipated problems. In 1981, the Air Force policy that denied sickle-cell carriers admission to the Academy was reversed under threat of a lawsuit. In other cases, carriers were reportedly turned down for insurance and employment, even though carriers have no inherent health problems.

Some sickle-cell screening programs were criticized for not maintaining confidentiality of records and not providing counseling to those identified as heterozygotes (carriers). In the late 1970s and early 1980s, many of these screening programs were cut back or reorganized, and only a few states currently offer, but don't mandate, testing for adults, although all 50 states and the District of Columbia screen newborns for sickle-cell anemia.

voluntary carrier screening programs were initiated in Ashkenazi populations in the United States. More than 300,000 individuals were tested in the first 10 years of screening, and 268 couples were identified in which both members were carriers and had not yet had a child. This test is not DNA based; rather, it tests for the level of an enzyme in the blood. These simple blood tests have been effectively used for years.

In the Ashkenazi community, Tay-Sachs screening programs are coupled with counseling sessions that provide information about the risks of having an affected child, the availability of prenatal testing, and reproductive options. These programs are voluntary, although some states require that couples be informed about Tay-Sachs testing. Before the screening programs began, each year between 50 and 100 children were born with Tay-Sachs disease in the United States. This number decreased to fewer than 10 such births each year after screening programs were implemented.

Another population screening program began when the U.S. Congress passed the National Sickle-Cell Anemia Control Act in 1972. This law was designed to establish a screening program to identify carriers of sickle-cell anemia (Figure 8.15). Each state received funds to set up a screening program. Some states established compulsory programs requiring that all African American children be tested before attending school; others required testing before obtaining a marriage license. Testing was also done on professional football players and applicants to the U.S. Air Force Academy, because carriers were not allowed to enter the academy.

Masterfile

8.6 Essentials

- **Genetic screening tests many individuals in a population.**

8.7 Which Genetic Conditions Can Be Treated?

Many of the genetic disorders we have discussed can be treated to alleviate symptoms, but none of them has a cure. Table 8.3 lists some conditions and their treatments, if available.

Table 8.3 **Treatments for Genetic Disorders**

Name of Condition	Is Treatment Available?	Treatment
PKU	Yes	Diet low in phenylalanine
Sickle-cell anemia	Yes	Bone marrow transplants, blood transfusions
Hemophilia	Yes	Recombinant DNA Factor VIII
Cystic fibrosis	Yes	Inhalers, antibiotics
Polycystic kidney disease	Yes	Kidney transplant, dialysis
Tay-Sachs disease	No	
Huntington disease	No	
Muscular dystrophy	No	
Trisomies (Down syndrome, trisomy 18, etc.)	No	

CASE B: Company Gives Big Party

A wonderful party was in full swing. Representative Prescott Smith had been to many parties since being elected to Congress from his state, but this one was the best. The party was hosted by MegaGene,

A company celebration.

© Ace Stock Limited/Alamy

a large pharmaceutical company. MegaGene had just developed a test to detect the mutant allele for a disease that was rare and serious but had no treatment or cure.

Finally, at the end of the evening, MegaGene came to the point. Company representatives suggested that Mr. Smith sponsor a bill the company had drafted to require that every baby born in the United States be tested for this gene.

1. **Why was MegaGene throwing this party? Give three reasons.**

2. **Why did the company draft the law? Give two reasons.**

3. **What should Congressman Smith do if asked to sponsor such a bill?**

4. **Today all babies born in hospitals are tested for several to dozens of genetic diseases by law (which condition is dependent on the state). Should parents be allowed to refuse this testing? Why or why not?**

5. **What do you think will happen, years from now, when we have tests for all the human genes?**

8.7 Essentials

- **A few genetic conditions can be treated but not cured.**

8.8 What Are the Legal and Ethical Issues Associated with Genetic Testing?

The use of genetic testing and screening has a number of accompanying ethical and legal problems. Because these methods examine and record a person's genetic makeup, privacy is extremely important. Both federal and state laws require that medical records be kept private and their contents not be revealed to anyone without the patient's explicit consent.

Information from these tests can affect a patient's health insurance, job security, and even personal life. For example, as discussed in Chapter 4, if an insurance company discovers that a family carries the gene for Huntington disease, it may not insure them. In another example, would a person be willing to marry someone who carries the gene for Huntington disease or the gene for polycystic kidney disease?

These questions are dealt with either personally, on a one-to-one basis, in the courts, or in the legislature. In some cases, laws have been passed prohibiting discrimination against someone because of his or her genetic makeup. During the 1970s, many people were discriminated against based on their carrier status for sickle-cell anemia, and many screening laws were repealed.

On another level, companies that work to create a test for a specific disease gene spend millions of dollars on research and development. For this reason, they are looking to sell this test to as many potential customers as possible. If, as Mega-Gene wanted, the company's test were used in a screening program of newborns, this would be one way to recover the money spent on research and development.

Table 8.4 addresses some of the ethical and legal questions in testing and screening.

8.8 Essentials

- **Genetic testing must be done under circumstances that ensure that the results will remain private.**

Table 8.4 **Legal and Ethical Questions about Genetic Testing**

Question	How Are These Questions Decided?	Related Case or Legal Issue
How should genetic test results be used?	Laws control the use of information, but insurance companies ask potential insurees to sign a release for their medical records.	The Kennedy-Kassebaum law stops insurance companies from discriminating against patients with genetic conditions.
Should we test for conditions that have no cure or treatment?	The individual makes the decision for this testing (except as mandated by screening laws) after meeting with a genetic counselor or a physician.	Newborns and children have no right to refuse testing; however, adults can refuse, and the law has upheld this right.
Should the government decide who should be tested?	Since the 1970s, the U.S. government has mandated testing for all newborns born in a hospital. Babies born at home may be missed. The numbers of tests is increasing.	Parents have sued to be excluded from the testing and have won on religious and moral grounds. But most parents don't even know the testing occurs.
Should insurance companies or employers be allowed to require genetic testing of their potential customers or employees?	This has happened only in a few cases. The argument is usually that it is for the customer's or employee's own good.	In *EEOC v. Burlington N. Santa Fe Railway Co.* (see Spotlight on Law, Chapter 5), the company wanted to test all employees for a gene that predisposes them to carpal tunnel syndrome. The court decided that the testing violated the Americans with Disabilities Act, and the company immediately stopped the testing.

Spotlight on Society

Iceland and Its Genetic Code

This spotlight will examine the nationwide genetic testing being done in Iceland. In 1996 an Icelandic scientist, Kari Stefansson, started a company called deCODE Genetics to find disease genes in the human genome. It seemed logical that if he could identify a large population of closely related people, any genetic differences related to disease would stand out.

Iceland has a relatively small, homogenous population of approximately 290,000. Few people have immigrated there, and even fewer have left. Iceland keeps meticulous records of families as well as comprehensive health records that date back to the early 1900s. Figure 8.16 shows some children from Iceland, all with similar phenotypes.

Figure 8.16 Members of this crowd in Iceland have very similar phenotypes, reflecting their genotypic similarities.

Stefansson thought that if he could get DNA samples from everyone in Iceland, as well as their health and genealogy records, he could identify disease genes more easily. In the beginning, he asked for volunteers to supply samples for his DNA database. However, to speed up the process of gene hunting, he approached the Icelandic government with an idea. He proposed that the government ask each citizen of Iceland to come forward and give a sample of his or her DNA for this database. After a great deal of debate, a law was written, and on December 17, 1998, the Health Sector Database Act was voted into law by the Icelandic parliament.

For a one-time fee of $950,000 paid to the Icelandic government, deCODE was given complete access to family histories, medical records, and genetic information for the entire population of Iceland. According to the law, deCODE will construct a database (at a cost of $10–20 million) of DNA samples from members of the population. Differences in DNA sequence would be used to search for disease-associated genes. In exchange, deCODE will receive and retain all profits derived from its operating license.

To ensure privacy of medical records, the law forbids deCODE from sharing medical information with anyone. In exchange, Icelanders have been promised free access to genetic tests and medications developed from this project. deCODE has signed drug development contracts with several pharmaceutical companies, including Merck, Roche Diagnostics, and Hoffmann–La Roche.

Some of deCODE's accomplishments include the discovery of a number of disease genes, including those linked to schizophrenia and heart disease, and clinical trials on drugs developed to treat these disorders are underway. In addition, deCODE has recently started a service called deCODEme. Using a DNA sample collected by swabbing the inside of the cheek, this online service offers a genome scan to identify genomic variations associated with risks for a number of genetic disorders. Results from the genome scan can also be used to trace an individual's genetic ancestry, compare genome data with other individuals, and make predictions about someone's physical attributes, based on the analysis of the genome scan.

Iceland has had a serious financial setback recently and is trying to recover. As a result, deCODE genetics has also met with some financial difficulties (Figure 8.17). The CEO of deCODE has suggested the personal testing wing of the company (deCODEme) will carry it through the next few years.

QUESTIONS

1. One of the interesting parts of Iceland's Health Sector Database Act is that it presumes consent of all the members of the population. What problems do you see with this?

2. Should Icelanders have the right to opt out of the testing? Explain.

Figure 8.17 The Icelandic headquarters of deCODE, where much of its work takes place.

Figure 8.18 A Page from an Icelandic Phone Book

Notice the similarities in names, and the number of people with the same name.

3. Why should deCODE have a monopoly on the data and ability to profit from such information? Give both pro and con arguments.

4. Will deCODE or other companies own part or all of an individual's genome? Explain.

5. deCODE has promised Icelanders free medications. Will [it be] liable for any unforeseen side effects? Explain.

6. Figure 8.18 shows a page from an Icelandic phone book. Notice how many of the first and last names are the same. What does this tell you about the genetics of the people [of] Iceland?

7. If deCODE genetics has to go out of business, what do you think might happen to all the samples taken from [the] citizens of Iceland?

The Essential 10

1 Genetic testing determines a person's genotype. (Section 8.1)

This information is used in various ways.

2 Genetic testing can be done on fetuses, newborns, children, and adults. (Section 8.1)

As more tests become available, more people can be tested for conditions.

3 Prenatal testing is usually done to detect genetic disorders in the fetus when there is a risk of such conditions. (Section 8.2)

Usually this testing is done early in a woman's pregnancy.

4 Preimplantation genetic diagnosis can detect genetic abnormalities *before* the embryo is implanted into the mother. (Section 8.2)

Children born after this procedure have no side effects.

5 Prenatal testing can determine the genotype of a fetus by examining the cells retrieved using any of several different procedures. (Section 8.3)

Results of these procedures are usually processed in accredited laboratories.

6 PKU is one of the few genetic conditions that can be treated. (Section 8.4)

The treatment is a diet low in phenylalanine.

7 Adults can be tested for many genetic conditions. A person may want to find out if he or she is a carrier of a condition before marriage. (Section 8.5)

Some genetic disorders are much more prevalent in certain ethnic groups.

8 Genetic screening tests many individuals in a population. (Section 8.6)

Screening is done most frequently on newborns to identify certain conditions.

9 A few genetic conditions can be treated, but not cured. (Section 8.7)

A question exists as to whether we should test for conditions that have no treatment.

10 The information gleaned from genetic testing must remain private. (Section 8.8)

Discrimination can occur against individuals with genetic conditions.

Review Questions

1. Who might be screened for a genetic disease?

2. Who might be tested for a genetic disease?

3. Draw a simple pedigree for a family that might have *one* of the tests listed in Section 8.7. Indicate the condition and those who would have it.

4. As we find more and more genes associated with genetic disorders, what might happen to genetic testing?

5. If parents do not want to have their child tested at birth, what legal action could they bring?

6. What is the difference between testing and screening?

7. List the three types of genetic testing.

8. What condition is prenatally tested for the most?

9. What are the main types of prenatal testing?

10. Explain the process of PGD.

11. List four states that have the fewest number of screening tests for newborns.

12. What is the full name of the *PAH* gene found on chromosome 12?

Application Questions

1. In Chapter 3, we discussed some prenatal testing and examined Martha Lawrence's case. After reading this chapter, have you changed your mind about any of your answers in her case?

2. Research the screening laws in your state and list the conditions for which screening is conducted. Are there serious conditions that should be added to the list of screened diseases? Why?

3. When a company spends a great deal of money developing a genetic test, what would be the best way for it to earn the money back?

4. Create an idea for a new prenatal test for pregnant women who wish to find out if they are carrying a fetus with a genetic disorder.

5. Very recently, small fragments of charred bone were found in a grave in Yekaterinburg, Russia. They were identified as two children of Czar Nicolas of Russia. Do some research and find out which children they were.

6. What would be the pros and cons of each of the tests listed in this chapter?

7. Explain why Iceland's economy would so seriously affect deCODE genetics. See if you can find deCODE's stock and track its price for the last year.

8. Plan a marketing strategy for a new genetic test.

9. Almost all obstetricians have ultrasound machines in their offices. They are expensive and usually a nurse is trained in their use. What might happen if the doctor does not see a problem in one of the ultrasounds?

. . . you knew there was a genetic test for breast cancer, which runs in your family?

. . . you could have your newborn genetically tested for breast cancer?

. . . the state you live in was considering a law requiring officials to collect DNA samples from every newborn, test those samples for more than 100 diseases, and then save them in a DNA data bank?

. . . in the future, a small company, ezPGD, has developed an easy test for PGD and wants to market it to doctors, and you were an advertising executive given the project?

. . . you and your partner were going to get married next year, and a new state law required all couples getting marriage licenses to be tested to see whether they were carriers of cystic fibrosis or sickle-cell anemia?

10. A few years ago there was a case of a child who was tested and found to carry the gene for adult polycystic kidney disease. Physicians asked for a blood sample from both the mother and the father for testing. The mother told them that her child was conceived with a sperm donor. When questioning the sperm bank about his donor, they admitted that they didn't release the information to the mother when she purchased the sperm. Design a legal case as if you were the mother's attorney. What would you do step by step and what would you ask for?

11. In some cases women who use PGD want the cell tested for a specific condition that runs in their family. After finding a healthy embryo, what might be done to the remaining embryos? List five options.

12. If a patient lives in a state that does not test for a specific condition and his or her child is born with that condition, can the parent sue the state? Why or why not?

Online Resources

Preparing for an exam? Log on at www.cengagebrain.com for study tools to help you assess your understanding. If assigned by your instructor, the Case A and Spotlight on Society activities for this chapter, "Hospital Tests Babies" and "Iceland and Its Genetic Code," will also be available.

In the Media

Los Angeles Times, March 9, 2011

FDA Panel Advises Caution on Personal GeneticTesting
Andrew Zajac

A Food and Drug Administration advisory panel, comprised predominantly of both physicians and academics, said Tuesday that genetic tests directly marketed to consumers should be allowed only under a doctor's supervision.

These tests can produce ambiguous or misleading results without proper analysis by a physician, panel members said.

"It's very dangerous to get a false reassurance when you don't know about environmental and other risk factors," said panel member George Netto of the Johns Hopkins School of Medicine.

The 21-member panel expressed general agreement that doctors should be in charge of ordering and interpreting the tests. In addition the panel suggested certain regulations be put in place. However, the panel's consensus on new regulations is not binding on the FDA, but the agency usually follows them.

Source: http://articles.latimes.com/2011/mar/09/nation/la-na-genetic-testing-20110309

QUESTIONS

1. Should the FDA put more or fewer regulations in place, in your opinion?

2. What might happen if the companies marketing these genetic tests were required to follow stricter regulations?

Political Gateway, April 16, 2007

Genetic Link to Crohn's Disease Found

Canadian scientists have identified seven genes associated with Crohn's disease. Crohn's disease is one of the most common forms of inflammatory bowel disease, affecting 100 to 150 people per 100,000 individuals of European ancestry.

The researchers conducting a search of the entire human genome found that three of the seven genes were previously known, but four are newly associated with the disease.

The researchers found that one gene, *ATG16L1*, may have a role in the degradation and processing of bacteria by the body.

To access this article online, go to www.cengagebrain.com.

QUESTION

Do some research on Crohn's disease and discuss how the genetic component of this disease makes a difference in its treatment.

CTV.com

Down Syndrome Families Worry about New Prenatal Tests

Some parents of children with Down syndrome worry about what a simpler, less invasive test might mean. The new test would be able to identify Down syndrome using only a sample of blood drawn from the mother's arm. Researchers recently reported that a clinical study had found the test highly accurate.

The test could be introduced to the market as early as a few years from now. What might it mean to prospective parents grappling with the decisions that come from these tests?

Many parents have expressed excitement about the new blood test, noting it would be much less invasive than the current amniocentesis test, which involves drawing a sample of amniotic fluid—a procedure that carries a 1% risk of sparking a miscarriage. In addition, the new test can also be performed early in the pregnancy, instead of the second trimester, when amniocentesis tests are typically performed.

To access this article online, go to www.cengagebrain.com.

QUESTIONS

1. Give one reason why the parents might be concerned about the new test; after all, they already have a child with Down syndrome.

2. How might this type of a test change wrongful birth lawsuits based on testing?

9

DNA Forensics

REVIEW

What are the chemical subunits of DNA? (BB2.2)

How is DNA replicated? (BB2.8)

CENTRAL POINTS

- DNA profiles can identify individuals.

- DNA forensic analysis is done in specialized laboratories.

- DNA forensics is used in courts of law.

- Ancestry can be determined by DNA testing.

- A person may refuse to give a DNA sample to the police.

CASE A: DNA Frightens Victim

Margaret Sackler was really frightened. It had been five years since William Bern attacked her, and now he was asking to be released from jail on new evidence his attorney had presented. She didn't really understand the evidence, but she had read about it in the paper. A number of other prisoners had been released using DNA evidence.

Margaret had been walking in a parking lot outside a mall near her home. It was dark and near closing time, so there weren't too many people around. A man with a gun came up to her and ordered her to hand over her money and jewelry. She had been told never to fool around by trying to protect her valuables, so she gave him everything. But then he started to beat her. She fought back, scratching and hitting him. His blood was all over her. Then a car passed by, and her assailant got scared and ran. She always wondered what he might have done to her if the car hadn't come by.

She identified him in a lineup and testified at the trial. She had seen his eyes clearly, but he had covered the rest of his face with a mask. She would never forget his eyes. He was found guilty and sentenced to 25 years to life for assault. She finally felt safe.

But, William Bern, the man convicted, had always claimed he was innocent, so when he heard about the use of DNA testing to exonerate people, he asked his attorney to arrange testing of the blood

on the blouse Margaret was wearing at the time of the attack.

The blood was 5 years old, but with a type of DNA testing called PCR, old DNA present in very small amounts can still be tested.

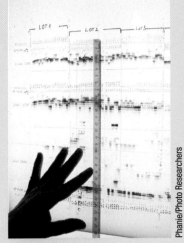

A lab tech examining a DNA profile.

Phanie/Photo Researchers

Margaret and William were asked to give a blood sample, and all the samples were tested. William's DNA did not match the blood on Margaret's shirt. Based on the results of the DNA testing, he was asking to be released from jail.

Margaret didn't care about the DNA test results—she was sure he was the one. She'd never forget his eyes.

Some questions come to mind when reading about this case. Before we can address those questions, let's look at the biology behind DNA forensics.

A Crime Scene Technician Gathers Evidence

Alamy Limited

9.1 How Is DNA Testing Done?

In 1975 an English scientist, Dr. Alec Jeffreys (Figure 9.1), developed the first method of preparing what we now call DNA profiles and used them to compare profiles from different people. He called this method **DNA fingerprinting**. Soon after the method was developed, he used it to solve two murders in northern England by comparing DNA fingerprints from a number of men to fingerprints prepared from DNA found at the crime scenes (see "Spotlight on Law: Narborough Village Murders"). This spectacular discovery has revolutionized criminal investigations all over the world.

John Smock/AP Photos

Figure 9.1 **Alec Jeffreys**
He developed DNA fingerprinting, and this discovery led to the development and widespread use of DNA forensics across the globe.

DNA fingerprinting a method of DNA forensics

- **DNA forensic testing began in England in 1975.**

9.2 What Is a DNA Profile?

Jeffreys's method of DNA fingerprinting was based on the discovery of small differences among people's DNA. These differences are present as variations in the length of certain DNA sequences called **minisatellites**. Minisatellites are repeated sequences of 10 to 100 base pairs (see "Biology Basics: DNA"). They are found at many different locations on all chromosomes and can be used in making DNA fingerprints because the repeats have different lengths in different people. Here is what one repeat of a minisatellite looks like:

```
...CCTGACTTAGGATTGCCA...
```

Later other, shorter sequences called **short tandem repeats (STRs)** were discovered. The repeats in STRs are much shorter than those in minisatellites and are only two to nine base pairs long. Clusters of STRs are found widely distributed on all human chromosomes. Variations in the number of repeats in STRs are used in preparing DNA profiles because they have different lengths in different people. Here is what an STR of the repeat TTCCC looks like:

```
...TTCCCTTCCCTTCCCTTCCCTTC...
```

In addition, **single nucleotide polymorphisms (SNPs),** pronounced "snips," are present in the genome. SNPs are single nucleotide differences in a DNA sequence. Because they consist of only *one* base pair, they can be specific to an individual. Here is what a SNP looks like compared with DNA sequences of others within a population:

minisatellites DNA sequences composed of 10 to 1000 base pairs that are scattered throughout the genome

short tandem repeats (STRs) DNA sequences that are two to nine base pairs long

single nucleotide polymorphisms (SNPs) DNA sequence variations that are only one base pair long

DNA profile the pattern of DNA fragments that can identify individuals

population frequency a measurement that determines how often specific combinations of alleles are present in a population

restriction fragment length polymorphism (RFLP) analysis one of the methods of DNA testing

polymerase chain reaction (PCR) a molecular technique that allows production of many copies of a specific DNA sequence

```
...ATCGTCGAGCCTAAATA...
...ATCGTCGAGCCTAAATA...
...ATCGTCGAGCCTAAATA...
...ATCGTCGTGCCTAAATA...
...ATCGTCGAGCCTAAATA...
...ATCGTCGAGCCTAAATA...
```

As you can see from these sequences, each taken from a different person, there is a difference (see red T) in person #4.

Each of these different types of DNA variation can be used to create a DNA profile and identify individuals. The profiles developed from this information can be used in numerous ways, including criminal cases, paternity lawsuits, studies of human evolution, and identification of bodies. STRs are now routinely used, and the term *DNA fingerprint* has been replaced by the term **DNA profile**.

How are STRs used to make a DNA profile of individuals in a population?

Let's look at the steps in preparing a DNA profile. Each STR is a DNA sequence composed of two to nine nucleotides. Each allele of an STR contains a different and unique number of repeated copies of these nucleotides. For example, an STR might contain different copies of the sequence AGA. An allele of this STR (allele 1) might contain one copy of AGA, another might have three (allele 2), and yet another (allele 3) might have five, as shown below:

```
Allele 1:  ...AGA...
Allele 2:  ...AGAAGAAGA...
Allele 3:  ...AGAAGAAGAAGAAGA...
```

In preparing a DNA profile, several different STR alleles obtained from an individual are analyzed and compared with information that shows how often a specific combination of these alleles is found among individuals in a population. Different populations may have different frequencies of these alleles. This analysis is done in a series of steps:

1. DNA samples from a population are analyzed to establish which STR alleles are present in members of the population.

2. The results are analyzed to see how often specific combinations of STR alleles are present in a population; this result is called the **population frequency** (discussed in more detail in Chapter 15).

3. The population frequencies for each STR allele are multiplied to estimate the probability that anyone who carries this specific combination of alleles is a match to the sample being tested (see Table 9.1).

The combined frequency shows that the combinations become more rare in the population as more alleles are analyzed. Later discussion will show how this applies to criminal cases.

Table 9.1 **Population Frequency Calculations**

STR	Allele	Frequency in Population	Combined Frequency
A	1	1 in 25	
B	2	1 in 100	A1 × B2 (1 in 25 × 1 in 100) = 1 in 2500
C	3	1 in 320	A1 × B2 × C3 (1 in 25 × 1 in 100 × 1 in 320) = 1 in 800,000
D	4	1 in 75	A1 × B2 × C3 × D4 (1 in 25 × 1 in 100 × 1 in 320 × 1 in 75) = 1 in 60 million

How are DNA profiles constructed? DNA profiles can be constructed using a number of methods. Two methods commonly used in DNA forensics are **restriction fragment length polymorphism (RFLP)** analysis (Figure 9.2) and the **polymerase chain reaction (PCR)**. RFLP analysis is used to detect differences in the number of copies of repeated DNA sequences.

How are RFLPs done? RFLP analysis uses blood samples or other relatively large amounts of biological material (biopsy specimens, semen, etc.). The process consists of the following steps (follow along using Figure 9.3, page 186):

1. DNA is extracted from cell samples. Each sample is from a different person.
2. The DNA is cut into fragments using restriction enzymes (see Chapter 10).
3. Each sample is loaded into a small slot, called a well, on a slab of a gelatin-like substance called agarose.
4. Electrical current is passed through the gel (which is immersed in a liquid). As the current moves through the gel, the DNA fragments migrate through the pores of the gel.
5. As they migrate through the gel, the DNA fragments separate by size. Smaller fragments move faster and further than the larger fragments, creating a distinctive pattern.
6. The resulting pattern of DNA fragments is visualized, photographed, and scanned into a computer. The patterns obtained from one individual are stored in a database and compared to patterns obtained from other individuals.

How does PCR work? PCR uses very small amounts of DNA to produce profiles, working in a way that is similar to DNA replication in a cell (see "Biology Basics: DNA"). During the process of PCR, DNA is heated and cooled. DNA responds to the temperature changes by opening up its helix; this creates two single-stranded molecules. After every round of replication, the amount of DNA is doubled. This process can be repeated many times. In this way, a single DNA fragment can be amplified to make millions of copies. These copies are separated by electrophoresis and used to prepare DNA profiles.

How was PCR discovered? Some think that because of its use in DNA forensics and biotechnology, PCR is *the* scientific breakthrough of the last 50 years. The process was discovered by Kary Mullis (Figure 9.4, page 186), who won the Nobel Prize in 1993 for his invention. He was working for Cetus (a biotech company) when he got his idea to work. He never owned the patent on the process, and Cetus paid him a $10,000 bonus for the invention. After winning the Nobel Prize, Mullis left science and spent

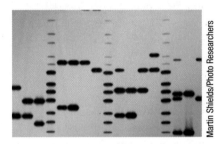

Figure 9.2 The RFLP test makes it possible to identify samples found at a crime scene and match them to suspects. In this gel, the band pattern is produced by DNA fragments containing different numbers of repeats. The three sets of bands that run from top to bottom are size standards.

Martin Shields/Photo Researchers

Figure 9.3 **RFLP Analysis** DNA is extracted from samples and cut into fragments with a restriction enzyme. The fragments from each sample are placed in different lanes on a gel and separated by size using an electric current. The various fragment sizes reflect differences in the number of copies of a DNA repeat.

1 DNA is extracted from cells belonging to three different people.

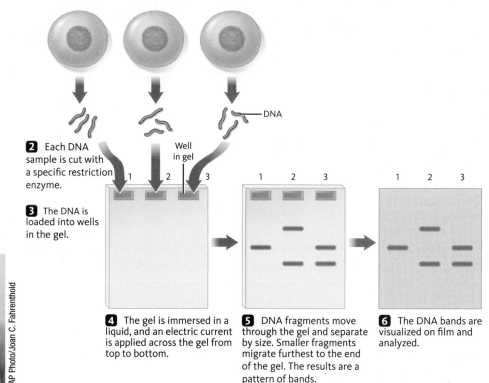

2 Each DNA sample is cut with a specific restriction enzyme.

3 The DNA is loaded into wells in the gel.

4 The gel is immersed in a liquid, and an electric current is applied across the gel from top to bottom.

5 DNA fragments move through the gel and separate by size. Smaller fragments migrate furthest to the end of the gel. The results are a pattern of bands.

6 The DNA bands are visualized on film and analyzed.

Figure 9.4 **Kary Mullis, the Inventor of the PCR Technique**

AP Photo/Joan C. Fahrenthold

years as a surfer, stating that he had nothing left to win in science. Later he developed a company that made jewelry that contains DNA of dead celebrities copied by PCR.

The PCR process consists of the following steps (follow along using Figure 9.5):

1. DNA is extracted from the sample and placed in a solution.

2. The DNA is heated and separated into two single strands.

3. The temperature of the solution is lowered, and short nucleotide sequences called **primers** are mixed with the DNA.

4. The primers find and pair with complementary regions on the single-stranded DNA. These primers act as the beginning of a new DNA strand.

This test tube contains DNA that has been collected from a crime scene or old evidence.

1 DNA is recovered for PCR and processed before being copied.

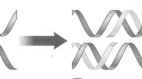

Figure 9.5 **The PCR Method**
This shows one round of DNA duplication. Most PCR reactions involve several hundred rounds.

2 DNA is heated and separated into single strands.

3 Primers are added and bind with complementary regions in the DNA.

4 Primers serve as starting points for synthesis of new DNA strands.

5 New double stranded DNA is formed.

primers short nucleotide sequences used in PCR

5. DNA polymerase (the enzyme involved in DNA replication) is added to the solution along with nucleotides (C, A, T, and G). The DNA polymerase uses these nucleotides to synthesize a double-stranded DNA molecule from each of the single-stranded DNA templates.

These steps make up one PCR cycle. The cycle can be repeated over and over, as shown in Figure 9.6. The amount of DNA present doubles with each PCR cycle. The power of PCR allows millions of copies to be made in hours from tiny amounts of DNA. These cycles are done quickly using specialized machines that are called PCR machines (Figure 9.7). Although they are expensive, most criminalists work in labs that have at least one.

Table 9.2 shows how the amount of DNA increases with each PCR cycle.

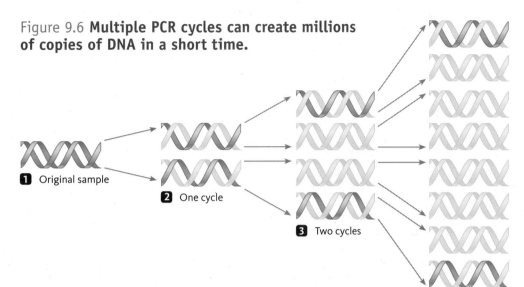

Figure 9.6 Multiple PCR cycles can create millions of copies of DNA in a short time.

1 Original sample

2 One cycle

3 Two cycles

4 Three cycles

Where do DNA samples come from? Unlike RFLP analysis, PCR can be used to prepare DNA profiles from very small DNA samples. Sources of DNA can include single hairs, licked envelope flaps, toothbrushes, cigarette butts, and dried saliva on envelope flaps (Figure 9.8, page 188).

Evidence from crimes including murder, rape, break-ins, and hit-and-run accidents can be used. Profiles can also be obtained from very old samples of DNA, increasing its usefulness in legal cases. In fact, DNA profiles have been prepared from mummies more than 2400 years old.

What are DNA microarrays? Many technologies first developed for laboratory use have been adapted to forensics. One of these is the use of DNA microarrays (also called DNA chips). Originally developed for quick DNA analysis by scientists, microarrays are now used in crime labs. Portable devices containing a DNA microarray allow small amounts of DNA at a crime scene to be analyzed quickly and efficiently, sometimes directly on the scene. Because these devices are small and able to process many DNA samples, police can carry them to crime scenes and work through evidence quickly. These and other handheld devices, including PCR machines (Figure 9.9, page 189), will revolutionize crime scene forensics. These devices also have a far-reaching future in other fields, especially in getting DNA sequencing out of research labs and into physician's offices where it can be used to tailor medical treatments to an individual's unique genetic makeup.

9.2 Essentials

- **Three types of DNA variations are used in DNA profiling: minisatellites, short tandem repeats, and single nucleotide polymorphisms.**
- **Two common methods used in DNA forensics are restriction fragment length polymorphism analysis and the polymerase chain reaction.**
- **DNA samples for profiles can come from any number of sources, including cheek swabs, semen, saliva, and blood.**
- **Once constructed, DNA profiles are compared with other profiles in forensic applications.**

Figure 9.7 **A Large-Scale PCR Machine**

Philippe Psaila/Photo Researchers, Inc.

Table 9.2 **Using PCR to Increase DNA**

Cycle	Number of Copies of DNA
0	1
1	2
5	32
10	1024
15	32,768
20	1,048,576
25	33,544,432
30	1,073,741,824

Figure 9.8 Sources of DNA Evidence (a) Skin cells deposited on the inside of the ski mask can be used to retrieve DNA. (b) When a suspect licks the back of the postage stamp, cells from the saliva serve as a source of DNA. (c) Saliva deposited while drinking from a soda can may also be used. (d) Saliva is also left on cigarette butts. (e) White blood cells can be recovered from blood samples, and DNA can be extracted from very small numbers of cells. (f) Skin cells inside a sock and other clothing can be used for DNA profiles. (g) Extraction of DNA from cells in hair roots can be used for nuclear DNA, but follicle cells also contain mitochondria, and that DNA can also be used. (h) Bone or bone fragments can be pulverized and exposed to chemicals that free the DNA from the calcium in bone. (i) Teeth can also serve as a source of DNA.

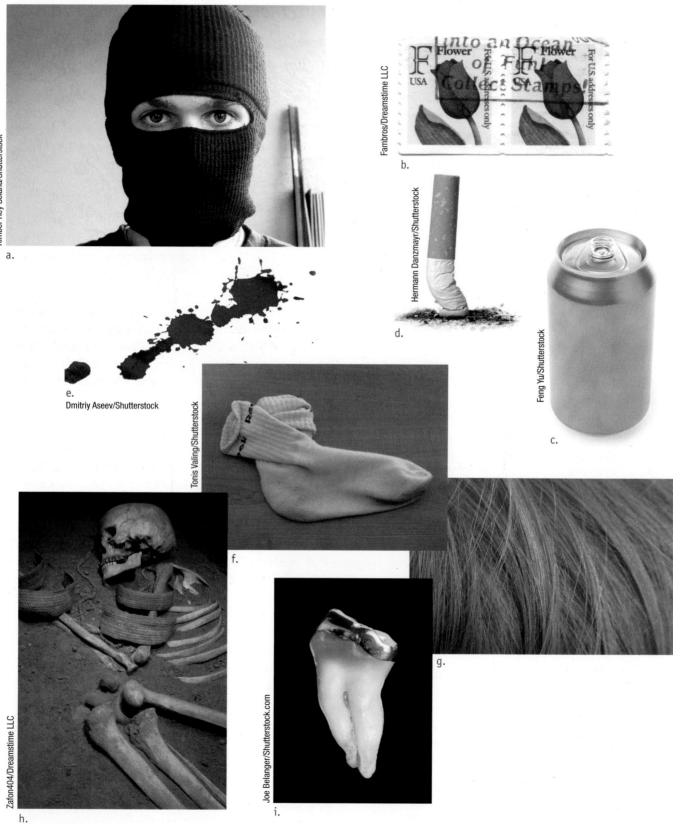

Kimber Rey Solana/Shutterstock

a.

Fambros/Dreamstime LLC

b.

Hermann Danzmayr/Shutterstock

d.

Feng Yu/Shutterstock

c.

Dmitriy Aseev/Shutterstock

e.

Tonis Valing/Shutterstock

f.

Jupiter Images

g.

Zafon404/Dreamstime LLC

h.

Joe Belanger/Shutterstock.com

i.

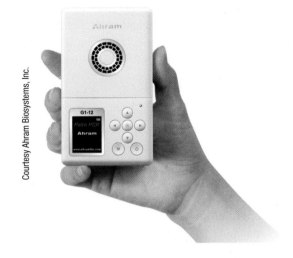

Figure 9.9 This handheld PCR machine will allow samples to be processed and analyzed at the crime scene.

Courtesy Ahram Biosystems, Inc.

posed that for more than 90 years would decide whether science was "good science" and not "junk science." The Frye test was applied to DNA forensics; it passed this test and has been accepted in all state and federal courts.

The first criminal case to use DNA forensics began in England in 1983 (see "Spotlight on Law: Narborough Village Murders"). It took several years before this type of evidence was used in the United States. The use of DNA evidence in the U.S. legal system was scrutinized as thoroughly as any other scientific method. Each state had to set its own standards for the use of DNA profiles using the *Frye* test, and the public was generally not aware of the use of DNA forensics for several years.

The first U.S. case to challenge the admissibility of a DNA profile was *People v. Castro*, 545 N.Y.S.2d 985 (S.Ct. 1989). Castro murdered a 20-year-old woman and her 2-year-old daughter. Blood was found on Castro's watchband.

In response to the challenge, the court decided that DNA identification procedures were generally accepted among the scientific community. This landmark case was just the beginning of DNA evidence in the U.S. courts. Since then it has been admitted in all types of cases including murder, divorce, paternity, and child support.

DNA profiling has had a significant impact on rape cases. Previously, rape cases were difficult to prosecute. It usually came down to one person's word against another's. When evidence could be obtained (semen or blood) and matched to a suspect, it gave the court solid evidence to use. Today victims are much more likely to report rapes, and more convictions are obtained.

How are DNA profiles used in the courtroom?
The use of scientific knowledge in civil and criminal law is called forensics. Forensic DNA analysis is usually performed in state and local police crime labs, private labs, and the Federal Bureau of Investigation (FBI) lab in Washington, D.C.

In criminal cases, DNA is often extracted from biological material left at a crime scene, which can include blood, tissue, hair, skin fragments, and semen. DNA profiles are prepared from evidence and compared with those of the victim and any suspects in the case.

Most forensic testing in the United States uses a panel of 13 different STRs (review preceding section) for preparing DNA profiles. The number 13 is not random. When the first DNA database was established in Virginia, it became the template for all others. The largest database is kept by the FBI and is called **Combined DNA Index System (CODIS) panel**. It is used by law enforcement and other government agencies to compare DNA profiles of convicted criminals at both the

---

CASE A: QUESTIONS

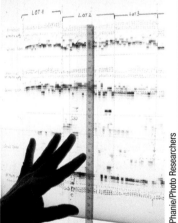

Phanie/Photo Researchers

Now that you understand how and why a DNA profile is constructed, let's look at some of the issues raised in Margaret Sackler's case.

1. If you were the judge, would you give William Bern a new trial? Why or why not?

2. If you were the prosecutor, what arguments would you give for not using the new DNA evidence?

3. If you were William's attorney, what arguments would you give to convince the judge that the DNA evidence should be submitted and a new trial granted?

4. Margaret is an eyewitness. Which is more reliable, an eyewitness or DNA testing? Why?

5. Should people who have already been convicted use DNA testing to get new trials? Why or why not?

6. In some cases, the DNA match is not perfect but is a 60% match. What could that mean?

9.3 When Did DNA Forensics Enter the U.S. Courts?

As discussed in Chapter 1, scientific evidence is not automatically admitted into a court of law. In the landmark case *Frye v. U.S.* (Spotlight on Law: Chapter 1), four questions were

---

Combined DNA Index System (CODIS) panel a national database of DNA profiles begun by the FBI

luchschen/Shutterstock.com

Figure 9.10 These DNA profiles were prepared from blood samples taken from a crime scene and samples from several suspects. Which of the suspects' profiles looks most similar to the crime scene profile?

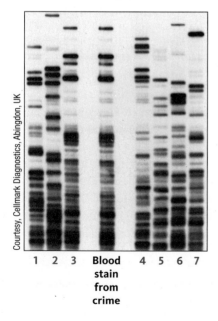

Courtesy, Cellmark Diagnostics, Abingdon, UK

1 2 3 **Blood stain from crime** 4 5 6 7

federal and state levels. There is no national database, but some legislation has been written to allow cross-state testing.

A DNA sample is tested using STRs from the CODIS panel, and a profile is compared to stored samples and to samples from the crime scene (see Figure 9.10). In a criminal case, if a suspect's profile does not match that of the evidence, he or she can be cleared, or excluded, as the criminal. About 30% of DNA profile results clear innocent people by exclusion.

What is a DNA database? The FBI began cataloging DNA profiles and DNA samples obtained from convicted felons in 1998, and that database now contains more than 1,700,000 profiles. At first, samples were collected and profiles kept only from convicted felons. However, many states are now collecting DNA samples from anyone arrested for a felony and entering their DNA profiles into a database. Other states keep the DNA of anyone arrested. As they accumulate large numbers of profiles, DNA databases are becoming more important tools in solving crimes. Over a three-year period in Virginia, for example, matching DNA profiles from evidence at crime scenes with those in the state database solved more than 1600 crimes in which there were no suspects.

In other countries, DNA data banks are regulated differently. In the United Kingdom, a law requires that, if asked, an individual must submit a sample and that sample can be kept forever. Other countries are moving to create larger and larger databases to make identification easier and faster. It has been suggested that this would be a good idea for the United States, but it has not happened yet. Most of the databases contain samples from people in jail or those awaiting trial. However, if asked for DNA samples, ordinary people are usually willing to volunteer.

9.3 Essentials

- **DNA forensics entered the U.S. court system in 1989 and is now used in courts all over the country.**
- **Many people wrongly accused have been freed by DNA evidence with the help of innocence projects all over the United States.**
- **DNA databases are used to identify perpetrators from DNA left at crime scenes.**

9.4 What Are Some Other Uses for DNA Profiles?

DNA profiles have many uses outside the courtroom. **Biohistorians** use DNA analysis to correctly identify bodies and body parts of famous and infamous people whose graves have been moved several times or whose graves have been newly discovered. For example, Czar Nicholas II of Russia (Figure 9.11) and his family members were killed in 1917 and buried in unmarked graves. In 1991, remains found in Yekaterinburg, Russia, were identified as those of the czar and some of his family. In 2007, bones found at a nearby site were identified as those of two more of his children, accounting for all family members. For years previously, a woman claimed to be Anastasia, one of the czar's daughters. DNA testing from a lock of hair established that this woman, Anna Anderson, who died in 1984, was not one of the czar's children.

DNA profiles are used to identify the remains of military personnel killed in action and were used to identify people killed in the September 11, 2001, attacks in the United States. In the wake of the devastation caused by Hurricane Katrina in the southern United States in 2005, many of the dead were identified by DNA profiles. In

Figure 9.11 **Czar Nicholas II of Russia** The Czar and his family were killed during the Russian Revolution. Years later their remains were identified using DNA forensics.

Hulton Archive/Getty Imagesv

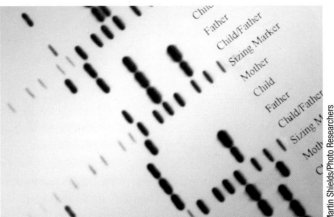

Figure 9.12 **This RFLP test is being used to determine paternity.** Although the entire test is not shown, the child and mother's DNA test are clearly visible.

Martin Shields/Photo Researchers

both the Katrina and the September 11 disasters, family members brought in hairbrushes, toothbrushes, pillows, and clothing for DNA testing and matching. After years of lab work, all remains from September 11 were identified.

Another important use of DNA profiling is paternity identification. Matching samples from the mother, father, and child allow 100% accuracy in identifying fathers. Use of DNA profiles in paternity testing ensures that many children will get the child support that they need. This is another area where DNA profiling has made a huge difference in legal proceedings. Before DNA testing, fathers could be eliminated in paternity claims only through the use of blood typing as an identifier (see Chapter 14). Today positive identification is done easily when samples can be obtained.

Paternity testing has become big business. Kits that collect DNA samples from the child, mother, and father are available over the counter in major drug stores. After collection, the samples are sent to a lab to be processed. Although the kits cost around 40 dollars, the test itself is much more expensive.

Several talk shows on television run episodes that help women find the fathers of their children. All the people involved come on the show and are tested. Then, on television, the truth is revealed!

Figure 9.12 is a photo of a paternity test that compares the mother's, father's, and child's DNA.

Can DNA profiles trace our ancestry? The

use of DNA in forensic identification in high-profile paternity cases and on television crime shows has raised awareness that DNA can be used to trace an individual's ancestry. Many companies now offer ancestry genetic testing on the Internet. Two types of testing are usually offered: mitochondrial DNA testing and **Y chromosome testing**.

Mitochondria are found only in the cytoplasm of a cell (see "Biology Review: Cells and Cell Structure"). Because the egg is the only gamete with cytoplasm, mitochondria are always passed from a mother (Figure 9.13) to all her children (males and females). Mitochondria contain small amounts

of DNA, which contains a small number of genes. Because sperm have no cytoplasm, men do not contribute **mitochondrial DNA (mtDNA)** to their children. Differences in mitochondrial DNA sequences provide a series of markers called **haplotypes.** Each person has the same haplotype as his or her mother, so comparison of haplotypes can help trace ancestry through the mother's line.

Similar haplotypes can be grouped together to form **haplogroups.** When collected from large population groups such as European, Asian, or African, they can be used to provide

Figure 9.13 **Inheritance of Mitochondrial DNA**
This pedigree shows the inheritance of mitochondrial DNA through a family. The red circles and squares show how the mitochondrial DNA from the woman in generation 1 is passed from one generation to the next.

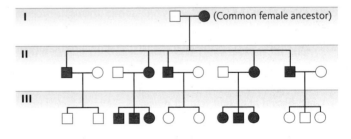

biohistorian a person who specializes in the history of biology

Y chromosome testing a process that maps sequence variations on the Y chromosome

mitochondrial DNA (mtDNA) DNA found in the mitochondria of a cell

haplotypes a series of markers that shows similarities and differences in the distribution of a specific DNA sequence

haplogroup groups of haplotypes from a specific region of the world

Figure 9.14 Inheritance of the Y Chromosome

This pedigree shows how the Y chromosome is passed through a family. The red squares show that all males in this family have the same Y chromosome.

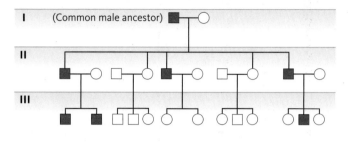

information about ancestry. When someone's haplotype is similar to that of a specific haplogroup, he or she is said to have ancestry within that group. Therefore, we can use mitochondrial DNA to trace our backgrounds.

The Y chromosome is found in the nucleus of a cell and is passed on by the sperm. Only males receive a Y chromosome (Figure 9.14), which comes from their father (the mother carries two X chromosomes and no Y chromosome). In addition, because the Y chromosome is small, very few changes in DNA occur, and most Y chromosomes are passed from father to son virtually unchanged. Therefore, it is easy to trace male lineages through haplotypes on the Y. The Y chromosomal haplotypes are used to trace paternal ancestry.

DNA testing using mtDNA or Y chromosomes can be used to search haplogroup databases to provide information about one's ancestry. One of the largest projects to trace ancestry and the patterns of human migration out of Africa and across the globe is being sponsored by the National Geographic Society. This effort is called the Genographic Project. Interested individuals must pay to participate in this project. Those enrolled receive a vial and a swab to collect cheek cells. DNA from the swab is analyzed, and individuals receive information about their ancestry. The Genographic Project uses the results anonymously to construct a detailed chart of human migrations over the last 50,000 to 100,000 years.

9.4 Essentials

- **DNA profiles can be used to trace ancestry through the maternal and paternal lines.**
- **DNA profiles have many other uses in many areas of our society.**

9.5 What Are the Legal and Ethical Issues Associated with DNA Forensics?

What are some of the unforeseen consequences of DNA forensics? In an interesting twist of fate, another huge change

CASE B:
Samples Asked of All

There were very few clues to the murders of two women in a small rural community. The police had little evidence, but they did have skin samples from the killer, found under the fingernails of the victims.

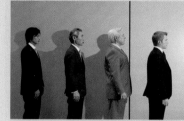

Men waiting in line to give DNA samples in an investigation.

From these samples a DNA profile was created, but there were no suspects. Because the crimes had occurred out in the forest, presumably at night, there were no eyewitnesses. No fingerprints, footprints, or pieces of clothing were recovered from the crime scene. If investigators could find someone whose DNA profile matched the sample, they might solve the case.

The investigator in charge of the case, Lt. DeMato, decided to try a new technique being used in the United Kingdom called a DNA dragnet. He planned to ask all the male members of the rural community over the age of 17 to give a cell sample (obtained with a cotton swab scraped along the inside of the cheek) and do DNA testing on all of them.

1. **What might happen if a man didn't want to come in and give a sample?**

2. **Suppose that after samples were taken from all area men over age 17, no match was found. What would you do next if you were Lt. DeMato?**

3. **Amazingly, a match was found in the samples taken in the dragnet. The district attorney questioned whether the dragnet would be accepted as evidence by the court during the trial. Give three arguments the prosecution could use to have the evidence from the dragnet admitted.**

4. **Give three arguments that the defense could use to keep the evidence from being admitted.**

5. **After Lt. DeMato sent out a letter to all men in the area, he began testing. The 23rd man on the list, Irving Tomston, didn't respond. A second letter was sent, and there was still no response. What should Lt. DeMato have done?**

in the criminal justice system has come about as a result of the use of DNA forensics. When DNA evidence was first used in U.S. courts, it seemed obvious to some that prisoners convicted before 1989 were not able to use this method to establish their innocence. When approved by a *Frye* hearing, should scientific discoveries be retroactive? In 1993, an attorney named Barry Scheck and his partner Peter Neufeld

The Innocence Project/NLM

Figure 9.16 Ronald Cotton was released after 10 years in prison for a crime he did not commit. DNA testing led to his release.

AP Photo/HO/Burlington Police Department

began the Innocence Project at the Cardoza School of Law at Yeshiva University (Figure 9.15) to address this issue.

Soon after the inception of the project, Scheck and Neufeld appeared on the "Phil Donahue Show." They talked about DNA forensics and the possibility of using it to reopen cases that had already been decided as a way to help free those who had been wrongly convicted. At the end of the interview, Scheck turned to the camera and said, "Anyone who thinks they are wrongfully convicted of a crime, write to us."

This was the first time that many members of the general public had heard of DNA forensics and how it might be applied to past cases. Since its inception, the Innocence Project has freed over 300 people, there are more appeals pending, and there are still more ready to be filed. In addition, other innocence projects have sprung up throughout the United States (there are over 100) at various law schools.

The Margaret Sackler and William Bern case described at the beginning of this chapter (Case A) was actually based on a real case, the case of Ronald Cotton (Figure 9.16). The Cotton case is one of the most interesting DNA exoneration cases because of how it was resolved.

Ronald Cotton heard that his name was on a list of men to be interviewed in a rape case. Instead of waiting for the police to come to him, he went in to tell them he was innocent. He was asked to do a lineup and was identified by the victim and arrested. Ronald Cotton was convicted of rape and assault of two victims, and this verdict was upheld in a number of retrials. All through these trials Cotton claimed his innocence. But, before the second trial was about to begin, an inmate in prison told another inmate that he had committed the rapes. The judge, however, refused to allow this evidence, and Cotton was convicted again.

In 1994, two new attorneys were appointed to represent Cotton, and they asked for the DNA collected at the time of the crime to be tested. The evidence had been saved, and they showed Cotton was not a match. He was released in 1995 and was given $5,000 by the state of North Carolina. After his release, Jennifer Thompson-Canino (Figure 9.17), the victim who identified Cotton, came forward and met with Cotton; together they now tour the country discussing the problem of wrongful conviction.

You can read more about Ronald and Jennifer in her book *Picking Cotton*, published in 2009.

One alarming aspect of these cases is that some of the prisoners released were on death row awaiting execution. One such case was that of Kirk Bloodsworth (Figure 9.18, page 194).

Figure 9.17 Jennifer Thompson-Canino was the woman who was raped and wrongly identified Ronald Cotton.

AP Photo/Chuck Burton

Kirk Bloodsworth
First capital conviction in the U.S. overturned in 1993 as a result of DNA testing.

After an anonymous tip and eyewitness testimony placed him near the crime scene, Kirk Bloodsworth was sentenced to death in Maryland for the 1984 sexual assault, murder, and mutilation of 9-year-old Dawn Hamilton. From the beginning, Bloodsworth insisted on his innocence. In prison, he learned about DNA profiling. Eventually his attorney, Bob Morin, with support from the Innocence Project, persuaded officials to compare Bloodsworth's DNA with the DNA of dried sperm found on the victim.

Testing on the DNA samples was done using PCR, which at the time was still a new forensic method. The results exonerated Bloodsworth. He was freed from prison in June 1993—the first death-row prisoner to be exonerated by post-conviction DNA testing.

In 2003, after much prodding from Bloodsworth and Innocence Project lawyers, Maryland authorities finally searched their DNA database for a "cold hit" match of the evidence in the Dawn Hamilton case. The search turned up Kimberley Shay Ruffner, a convicted rapist whom Bloodsworth had known in prison; Ruffner was then tried and found guilty of the 1984 murder.

The Bloodsworth case and other death-row exoneration cases have many states rethinking the death penalty. In 2000, Illinois Governor George Ryan stopped executions in his state because of the possibility that some prisoners were wrongfully convicted. In 2003, he commuted the death sentences of all 156 prisoners awaiting execution, and in 2011, Governor Pat Quinn signed legislation ending the death penalty in Illinois.

Another question that comes from these cases is this: have we already executed innocent prisoners who did not have the opportunity to use DNA forensics? The answer seems to be yes, but we cannot answer this question directly because of the difficulty of postexecution DNA testing, given that one major party of the case has already died.

How many wrongfully accused men and women are still in prison? It is difficult to ascertain. DNA forensic testing is expensive, and funding problems do not allow states to pay for it. Therefore, private funding, like that of the Innocence Project, is necessary to keep these cases going. In addition, when someone is convicted by testimony from an eyewitness, it is difficult for victims, juries, and the public to accept that these criminals were wrongly convicted, and that the real criminal went unpunished. To review the current situation, cases are available on the Innocence Project Web site at www.innocenceproject.org.

The results of DNA forensic testing are also being used with something called a John Doe warrant. When someone is suspected of a crime, a warrant may be issued for his or her arrest. This warrant would include a physical description of the person and other pertinent information. Beginning with New York, many states have allowed warrants to be issued when only a DNA profile is available. In this situation, the DNA profile replaces the physical description and other details usually included in an arrest warrant. This practice allows police more time to find a perpetrator.

Table 9.3 addresses some of the ethical and legal questions in DNA profiling.

9.5 Essentials

- **DNA profiling has been used to secure the release of hundreds of wrongfully convicted prisoners.**
- **Most states and the federal government have constructed DNA databases for use in solving crimes.**

1. DNA forensic testing began in England in 1975. (Section 9.1)

   Alec Jeffreys first developed a way to do this testing on blood samples.

2. DNA samples for these tests can come from any number of sources, including cheek swabs, semen, saliva, and blood. (Section 9.2)

   Any tissue from a person's body can be used to extract cells for testing.

3. Three types of DNA variations are used in DNA profiling: minisatellites, short tandem repeats, and single nucleotide polymorphisms. (Section 9.2)

   These variations occur in all individuals.

4. Two common methods in DNA forensics are restriction fragment length polymorphism (RFLP) analysis and the polymerase chain reaction (PCR). (Section 9.2)

   Both methods are used in criminal trials.

5. Each of these methods can be used to analyze DNA patterns and then make a comparison with other samples. (Section 9.2)

   The comparisons are often done with samples found in crime scenes.

6. DNA forensics entered the U.S. courts in 1989 and is now used in courts all over the country. (Section 9.3)

   This type of testing has revolutionized personal identification.

7. DNA profiles can be used to trace ancestry through both the maternal and paternal lines. (Section 9.4)

   Ancestry testing has become popular but is still in its infancy.

8. DNA profiles have many other uses in many areas of our society. (Section 9.4)

   Many think that profiling will be the way to identify people in the future.

9. DNA profiling has been used to secure the release of hundreds of wrongfully convicted prisoners. (Section 9.5)

   More prisoners are being released every day.

10. Most states and the federal government have constructed DNA databases for use in solving crimes. (Section 9.5)

    These databases contain valuable information that should be kept private.

Review Questions

1. The term *DNA fingerprinting* was developed in the infancy of the technology. Make a chart comparing and contrasting DNA profiling and the older technology, DNA fingerprinting.

2. What are the differences among minisatellites, STRs, and SNPs?

3. In some cases, why is PCR more desirable in testing than RFLP?

4. What is the meaning of the name *restriction length polymorphism* and where does it come from?

5. Who invented the PCR technique? Research his background.

6. What moves the DNA through the gel in RFLP testing?

7. What two benefits do DNA profiles have in the courts?

8. Do some research and find out which state has the largest DNA database.

9. At some point, RFLP testing and PCR testing merged. How did this happen?

10. Why is it useful to be able to test smaller and smaller amounts of DNA?

11. Can someone be forced to donate DNA for a database in the United States?

12. What does CODIS stand for?

Application Questions

1. Research the laws in the United Kingdom that control DNA dragnets, and write a short report.

2. Read up on the DNA microarray, and discuss how it will change criminal detection.

3. If an allele is prevalent in a certain group of people, what does this probably mean?

4. DNA forensics took a number of years to enter the U.S. court system. Why do you think this is?

5. If a person is questioned about a crime, should his or her DNA be taken? Should a profile be made and entered into a database? Give your reasons.

6. Go back to Chapter 1 and re-read *Frye v. U.S.* How does this case apply to the use of DNA forensics in the U.S. courts?

7. As PCR allows us to test smaller and smaller samples of DNA, how do you think this will change the way police and forensics experts treat a crime scene? List three ways.

8. DNA testing is also done to identify purebred dogs for dog shows. If this is done all the time, what effects might it have on the dog breeds?

9. Some think that a huge DNA database should be assembled from DNA taken from all newborns, so that eventually, the whole population is represented in the database. What do you think? Why?

10. In an interesting case in New York State, no perfect match was found for DNA taken from a murder scene. However, a 60% match was found from a man already in jail. This meant that the perpetrator was somehow related to this man. Police began to investigate the man's son and found evidence from the crime in his home. Should this evidence be used at his trial? If a DNA profile from the son matches the profile from the murder scene, should this evidence be admitted in court?

. . . you were asked to participate in a DNA dragnet?

. . . you were asked to vote on a referendum that would create a DNA database for all newborns?

. . . you were on a jury that was asked to release a felon from prison after 20 years because of DNA evidence?

. . . you were on the Sackler jury?

. . . you were given the opportunity to participate in an ancestry test?

11. In 1998, when the entertainer and politician Sonny Bono died, a man came forward claiming to be his son. The birth certificate issued had the name "Salvatore Bono" in the father's slot. But because he was a celebrity, the court wanted more evidence than that. They asked for a DNA test of Sonny Bono's remains, the man's mother, and the prospective son. Should this be allowed? How would they get the DNA from someone who has died? Do some research and see how this was resolved.

12. List several pros and cons of DNA databases.

13. Compile a list of celebrities who have been sued for paternity.

Online Resources

Preparing for an exam? Log on at www.cengagebrain.com for study tools to help you assess your understanding. If assigned by your instructor, the Case B and Spotlight on Law activities for this chapter, "Samples Asked of All" and "Narborough Village Murders," will also be available.

Learn by Writing

In Chapter 1 we suggested that you start a blog with members of your class or others who are interested in the impact of genetics on individuals and society. Now is the time to revisit your blog and consider some of the questions in Chapters 5–9. E-mail others whom you think might be interested, and invite them to contribute to the blog.

Here are some ideas to address in your blog:

• Is a single gene defect such as sickle-cell anemia more serious than an extra or missing chromosome?

• What compensation should be given to people who are released after years in jail due to a wrongful conviction?

• It look many years to identify all the body parts taken from the 9/11 destruction of the Twin Towers in New York City, but DNA matching hasn't been as successful in identifying bodies found in the aftermath of Hurricane Katrina. Why do you think that is?

• Some have suggested making a DNA database by taking DNA samples from all newborns. Is this a good idea?

• Discuss any other questions or comments you want to make or address with your fellow students.

In the Media

60 Minutes (CBS), July 15, 2007

DNA: Going Too Far?

An episode of *60 Minutes* discussed the use of DNA profiles to search databases to find relatives of suspects. In the U.K. "shoe rapist" case, it was very effective. Using a DNA match that was not 100%, police tracked down a woman whose DNA profile was in the database. They asked her if she had a brother, and when they located him they found hundreds of shoes under the floor of his home. Subsequent DNA testing showed that he had raped a large number of women.

To access this episode content online, go to www.cengagebrain.com.

QUESTIONS

1. Can you see why some people might be against this use of DNA? Why?

2. In the United Kingdom, as discussed in this chapter, many laws control one's ability to refuse DNA testing. Do you think these laws could be passed in the United States? Why or why not?

DNA Diagnostic Center, April 8, 2011

18-Year Prisoner Freed from Life Sentence Based on DNA Test Performed by a Private Company

In a landmark case, a prisoner with a life sentence was freed after serving 18 years in a Florida prison, based on DNA findings presented by DNA Diagnostics Center, a private company. The evidence was presented by Dr. Julie Heinig, DDC's DNA Forensics Assistant Laboratory Director. DDC is the largest provider of private DNA paternity and other DNA testing in the United States and worldwide. The recent DNA tests were conducted for the Innocence Project of Florida.

Dr. Heinig presented DNA test results in the evidentiary hearing of Derrick Williams, a Palmetto, Fla., man who was convicted of kidnapping and rape in 1993. She isolated DNA from the shirt worn and abandoned by the true perpetrator 19 years ago. DNA results excluded Williams as a contributor to DNA found on the shirt.

To access this product site online, go to www.cengagebrain.com.

QUESTION

Some people think that heavy coverage of DNA testing in the press increases the public's knowledge of science. Do you agree? Why or why not?

Identigene Product Site, launched June 2007

DNA Paternity Test: At Home

With Identigene's home paternity test, you collect DNA with a cheek swab and send the DNA collection kit back to Identigene's DNA testing laboratory, and in a matter of days you receive the DNA test results. Results that are 100% accurate are necessary to exclude someone; those that are 99% accurate or better are sufficient for identifying the father. Identigene claims that these are the most accurate DNA test results available today. It also claims that its results are admissible in most courts of law. Although Identigene adheres to the most stringent standards for testing, some states may differ in their requirements.

One man who testified said, "A girl I dated off and on informed me that I was a dad. What to do?!! I had to know if the child was actually mine. I was sent a kit, I followed the simple instructions, sent the lab samples back and days later, received the DNA results telling me that the child was mine. I can finally look into her eyes and feel at ease knowing I did the right thing."

To access this product site online, go to www.cengagebrain.com.

QUESTIONS

1. Making paternity testing easier and more accessible can cause some problems. What problems might arise if it were cheap and readily available?

2. Suppose a woman thinks her old boyfriend is the father of her child. She meets him for coffee, takes his DNA from his coffee cup, and sends it to be analyzed. Do you think this could be used in court? Why or why not?

10

Genomics

REVIEW

How are traits transmitted from generation to generation? (4.1)

What is recombinant DNA technology? (7.2)

CENTRAL POINTS

- A large, international project is analyzing the human genome.

- Researchers are obtaining information by sequencing and mapping all the human genes.

- There have been a number of surprises as the sequence of the human genome was analyzed.

- Scientists are applying information from the Human Genome Project (HGP) to create new methods of medical diagnosis and treatment.

- In the future, physicians will use your genome sequence to provide you with better medical care.

- Gene therapy is one of the applications of genomics.

- Some ethical and legal aspects of the HGP are still being discussed.

CASE A: The Future Tells All

Natalie and Ben Coleman had already been to the obstetrician a number of times. They had been tested to find out their carrier status for many conditions. Natalie gave a blood sample so that the fetal cells present in her blood could be isolated and the DNA sequenced to analyze all the genes in its genome. This analysis would provide detailed information about their child, including risks for genetic diseases, and other characteristics such as intelligence, looks, and height, just to name a few.

All of their friends had decided to have this fetal genome sequencing when they had children and had told them how exciting it was to find out about their baby before it was born. Many said it helped them "know" their baby way before they could actually interact with it.

Natalie's baby was due to be born in three months, and her obstetrician had already set a date for its delivery: May 3, 2025. He said it was simple to time births down to the minute.

Sitting in the waiting room, Natalie and Ben were nervous. What would they find out? Who would their child be?

Soon they were called in to Dr. Rudin's office. He was sitting at his desk with a thick ream of paper in front of him. He put his hand on the papers and said, "Well, let's see what we have here."

Some questions may come to mind when reading about this futuristic case. Before we can address those questions, let's look at what a genome is and what we can learn from studying a genome to help make this story a reality.

Parents are always concerned about changes in their new baby.

Scientists Building a Model of DNA

10.1 What Is Genomics?

Simply put, **genomics** is the study of genomes. A **genome** is composed of the string of A, T, C, and G bases carried on an organism's chromosome set. Figure 10.1 (page 202) shows what part of a genome sequence looks like.

The science of genomics is based on information gleaned from recombinant DNA technology but has gone far beyond that, leading to the development of several new scientific fields. Because genome sequencing generates huge amounts of information, a new field called **bioinformatics** emerged to develop computer software to collect, store, analyze, and display this information. As genes are identified, attention is turning to the set of proteins that can be encoded by a genome. This new field is called **proteomics**. These are just a few of the fields, grouped under the term genomics, that contribute to the study of genomes. Scientists study our genome as well as the genomes of organisms that cause disease, plants and animals we use for food, and even our pets.

Through genomics, scientists hope to learn about all the genetic disorders humans have. They want to develop treatments for those diseases, but we also need to know how many genes humans carry and where they are located on our chromosomes. We need to have ways of studying what these genes do.

genome set of DNA sequences carried by an individual

genomics study of the organization, function, and evolution of genomes

bioinformatics a field of genomics that uses computers and software to collect, analyze, and store genomic sequence information

proteomics the study of the functions and interactions of the set of proteins in a cell

Figure 10.1 Short Segment of DNA Sequence

```
GCAAAAATACAAAAAATCTTGGATTCTATCGATAACAGCCGAGGTGCCAATCCATATGC
TACAAATAAAAAGCTTACTTTGGATACTTTGACAGGTGGACACTCAAAAGAATCTTATT
TGCGAAGTTATATTAATGGCAAACGTATTCCTGAGACTGCCAGAGCTGTAATCGAACCC
TCTATGAATAAAACTGGCTTTATTGAAGTACCATCTTACATTTTAAACAAGTTAAGAGA
TGTTGTCTTTTATAATCACGTTACGAAAGATAACATACTCAAAAGTCTTCAAAACGAAC
AAGCTTTTCTAACATATATCAAAAGTGATCATAATTCTGAAAATCCTTATATGGTTTAT
```

What is the Human Genome Project? The **Human Genome Project (HGP)** is an interdisciplinary program launched in 1990 with several long-term goals:

1. to create *maps* of the human genome and the genomes of other organisms

2. to find the location of all genes in our genome and locate each gene on a map of our chromosomes

3. to compile lists of **expressed genes** and **nonexpressed sequences**

4. to discover the function of all genes

5. to identify all proteins encoded by genes and their functions

6. to compare genes and their proteins among species

7. to analyze DNA differences between genomes

8. to set up and manage databases based on the genomes discovered

The history and timeline of the Human Genome Project is shown in Figure 10.2.

The human genome contains about 3.2 billion nucleotides (See "Biology Basics: DNA and RNA").

Human Genome Project (HGP) the program to sequence and analyze all the genes in the human genome

expressed genes the set of genes that are active in transcription

nonexpressed sequences the set of genes that are transcriptionally inactive

linkage two or more genes on the same chromosome that tend to be inherited together

nail-patella syndrome a dominantly inherited disorder with malformations of the nails and the kneecap

linkage map map derived from crossing-over studies showing the order of and distance between genes on the same chromosome

genetic map diagram showing the order of and distance between genes on a chromosome

centimorgans (cM) unit that measures how far apart genes are on chromosomes

The portion of the human genome that carries genes was completely sequenced in 2003. Efforts are now directed at identifying all genes, mapping their locations, and establishing their functions.

10.1 Essentials

- **The Human Genome Project, which seeks to sequence and map the human genome, began in 1990.**

- **The project will also identify all the proteins encoded by human genes, and their functions.**

Figure 10.2 Genome Project Timeline This timeline shows the major developments in the Human Genome Project and other genome projects.

Genome Project Timeline	
1984	Discussion, debate in scientific community
1990	Human Genome Project (HGP) begins on October 1
1992	First genetic map of human genome
1993	Revised goals call for sequencing genome by 2005
1994	High-resolution genetic map
1995	First physical map of genome
1996	16,000 human genes cataloged
1997	National Human Genome Research Institute (NHGRI) created
1998	Celera Corporation announces plans to sequence the human genome
1999	Full-scale sequencing begins in HGP
2000	HGP and Celera jointly announce draft sequence of genome
2001	Working draft of genome published
2002	Mouse genome sequenced
2003	Sequence of gene-coding portion of human genome finished
2004	Rat and chicken genomes sequenced
2005	Chimpanzee genome sequenced
2006	Rhesus monkey genome sequenced
2007	J. Craig Venter sequences his own genome
2008	Platypus genome sequenced

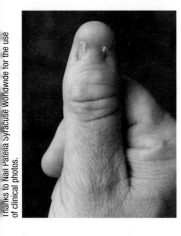

10.2 How Are Genes Mapped?

There are several ways to map genes. Before the development of genomics, scientists in the 1930s began searching for genes located near each other on the same chromosome. Genes close together on the same chromosome are said to show **linkage** because they tend to be inherited together. The closer the genes are on the same chromosome, the more likely they are to be inherited together. In 1936, using pedigree analysis, scientists showed that the genes for hemophilia and color blindness were both on the X chromosome as shown at right.

Because of their distinctive pattern of inheritance, it is easier to assign genes to the X or Y chromosome than to individual autosomes (chromosomes 1–22). Mapping genes to autosomes requires the use of very large families that carry two different genetic traits. These families are very rare, making the task very difficult.

However, in 1955, researchers did find linkage between the *B* allele of the ABO blood type and an autosomal dominant condition called **nail-patella syndrome**, which causes deformities in the nails and kneecaps (Figure 10.3). A pedigree of a family with this linkage is shown in Figure 10.4. As you examine this pedigree, notice how the two traits are inherited together (that is, they are linked) in most cases. However, the pedigree also shows that individuals II-8 and III-3 (starred) inherited only the nail-patella allele *or* the *B* blood type allele, but not both. This shows that the two alleles can separate from each other and be inherited independently.

When the two alleles separate from each other, some people in the family inherit only the *B* blood type or the nail-patella trait. This happens as a result of *crossing over* between the two genes on chromosome 9. Crossing over is an event that takes place during meiosis and occurs randomly only a few times on any one pair of chromosomes (Figure 10.5, page 204). The closer two genes are on a chromosome, the less likely it is there will be crossing over between them and the more likely they will inherited together.

Because the frequency of crossing over is related to the distance between the two genes, one can tell how close these genes are by looking at how often they are inherited together and how often they are inherited separately. This information is used to construct a **linkage map** of a chromosome.

A linkage map (also called a **genetic map**) shows the order of genes on a chromosome and the distance between them. The units of distance are expressed as a percentage of the number of times crossing over occurs between two genes. This means that two genes that undergo crossing over 10% of the time are 10 map units apart; genes that undergo crossing over 1% of the time are 1 map unit apart (Figure 10.6, page 204). These map units are called **centimorgans (cM)**.

Location of hemophilia A gene

Location of color blindness gene

X Chromosome

X chromosome showing the area where the hemophilia A gene and the color blindness gene are located

Figure 10.4 Linkage between the gene for nail-patella syndrome and the gene for the ABO blood group is shown in this pedigree. Shaded symbols represent members with nail-patella syndrome, an autosomal dominant trait. Genotypes for the ABO alleles are shown below each symbol. Nail-patella syndrome and the *B* allele of the ABO gene are linked and tend to be inherited together. Individuals marked with an asterisk (II-8 and III-3) have either nail-patella syndrome or the *B* allele, but not both. The separation of these alleles occurred by crossing over.

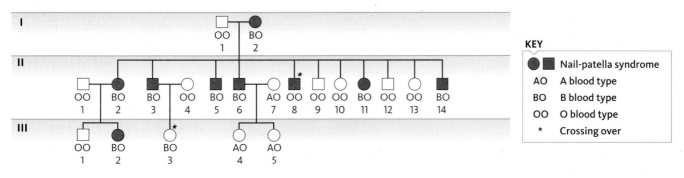

KEY

- ● ■ Nail-patella syndrome
- AO A blood type
- BO B blood type
- OO O blood type
- * Crossing over

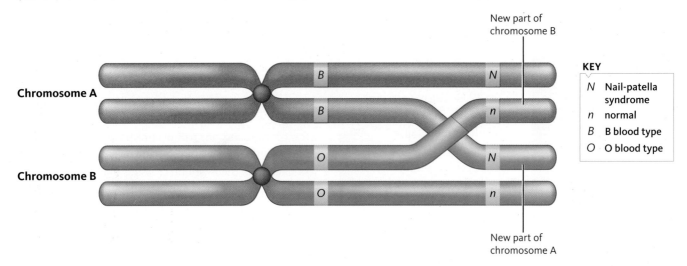

Figure 10.5 Crossing over between homologous chromosomes during meiosis involves exchange of chromosome parts. In this case, crossing over between the *B* and *O* alleles of the ABO blood type gene and the *N* and *n* alleles of the gene for nail-patella syndrome (alleles *B* and *O*) separate the mutant nail-patella allele (*N*) from the *B* allele.

New part of
chromosome B

Chromosome A

KEY

N Nail-patella
syndrome

n normal

B B blood type

O O blood type

Chromosome B

New part of
chromosome A

We can use the frequency of crossing over to calculate the distance between the gene for nail-patella syndrome and the gene for the ABO blood type. Looking at the pedigree in Figure 10.4, we can see that two family members (II-8 and III-3) out of sixteen inherited only one of the alleles in question. In other words, the frequency of crossing over between the gene for ABO blood type and nail-patella syndrome is 2/16, or 12.5%. From this pedigree, we can calculate that the distance between these two genes is 12.5 cM.

To refine this measurement, many pedigrees of families with nail-patella syndrome have been studied and combined; this gives a more accurate estimate of the distance between these two genes—10 cM.

It took about 20 years to find the first example of linkage between autosomal genes, and by 1969 only five cases of linkage had been discovered. It was apparent that the effort to map all human genes by linkage analysis using pedigrees was progressing slowly.

What changed genetic mapping? In the 1980s, a method called **positional cloning** based on recombinant

DNA technology was used to directly map the genes for cystic fibrosis, neurofibromatosis, Huntington disease, and dozens of other genetic disorders. Some of these conditions were discussed in Chapter 4. Positional cloning was also used to construct genetic maps of most human chromosomes.

In positional cloning, the first step is to identify **markers**. These markers do not have to be genes; they can be differences in restriction enzyme cutting sites (see Chapter 7) or differences in the number of repeated DNA sequences (such as short tandem repeats). Once identified, other methods are used to assign these molecular markers to specific chromosomes. Then, as was done with ABO blood type and nail-patella syndrome, pedigrees can be constructed that trace the inheritance of a marker and a genetic disorder through a family, establishing linkage between the marker and the genetic disorder. Once a genetic disorder has been identified as being on a specific chromosome, and located in a specific region on that chromosome, recombinant DNA methods can be used to isolate the gene.

Although positional cloning identifies one gene at a time, by the late 1980s more than 3500 genes and markers had been assigned to human chromosomes. Some of the genes mapped by positional cloning are listed in Table 10.1.

How is mapping done today? The development of methods to rapidly sequence DNA (Figure 10.7) and the creation of computer software to analyze that sequence information were a springboard for launching the Human Genome Project. The idea behind the project was that instead of finding and mapping markers and disease genes one at a time, the HGP would sequence and map the entire human genome using high-speed DNA sequencers and computers.

Figure 10.6 **Human Gene Linkage Map** This chromosome shows the location of three genes: *a*, *b*, and *c*. Notice that each gene is a different distance from the others. The distance indicates how often crossing over occurs among these genes.

More crossing over would occur
here because there is a greater
distance between *a* and *b*.

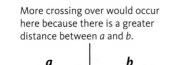

a   *b*   *c*

10.1 cM   6.2 cM

10.2 Essentials

- **Genomics is an extension of earlier methods of gene mapping, such as linkage studies.**
- **Efforts are now focused on identifying all the human genes and their functions.**

10.3 How Are Genomes Sequenced?

Geneticists developed several methods for determining the DNA **sequence** of a genome. First, DNA from an organism is cut with restriction enzymes to construct a collection of DNA fragments that contains all the sequences in a genome. This collection is called a **genomic library**. Individual fragments selected from the library are analyzed by a **DNA sequencer**, which generates the DNA sequence in each fragment. Using specialized software called **assemblers**, sequences of the fragments are compiled to eventually generate the sequence of the entire genome.

In 2003, the sequence of the gene-coding region of the human genome was published. The sequence is stored in a public database so it can be accessed and studied by scientists. New advances in this technology are being developed to sequence the remaining portion of our genome. It will be several years before the rest of the human genome is sequenced.

What happens after a genome is sequenced? After a genome is sequenced, the information is analyzed in several ways. First, the sequence is organized and checked for accuracy. Next, the sequence is analyzed to find all the genes in the sequence that encode for proteins. Although that sounds straightforward, it is not immediately obvious from looking at a DNA sequence (Figure 10.8) whether or not it contains genes (only about 1.5% of human DNA actually contains genes).

positional cloning recombinant DNA-based method of identifying and isolating genes

markers alleles, SNPs, or other molecular characteristics that can be used in linkage and other studies

sequence the order of nucleotides in a DNA molecule

genomic library collection of DNA sequences that contains an individual's genome

DNA sequencer a machine that analyzes the nucleotide sequence of DNA fragments

assembler software used in genomic research to assemble sequenced segments of DNA into the sequence of the whole genome

Figure 10.7 **DNA Sequencing Machines** These sequencers are linked to high-speed computers, allowing rapid genome sequencing and analysis of the results.

Sam Ogden/Photo Researchers

Table 10.1 **Some Genes Mapped by Positional Cloning**

Gene	Chromosome Number
Huntington disease	4
Familial polyposis	5
Cystic fibrosis	7
Wilm's tumor	11
Retinoblastoma	16
Breast cancer	17
Myotonic dystrophy	19
Amyotrophic lateral sclerosis	21

Figure 10.8 **A DNA Sequence** Within this sequence, it is difficult to tell where a gene might be located, where it begins, and where it ends. This DNA sequence shows 720 nucleotides; imagine looking through 3.2 billion nucleotides to identify genes, and you have an idea of the dimension of the problem of finding genes.

```
AAGTCTCAGG ATCGTTTTAG TTTCTTTTAT TTGCTGTTCA TAACAATTGT TTTCTTTTGT
TTAATTCTTG CTTTCTTTTT TTTTCTTCTC CGCAATTTTT ACTATTATAC TTAATGCCTT
AACATTGTGT ATAACAAAAG GAAATATCTC TGAGATACAT TAAGTAACTT AAAAAAAAAC
TTTACACAGT CTGCCTAGTA CATTACTATT TGGAATATAT GTGTGCTTAT TTGCATATTC
ATAATCTCCC TACTTTATTT TCTTTTATTT TTAATTGATA CATAATCATT ATACATATTT
ATGGGTTAAA GTGTAATGTT TTAATATGTG TACACATATT GACCAAATCA GGGTAATTTT
GCATTTGTAA TTTTAAAAAA TGCTTTCTTC TTTTAATATA CTTTTTTGTT TATCTTATTT
CTAATACTTT CCCTAATCTC TTTCTTTCAG GGCAATAATG ATACAATGTA TCATGCCTCT
TTGCACCATT CTAAAGAATA ACAGTGATAA TTTCTGGGTT AAGGCAATAG CAATATTTCT
GCATATAAAT ATTTCTGCAT ATAAATTGTA ACTGATGTAA GAGGTTTCAT ATTGCTAATA
GCAGCTACAA TCCAGCTACC ATTCTGCTTT TATTTTATGG TTGGGATAAG GCTGGATTAT
TCTGAGTCCA AGCTAGGCCC TTTTGCTAAT CATGTTCATA CCTCTTATCT TCCTCCCACA
```

exome the fraction of the genome that contains only exons

repetitive DNA regions of the genome that contain DNA sequences present in many copies

transposons DNA sequences that can move from location to location in the genome

Alu sequences a highly repetitive sequence that makes up 10% of the human genome

alternative splicing incorporating different combinations of exons into mRNA during splicing of pre-mRNA, creating a number of different proteins from the same genetic information

How are genes identified in a DNA sequence? Close examination reveals that genes leave several clues that can be used to locate them among the A, T, C, and Gs of a sequence. DNA sequences that encode proteins have a nucleotide sequence called an open reading frame (ORF) that encodes the amino acid sequence of a gene. The beginning of genes are marked by promoter regions (see Chapter 5), and the end of the coding sequences are marked by stop codons. These recognizable sequences provide giant clues about the location of genes.

Software programs scan sequence data searching for promoters, ORFs, stop codons, and other sequences that identify genes. After a gene has been identified, each of the nucleotide triplets is analyzed and matched to its corresponding amino acid. This analysis reveals the amino acid sequence of the protein is encoded by the gene in question. The amino acid sequence can be compared with those in protein databases. If the amino acid sequence matches that of a known protein, the identity of the gene and its protein is confirmed. In many cases, the amino acid sequence is known from studying another species, and it is the same or similar in humans. This confirms the evolutionary relationship that exists among all organisms.

10.3 Essentials

- **The development of new technology such as sequencers and software helped the HGP move forward.**

10.4 What Have We Learned about the Human Genome?

The content of the human genome has proven to be a big surprise. In some ways, the genome is like many basements or attics, filled with old things having no apparent value, mixed in with small clusters of useful items.

In the genome, these useful items include the genes that code for proteins. Much of the rest of the DNA represents remnants of our genome's evolutionary history.

Although the human genome contains more than 3 billion nucleotides of DNA, only about 1.5% of this DNA contains information for proteins. If we omit the introns (see Chapter 5) and include only the DNA present in exons (the DNA sequences that code for amino acids), the coding regions make up only about 1% of the genome. This fraction of the genome is called the **exome**.

The rest of our genome is noncoding DNA, meaning it does not code for proteins or RNA products. About half of this noncoding DNA, called **repetitive DNA**, is composed of sequences repeated thousands or millions of times.

Not all the genes in the human genome have been identified, but humans carry about 20,000 to 25,000 genes, far fewer than the 80,000 to 1 million originally predicted. Functions have been assigned to about 60% of our genes, but the functions of the rest are still unknown (Figure 10.9).

What are the repeats and where did they come from? There are several different types of repeats in our genome. One type, **transposons,** makes up 45% of our genome. If transposons replicate, the new copies may move (or transpose) and insert themselves into another region of the genome. Fortunately, most transposons are unable to replicate and move around.

A second type of repeat is called a LINE 1 sequence; copies of this sequence make up 17% of the genome. A third type, the **Alu sequence,** contributes 10% of our genome. Alu repeats first appeared in primate genomes about 65 million years ago and have played important roles in the evolution of our own genome.

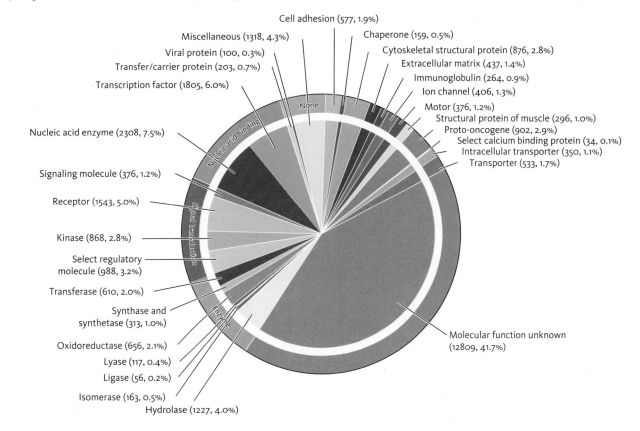

Figure 10.9 **Gene Functions** This pie chart shows the specific gene functions in the human genome. As you can see, the largest group of genes has no known function, emphasizing the work that must be completed before we can fully understand our genome.

Cell adhesion (577, 1.9%)
Miscellaneous (1318, 4.3%)
Chaperone (159, 0.5%)
Viral protein (100, 0.3%)
Cytoskeletal structural protein (876, 2.8%)
Transfer/carrier protein (203, 0.7%)
Extracellular matrix (437, 1.4%)
Transcription factor (1805, 6.0%)
Immunoglobulin (264, 0.9%)
Ion channel (406, 1.3%)
Motor (376, 1.2%)
Structural protein of muscle (296, 1.0%)
Nucleic acid enzyme (2308, 7.5%)
Proto-oncogene (902, 2.9%)
Select calcium binding protein (34, 0.1%)
Intracellular transporter (350, 1.1%)
Transporter (533, 1.7%)
Signaling molecule (376, 1.2%)
Receptor (1543, 5.0%)
Kinase (868, 2.8%)
Select regulatory molecule (988, 3.2%)
Transferase (610, 2.0%)
Synthase and synthetase (313, 1.0%)
Molecular function unknown (12809, 41.7%)
Oxidoreductase (656, 2.1%)
Lyase (117, 0.4%)
Ligase (56, 0.2%)
Isomerase (163, 0.5%)
Hydrolase (1227, 4.0%)

None
Nucleic acid binding
Signal transduction
Enzyme

Can any of these types of repeats produce a genetic disorder or inactivate a gene? Yes, in fact, copies of Alu sequences inserted in or near genes account for about 0.1% of all known genetic disorders, including some mutant alleles of hemophilia, neurofibromatosis, and some forms of breast cancer.

Although Alu sequences may pose a threat to our genetic health by creating mutant alleles, they may also have played an important role in human evolution. About 2.8 million years ago, an Alu sequence inserted itself and turned off a gene that limits cell growth in one of our ancestral species. At about the same time, brain size began to increase in the line leading to our species. Scientists think that switching this gene off by an Alu sequence insertion may be related to the large brain size that is characteristic of our species.

Other repetitive sequences are found in clusters throughout the genome. Some of these include the short tandem repeats (STRs) now used to construct DNA profiles (discussed in Chapter 9).

Are there other surprising findings about our genome?
While cataloging the proteins found in human cells, scientists were surprised to find that the number of proteins our cells can make far outnumbers the protein-coding genes we carry. It is estimated that our genome has about 20,000 to 25,000 genes, but the catalog of human proteins now includes over 500,000 and may exceed 2 million.

How can 20,000 to 25,000 genes encode the information for so many proteins? The difference is the result of several mechanisms that can work independently or together.

During processing of pre-mRNA in the nucleus (see Chapter 5 to review), exons can be retained or removed, allowing the finished mRNA molecules to contain different combinations of exons. As a result of this process, called **alternative splicing** (Figure 10.10, page 208), one gene can encode the information for several different forms of a protein (A, B, and C in Figure 10.10). It is estimated that alternative splicing of mRNA occurs in more than 95% of all human genes, greatly increasing the number of proteins encoded by our genome. In addition, once they are made, proteins can be chemically modified. These modifications can change the structure and function of the protein, again multiplying the number of different proteins that can be derived from a single gene. An international effort called the Human Proteome Project (HUPO) is working to identify all the proteins humans can make and to establish their functions and interactions with other proteins.

10.4 Essentials

- **While completing the HGP, scientists found a number of surprises.**

Figure 10.10 **Exon Splicing in a Gene with Four Exons** Normally, the pre-mRNA is processed to include all the exons (protein A). In alternative splicing, some exons can be removed during processing, resulting in the production of different proteins (proteins B and C).

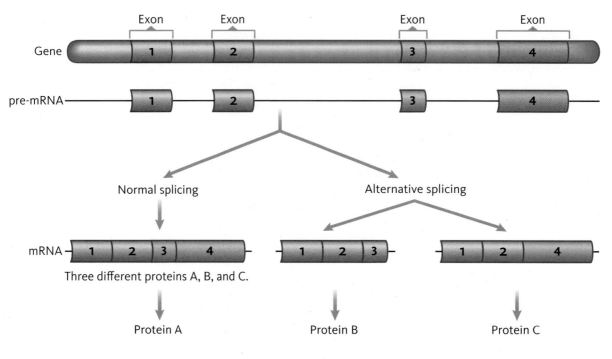

10.5 **How Is Genomic Information Being Used?**

Because of the Human Genome Project, genomics has become an important component of genetic research and health care. Its impact will continue to grow as information from genomics is analyzed and applied to the diagnosis and treatment of human diseases. More than 10 million children and adults in the United States have a genetic disorder, and a newborn has a 3% chance of having a genetic disorder. These numbers underscore the need to identify all the genes responsible for diseases and to develop ways to diagnose heritable diseases at all stages of life.

How can genomics be used to search for new genes? As genomes from more people were sequenced and analyzed, the results revealed a surprising amount of variation in the sequence of nucleotides in the genome of different people. One of the simplest types of variation is a single nucleotide change, called a single nucleotide polymorphism (Figure 10.11) or SNP (pronounced "snip"). Because these nucleotide variations are common, they are not regarded as mutations, but instead are designated as variations.

SNPs located close together on a chromosome tend to be inherited together, and a cluster of SNPs is called a haplotype.

genome-wide association studies (GWAS) large-scale studies of populations to find common haplotypes

Because SNP haplotypes are widely distributed in the genome, they can be used as markers in linkage studies to identify the location of disease-causing alleles. For example, if a SNP haplotype is linked to (or near) a mutant allele causing a genetic disorder, people carrying this mutant allele will tend to have the same SNP haplotype. Comparison of haplotypes between those affected with a genetic disorder and those who are not affected will show different SNP haplotypes, and this information can be used to locate and identify the mutant allele associated with this genetic disorder.

These studies, which can be administered to large numbers of people, are known as **genome-wide association studies (GWAS)**. They have been used to identify genes associated with type II diabetes, various cancers, cardiovascular diseases, and neurodegenerative disorders including Parkinson's disease and Alzheimer disease. The use of SNPs and GWAS is helping unravel the number and identity of genes associated with complex diseases and is rapidly changing human

Figure 10.11 **DNA Sequences from Four Individuals** Single nucleotide polymorphisms are shown in red.

	SNP SNP	SNP SNP	
Person 1	...AA**C**TTCGCC....	...TTGA GGCATC...	Haplotype 1
Person 2	...AAGCTTCGCC....	...T**A**GA GGCATC...	Haplotype 2
Person 3	...AAGCTT**C**CC....	...TTGA GGCATC...	Haplotype 3
Person 4	...AAGCTT**CT**CC....	...TTGA GGCA**A**C...	Haplotype 4

genetics. In Chapter 13 we will see how GWAS are used to study and diagnose behavioral disorders such as schizophrenia and bipolar disorder.

How is genomics being used to diagnose genetic disorders?

Genomics also provides a new and faster approach to the diagnosis of rare genetic disorders. Let's look at an example. A 5-month-old baby was diagnosed with Bartter syndrome, a rare genetic disorder affecting the kidneys. To confirm the diagnosis, a blood sample was sent to a lab at Yale University where pioneering studies are being done in whole exome sequencing. Rather than sequencing entire genomes, this method focuses on the 1% of the genome that actually encodes the information for proteins. Sequencing the baby's exome revealed that this infant did not have Bartter syndrome, but carried two mutant alleles of a gene responsible for another rare genetic condition, called congenital chloride diarrhea. Once a correct diagnosis was made, the infant's physicians were able to develop a program of treatment for this condition.

10.5 Essentials

- **Information from the HGP is being used to diagnose genetic disorders.**

10.6 What Is the Future of Genome Sequencing?

Currently, genome sequencing is expensive and time-consuming. However, newly developed technologies are reducing both the cost and the time needed to sequence areas of the genome, or an entire genome. These new methods may make sequencing a routine part of medical care, and many couples will soon be able to have the experience that the Colemans are having at Dr. Rudin's office. Eventually, genome sequencing will cost $1000 or less and take under a day. Having the sequence of your genome will make it possible for physicians to monitor your health, provide information on how to reduce risks for certain diseases, and make an early diagnosis of conditions such as cancer, when the disease is more treatable.

10.6 Essentials

- **In the near future, genome sequencing will probably be a routine part of medical care.**

10.7 Can We Correct Genetic Disorders?

As we saw in Chapter 7, using recombinant DNA techniques, scientists can transfer genes from human cells to bacteria to create drugs and other human proteins. Is it possible to

transfer genes *into* humans? Imagine a mutant gene that is making an incorrect protein. If the normal allele were transferred into a person with this genetic disease, the correct protein could be produced and the condition cured.

Before 1990, researchers were working to use recombinant DNA technology to develop **gene therapy.** The idea behind gene therapy is the transfer of copies of normal genes into cells (or people) that carry defective copies of these genes. Once transferred, the normal genes should direct the synthesis of the normal gene product, which in turn produces a normal phenotype.

How are genes transferred to people in gene therapy?

In most cases, cells are removed from the body of the person affected with a specific genetic disorder and modified in the lab (Figure 10.12, page 210). Normal copies of the gene are inserted into these cells using a

gene therapy the transfer of a normal allele of a gene to treat a genetic disease

Figure 10.12 **The Strategy for Gene Transfer** In this example of gene transfer, a virus is used as a vector and has a normal copy of a gene inserted. The vector/gene combination is transferred to white blood cells of a patient with a genetic disorder. Before inserting the modified cells, scientists check that the normal gene is active, then the patient receives the modified cells. The protein synthesized by the normal allele reverses the disease.

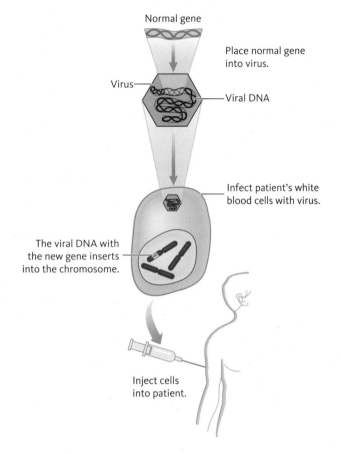

Normal gene

Place normal gene into virus.

Virus

Viral DNA

Infect patient's white blood cells with virus.

The viral DNA with the new gene inserts into the chromosome.

Inject cells into patient.

genetically modified virus, called a **vector**, to carry these genes. The virus, and the human gene it carries, enters the cell and is inserted into a human chromosome. The gene then becomes part of the genetic information carried by that cell. After gene transfer, the cells are grown in the laboratory and checked to ensure that the normal gene is active and making the normal protein. Then the cells are transferred back into the body.

Have genetic diseases been treated by gene therapy? In 1990, a young girl with a genetic disorder called **severe combined immunodeficiency disorder**

vector a self-replicating DNA molecule used to transfer foreign DNA segments between cells

severe combined immunodeficiency disorder (SCID) a complex genetic disorder in which there is no functional immune system

adenosine deaminase (ADA) the gene that is mutated in one form of severe combined immunodeficiency disorder

(SCID) was treated with gene therapy. Children with this disorder have no functional immune system and usually die from an infection that would not be serious in most people. White blood cells were removed from her body, and a vector carrying the human gene for the enzyme **adenosine deaminase (ADA)** was inserted into them. After the treated cells were checked to make sure they were making ADA, they were injected into her body. After this gene transfer, the girl went on to develop a functional immune system and is leading a normal life.

CASE B: Ownership of Gene Questioned

Technogene has been in the biotech business for quite a long time. During those years, it has worked in the secondary market supplying materials (DNA and cells) to other biotech companies and has never patented anything.

While looking for a gene for hypertension, Technogene's scientists found an interesting area of the genome that they thought might contain this important gene. They searched databases for the sequence of this gene and discovered that another company, Markogene, had already patented part of the gene.

Technogene's head scientist, Julie Mersiant, was shocked. How could Markogene patent a section of a gene? At the time it was patented, no one even knew what the function of this sequence was or even if it was useful.

Dr. Mersiant wanted to use this sequence in her experiments, but she knew that Technogene would have to pay a substantial royalty to have the right to use it. This would increase the cost of any discovery she made by millions of dollars.

1. **What might Dr. Mersiant or Technogene do now?**

2. **Many companies have patented human genes, parts of genes, their products, and the methods by which they were identified. Do you think this helps or hinders scientific knowledge? Why or why not?**

3. **If you were the attorney for Technogene, how would you argue for the right to use the information?**

4. **If you were the attorney for Markogene, how would you argue for the right to use the information?**

5. **Many people think that all information gleaned from the Human Genome Project should be open, and everyone should be allowed access. Do you agree? Why or why not?**

Are there any problems with gene therapy?

In the early and mid-1990s, gene therapy was used to treat a number of genetic disorders. Over a 10-year period, more than 4000 people received gene therapy. Unfortunately, those trials were largely failures. In a few cases, patients developed leukemia after gene therapy, and at least two people have died as a result of gene therapy.

In 1999, an 18-year-old, Jesse Gelsinger, died during a gene therapy trial designed to treat an inherited metabolic disorder. His death was triggered by a massive immune response to the vector, which was a modified virus. In 2000, two French children who underwent apparently successful gene therapy to treat an X-linked form of SCID later developed leukemia. In both children, the virus carrying the normal gene inserted itself into a gene that regulates cell division. The insertion activated the gene and caused uncontrolled production of white blood cells and the symptoms of leukemia.

In 2008 new methods began to raise hopes for gene therapy after successful gene transfer to treat a genetic disorder called Leber's congenital amaurosis. Leber's is a rare, inherited eye disease. Affected individuals develop loss of vision and often become blind during infancy or early childhood. Three young adults who were treated by transferring a normal copy of the *RPE65* allele reported improvements in vision that were maintained a year after treatment. Scientists continue to work on correcting problems with gene therapy and to develop new approaches to using genes as a treatment for genetic diseases.

10.7 Essentials
- **Gene therapy is available only experimentally.**

10.8 What Are the Legal and Ethical Issues Associated with Genomics?

Many of the legal and ethical questions regarding genomics were addressed as the Human Genome Project began, and as new technologies became available. The government-funded part of the HGP foresaw some of these problems and allotted a percentage of its funding to study the ethical, legal, and social issues (ELSI) related to the human genome sequence and its many applications. Scientific progress will undoubtedly be accompanied by new issues, such as those surrounding the testing of the Colemans' fetus.

Table 10.2 addresses some of the ethical and legal questions generated by the HGP and subsequent developments in genomics.

10.8 Essentials
- **Some of the legal and ethical questions surrounding the HGP are being addressed.**

Table 10.2 Ethical and Legal Issues Associated with Genomics

Question	How Are These Questions Decided Today?	How Might These Questions Be Decided in the Future?
Who will have control of the information gleaned from the HGP?	Confidentiality of medical records is controlled by law. Some anonymous genome sequence information has been made available to labs all over the world through computer databases.	If there were more understanding of genetics and its importance, courts could decide to diminish the right to confidentiality. This could make it easier for scientists to see records and use the material in them.
Will we have more genetic counselors as more information becomes available?	The academic community is developing more programs in genetic counseling.	Physicians will look for experts to work with their patients to explain genetic testing and the results that follow.
Will the use of this information lead to engineering the genetics of our children?	The first steps in genetic engineering are already happening in the use of PGD and other genetic testing. Many people are afraid that if this technology becomes commonplace, problems may arise.	The evolution of humans may be affected by genetic engineering, and those with disabilities may be selected against.
Could employers use genetic information to deny people jobs?	Employers are kept from discriminating against employees or job seekers by legislation.	If more genetic information is available, employers may require prospective employees to have genetic tests.
Can people patent human genes?	Currently, DNA sequences, genes, and entire genomes can be patented in the United States and other countries. These patents allow companies to work on areas of the genome without another lab doing the same research.	Almost all scientists feel that the information from the HGP should be made public so anyone can access it.

Spotlight on Law

Moore v. Regents of the University of California

(249 Cal.Rep. 494 [1988], 793 P.2d 479 Cal.Lex. 2858 [1990])

In 1976, John Moore entered the University of California Hospital in Los Angeles, California, for treatment of **hairy cell leukemia**, a cancer of white blood cells that invade the spleen. A *splenectomy* (removal of the spleen) is the generally accepted initial treatment of hairy cell leukemia; improvement often lasts months or even several years.

Moore's spleen was removed in October 1976. He went into remission and was thrilled with the cure. Moore returned several times from Washington State to California to provide blood samples, bone marrow, and sperm. He was told that these procedures were part of his treatment, and all of his expenses were paid by UCLA. On April 11, 1983, Moore was asked to sign a consent form to allow UCLA to conduct research on his cells. Moore signed the form.

Researchers at UCLA found that Moore's spleen cells were unique because they made a protein called a **lymphokine** that had been used to treat certain cancers. The researchers created a cell line from Moore's spleen, which produced a pharmaceutical product with medical uses and had tremendous commercial value. Moore was not informed about the use of his cells, the cancer-fighting product derived from them, or their potential commercial value.

On March 20, 1984, the U.S. Patent Office issued a patent for the cell lines to Moore's physicians, with Dr. David Golde and Dr. Shirley Quan listed as inventors. The cell line was originally named the "Moore Cell Line," but the name was later changed to "RLC Cell Line." Dr. Golde and Dr. Quan signed agreements with Sandoz Corporation and Genetics Institute to commercially develop and investigate this cell line. For this, they received $330,000 cash and the option to buy 75,000 shares of Genetics Institute stock at a reduced price.

On September 9, 1984, John Moore filed a lawsuit charging Dr. Golde, Dr. Quan, the Regents of the University of California, Sandoz Corporation, and Genetics Institute with **conversion,** the removal of another's property. As strange as it might seem, to win the case, Moore's attorneys had to prove that his cells and their products were his property.

ISSUES

Three major issues were involved in this case: informed consent, conversion, and the doctor-patient relationship. Informed consent asks: should a patient be informed of

hairy cell leukemia a cancer of white blood cells; the cancerous cells have hair-like projections from their surface

lymphokine a chemical in the body produced by cells of the immune system

conversion legal term meaning removal of someone's property

everything that might be done? The doctor-patient relationship asks: was Moore's relationship with his doctor breached Finally, conversion asks: were the defendants taking his property when they removed his cells?

RESULTS

Trial Court: The trial court did not allow the claim of conversion because Moore did not specifically state at the time of his surgery that he did not want his tissue used for research and he failed to attach a copy of the release for the splenectomy to the court documents. John Moore appealed this ruling to the California Court of Appeals.

Appellate Court: The California Court of Appeals stated in a two-to-one decision that Moore's allegation of a property right to his own tissues "is sufficient as a matter of law," and that he owned his own cells. Because the rights to the cell line had already been sold, the court could decide who should share in the proceeds. The defendants appealed to the Supreme Court of California.

State Supreme Court: On July 1, 1990, the Supreme Court California reversed the appellate court's decision and sent case back for a new trial on the grounds of breach of doctor patient relationship, because Dr. Golde should have informed Moore of the possible future use of his cells.

In considering the issue of ownership, the court stated that allowing Moore an ownership interest in his cells could impose "a duty on scientists to investigate the consensual pedigree of each human cell sample used in research."

The court did not allow Moore to stop the companies from using his cells or to have a part in the profits. But the quality of the doctor-patient relationship was important; the court said Moore could go back to court and sue the physicians because they did not inform him that his cells were being used for this purpose.

No such lawsuit has been filed on behalf of Moore, as suggested by the court. Therefore, it is assumed that a settlement was reached for an undisclosed amount, probably with a gag order

QUESTIONS

1. Do you think the physicians owed some of their earnings to John Moore? If so, why and how much? If not, why?

2. If you were Moore's attorney, what arguments would you to prove that his cells were his property? List three or more.

3. If you were the defendants' attorney, what arguments would you use to plead your case? List three or more.

4. Did the physicians do anything wrong? If so, what?

5. Do you agree with the California Supreme Court's ruling Why or why not?

6. Ethically, one of our most important concerns is autonomy the right to control our destiny and what happens to us How does autonomy play a part in the *Moore* case? How would you use an autonomy argument if you were one of Moore's attorneys?

1. The Human Genome Project, which seeks to sequence and map the human genome, began in 1990. (Section 10.1)
Scientists are working to identify all human genes and their functions.

2. The HGP will also identify all the proteins encoded by human genes. (Section 10.1)
This new field is called proteomics.

3. Genomics is an extension of earlier methods of gene mapping, such as linkage studies. (Section 10.2)
Mapping genes using older methods has been replaced by genomic techniques.
As scientists have found faster and more sophisticated ways to work with DNA, genomics has advanced rapidly.

4. The development of new technology such as sequencers and computer software helped genomics move quickly. (Section 10.3)
Sequencers are coupled to high-speed computers that analyze and store the data.

5. After a genome is sequenced, several steps are needed to analyze the information. (Section 10.3)
Finding open reading frames is one of the first steps in analysis of genome sequence.

6. We have been surprised by the results of sequencing the human genome. (Section 10.4)
There are many more proteins produced than there are genes in the genome.

7. Information from the HGP is being used to find genes associated with genetic disorders. (Section 10.5)
One of the main reason to decode the human genome is to help treat and possibly cure these disorders.

8. In the near future, genome sequencing will probably be a routine part of medical care. (Section 10.6)
Scientists working on the HGP see changes in health care as one of the ultimate uses of their work.

9. Gene therapy is available only experimentally. (Section 10.7)
Research is directed at making gene therapy a safe and effective procedure to treat genetic disorders.

10. Many of the legal and ethical questions surrounding the HGP have been and are being addressed. (Section 10.8)
These legal and ethical questions are discussed by the Ethical, Legal, and Social Issues (ELSI) arm of the HGP.

Review Questions

1. What is the difference between a DNA sequence and a gene map?

2. If a sequence is unexpressed, what does this mean in terms of a person's phenotype?

3. What is the combination of software, computers, and genetics called?

4. Why is crossing over important in gene mapping?

5. How is a linkage map constructed?

6. What is the next step after a gene for a genetic condition is located and isolated?

7. What is positional cloning?

8. List three possible consequences of finishing the HGP.

9. What are the goals of the Human Genome Project?

10. How are genes transferred in gene therapy?

11. If the human genome contains 3.2 billion nucleotides, and the exome is 1% of the genome, what is the size of the exome?

Application Questions

1. Some have compared the Human Genome Project to the development of the atomic bomb. They call it "big science." Research the atomic bomb project and list three similarities and three differences between the HGP and that project.

2. As the HGP progresses, large numbers of jobs will be created. What types of jobs are they? List three.

3. How are the science of the HGP and of DNA forensics similar? Go back to Chapter 8 and reread the sections on how forensic testing is done.

4. After all the human genes are mapped and cloned, we will have a great deal of information about ourselves, our children, and where we come from. Do we want to know this? Compare your answers to those of your classmates.

5. Research the ELSI program and see what progress it is making in studying ethical, legal, and social issues.

6. One of the most interesting cases involving gene therapy was the Jesse Gelsinger case. Jesse was a member of a clinical trial for gene therapy and died. Research this case and discuss what happened as a result of his death.

7. If there are so many areas of the genome that do not function, why do you think they are still in the human genome?

8. Research some of the animal genomes that have been mapped and sequenced. Why are they important?

Read the following three cases, and using the decisions in the *Moore* case as a precedent, pretend you are the judge in each case and write a ruling for each.

9. Recently in Illinois, a man was diagnosed with leukemia and was told he needed a bone marrow transplant. A computer found a perfect match for his rare tissue type. The woman, a stranger,

. . . you were deciding how far the Human Genome Project should go in collecting and publishing sequence information: Should the HGP publish the names of those individuals whose genomes have been sequenced? Why or why not?

. . . your state wanted to collect samples of DNA from everyone to use as a data bank for research and forensic purposes?

. . . you were asked to vote to allot more funds to work on the Human Genome Project?

was tested while in the hospital for other treatment. She refused to donate her bone marrow, even when the seriousness of the illness was explained and the leukemia patient called her and pleaded. The patient took her to court to get an injunction to force her to give him the bone marrow, but the judge told the man he could no longer bother her. They were *her* bone marrow cells. What would you decide using *Moore*?

10. A woman entered the hospital to have an appendectomy. She asked the surgeon for her appendix back. He said no: it was going to be examined in pathology and then destroyed. She went to court to try to force the hospital to release her appendix to her. What would you decide using *Moore*?

11. Recently a California couple had a second child in order to obtain bone marrow for their older child, who had leukemia. An organization tried to stop this use of the infant's cells by asking the court to stop the transplantation. What would you decide using *Moore*?

12. As more information becomes available about our genetics and testing becomes available, we can test our unborn children, our families, and ourselves. If we do not own our cells, as the California Supreme Court has ruled, do we own the results of tests on those cells? What is your opinion?

13. People are allowed to donate organs for transplantation. Courts have allowed some body parts (such as sperm and eyes) to be willed to someone. Does this mean that these parts are property? Would this argument have changed the results of *Moore*?

14. In Texas, a woman's eye had been removed because of disease and was lost down the drain. She sued for emotional trauma because her parts were treated so badly. How could you use the *Moore* case here?

15. In 2010, the University of California at Berkeley and Stanford University announced programs to collect and analyze DNA samples from incoming students. The information would be used in classroom discussions about human genetics. The tests would be voluntary, and samples would be tested only for three genes: lactose tolerance, flush reaction in response to alcohol intake, and folate metabolism. Results would be available only to the student who provided the DNA sample. Would you participate or not?

Online Resources

Preparing for an exam? Log on at www.cengagebrain.com for study tools to help you assess your understanding. If assigned by your instructor, the Case A and Spotlight on Law activities for this chapter, "The Future Tells All" and "*Moore v. Regents of University of California*," will also be available.

In the Media

7th Space Interactive, **April 18, 2011**

Exploring Single Nucleotide Variation in Entire Human Genomes
Jorge Salas

Ultra-sequencing technologies are starting to produce extensive quantities of data from entire human genome or exome sequences. Researchers in the field of bioinformatics present and analyze this vast amount of information. For example, the 1000 Genomes project has recently released raw data for 629 complete genomes representing several human populations. This is phase 1 of the project, which will sequence 1000 human genomes.

To access this article online, go to www.cengagebrain.com.

QUESTIONS

1. Where are these genomes coming from? Volunteers? Why?

2. What permissions will this company need to reveal someone's complete genome?

Pharma Biz.com, April 18, 2011

NIH Researchers Complete Whole-Exome Sequencing of Skin Cancer

Melanoma is the most serious form of skin cancer and its incidence is increasing faster than any other cancer. A major cause is thought to be overexposure to the sun.

Studying the genome of this form of cancer has shown scientists where mutations form in the tumor cells. The mutations studied in 14 melanoma tumor samples showed that some of the mutations were similar in many patients. The NIH researchers see this as a way to understand the genetics of tumors and eventually identify new treatments.

To access this article online, go to www.cengagebrain.com.

QUESTIONS

1. How can knowing the genome of a tumor cure patients with melanoma?

2. How can knowing about this gene help patients who don't have melanoma yet?

National Genome Research Institute Web site, February 2011

The Road to the $1000 Genome
Geoff Spencer, NHGRI Staff Writer

In 2004 National Human Genome Research Institute (NHGRI) asked its grantees to achieve high-quality human genome sequencing for $1000 or less. NHGRI launched this effort in 2004 with the award of the first grants from its Advanced DNA Sequencing Technology Program. At the time sequencing could produce a draft of a human genome for $20 million in three to four months.

According to Jeffrey Schloss, director of NHGRI, the promise of obtaining very-low-cost genomes would allow DNA sequencing to become a routine clinical test, much like blood tests that are conducted as a regular part of healthcare today. And for biomedical researchers, the low cost would allow genome sequences from thousands of people with diseases to be compared to genome sequences from healthy individuals.

Recently, James Watson (of DNA fame) became one of the first people to have his genome sequenced.

The ability to sequence genomes at this scale would allow an unprecedented understanding of a variety of diseases such as diabetes, mental health disorders, and heart disease. Today one machine can sequence about a billion samples a run and have results in a week. The cost has been reduced to $20,000 through the use of high-speed computers.

The question asked by the press and medicine alike is whether the $1000 genome is inevitable.

To access this article online, go to www.cengagebrain.com.

QUESTIONS

1. Do you think people will have their genomes sequenced if it costs only $1000? Why or why not?

2. Will insurance companies pay for genome sequencing? Why or why not?

New York Times, June 1, 2007

Genome of DNA Discoverer Is Deciphered

Nicholas Wade

The full genome of James D. Watson, who jointly discovered the structure of DNA in 1953, has been deciphered, marking what some scientists believe is the gateway to an impending era of personalized genomic medicine.

A copy of his genome, recorded on two DVDs, was presented to Dr. Watson yesterday in a ceremony in Houston by Richard A. Gibbs, director of the Human Genome Sequencing Center at the Baylor College of Medicine, and by Jonathan M. Rothberg, founder of the company 454 Life Sciences.

"I am thrilled to see my genome," Dr. Watson said.

To access this article online, go to www.cengagebrain.com.

QUESTIONS

1. What do you think the reason was for picking Dr. Watson?

2. Dr. Watson is a winner of the Nobel Prize; will his genome be different from that of an ordinary person? Explain your answer.

A Population of Young South Asians

Tim Graham/Getty Images

Genes, Populations, and the Environment

BB3.1 How Do Genes and the Environment Interact?

Our phenotypes are affected by more than our genomes. Our environment also plays a part in who we are. Environment can include things we don't usually think about, such as our prenatal environment, what we eat, what we breathe, where we live, and even our social environment. If someone near us has a cold, for example, he or she may pass it on to us and our immune system may fight it off. The outcome of interactions between our genomes and the environment make us what we are. Traits that are affected by both genetics and the environment are called **multifactorial traits**. We will discuss these in detail in Chapter 11.

multifactorial traits traits that are caused by the interaction of genes and environmental factors

BB3.2 Does Genetics Affect Our Perception of and Response to the Environment?

Beginning in the 1930s, researchers observed that people react differently to things in the environment. In 1933, a chemist found significant differences in how people taste a chemical called **phenylthiocarbamide (PTC)**. Some people are tasters, and others are not. Tasters find PTC has a bitter and very strong taste, whereas nontasters cannot taste it at all. Later, it was shown that the ability to taste PTC has a genetic basis. Nontasters are homozygous for a recessive allele (*aa*), and tasters are heterozygous or homozygous for a tasting allele (*Aa* or *AA*). There is a differing distribution of tasters around the world. In the map shown in Figure BB3.1, shading shows the geographic distribution of the nontaster allele. What do you think might cause this distribution?

Beginning in the 1970s, Linda Bartoshuk and her colleagues at Yale University began to study whether the genetics of tasting might have an impact on diet and food choices. She discovered that tasters can be divided into two groups: tasters and supertasters, who have intensely negative reactions to PTC. In her surveys she found that about 25% of people are nontasters, 50% are tasters, and 25% are supertasters. This seems to correlate with the *aa*, *Aa*, and *AA* genotypes. She also discovered that supertasters can have more than ten times as many taste buds as nontasters and are very sensitive to tastes in general.

It seems as though food preferences might be related to taster status. Supertasters often find that high-sugar foods are too sweet, coffee is too bitter, and hot peppers and spices have a more intense and often unpleasant taste. As a result of having more and highly sensitive taste buds, supertasters are less likely to include certain foods and beverages in their diet,

phenylthiocarbamide (PTC) chemical that only certain members of a population can taste

Figure BB3.1 **The Global Distribution of PTC Tasters** The darker the color, the higher the percentage of nontasters (up to 100%).

Source: From *Phenylthiocarbamide: A 75-Year Adventure in Genetics and Natural Selection* by Stephen Wooding from Genetics, Vol. 172, 2015-2023, April 2006, Copyright © 2006. Reprinted by permission of Genetics Society of America.

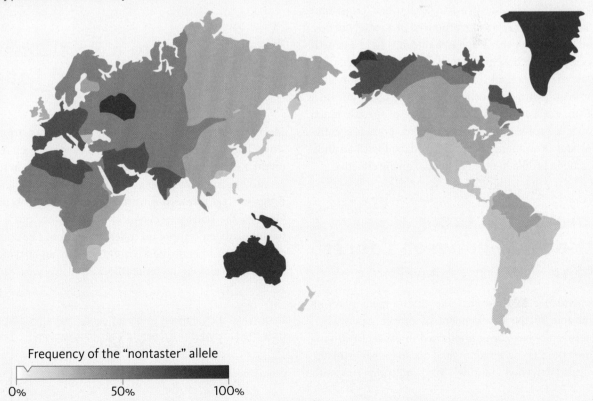

Frequency of the "nontaster" allele

0%   50%   100%

including Brussels sprouts, cabbage, spinach, soy products, grapefruit juice, and alcohol.

These findings raise questions about the relationships among our genotypes, taste preferences, and diet. Nutritionists are interested in studying supertasters to see if they can answer some of these questions: How does our taster status affect our food choices? Can food additives designed to block the sensation of bitterness in tasters and supertasters be used to improve nutrition? Do supertasters choose fruits and vegetables that are higher in cancer-fighting compounds?

BB3.3 Are There Population Variations in Responses to Drugs and Medicines?

In recent years, thousands of new drugs have been developed and are used to treat diseases. Some of these drugs produce a wide range of reactions that may be genetically influenced. For example, some people break down certain drugs very rapidly, making them less effective unless a higher dose is used. Others metabolize the same drug more slowly, and high levels of the drug remain in the body, which can prove toxic. Drug companies have responded to this problem in several ways. Some drugs are put into capsules that prolong the release of the drug, spreading the effective drug level over a longer time period, whereas other drugs are produced in a variety of dosages. The recognition that responses to drugs have a genetic basis has led to the development of a field called personalized medicine (also called stratified medicine). Personalized medicine begins with diagnostic molecular tests that measure metabolism, allele status, and the presence of specific mutations. Once a profile has been assembled, both preventive measures and drug therapies can be designed with an individual's genetic profile in mind to be maximally effective.

BB3.4 What Is the Relationship between Cancer and the Environment?

There is evidence to show that the environment plays an important role in the development of cancer. Many environmental factors have been implicated in cancer, including natural radiation (x-rays, radon gas), occupational exposure to chemicals (polyvinyl chloride), viral infections (hepatitis

Table BB3.1 Viruses and Cancer

Virus	Associated Cancer(s)
Epstein-Barr virus	Burkitt's lymphoma, nasopharyngeal cancer
Hepatitis B virus	Liver cancer
Human herpes virus 8	AIDS-related Kaposi sarcoma
Human papillomavirus	Cervical cancer

B, human papillomavirus), and personal choices such as exposure to ultraviolet light (sunlight and/or tanning lamps), smoking, and diet. Some of this evidence comes from studying workers with years of on-the-job exposure to synthetic pesticides, asbestos, and industrial chemicals such as benzene and polyvinyl chloride.

Naturally occurring environmental agents may also play a role in cancer. These include viral infections, which account for almost 15% of cancers worldwide. Table BB3.1 lists some human viruses that are associated with cancer. In most cases viral infection alone is not enough to convert a normal cell to a cancer cell; other factors are involved. Some of these factors include DNA damage, viral infection of actively growing cells, and the ability of the virus to disrupt cell cycle control.

BB3.5 Is There a Relationship between Genetics and Cancer Therapy?

More than 70% of breast cancers are classified as *estrogen sensitive*. This means that the tumor cells have estrogen receptors on their surface, and that the presence of estrogen in the body promotes growth of the cancer. One of the most widely used drugs to treat estrogen-sensitive breast cancer is **tamoxifen**. Once in the body, the chemical is converted into a powerful antiestrogen compound called **endoxifen**. The conversion of tamoxifen to endoxifen is controlled by an enzyme called CYP2D6, as shown in Figure BB3.2.

tamoxifen a drug given to breast cancer patients that controls the amount of estrogen in the bloodstream

endoxifen antiestrogen compound used in cancer chemotherapy

Figure BB3.2 **Tamoxifen** The enzyme CYP2D6 converts tamoxifen into a powerful anticancer drug, endoxifen. Alleles of the *CYP2D* gene produce enzymes that convert tamoxifen into endoxifen at different rates. This genetic variation causes variable effects of the drug as a cancer fighter.

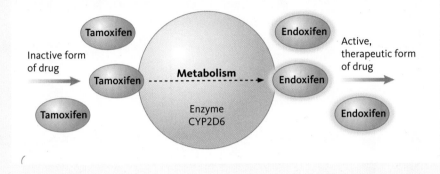

The gene that encodes for this enzyme has several alleles. Four groups of women have been identified: one group metabolizes tamoxifen very slowly, whereas members of the other three groups rapidly convert it to endoxifen. Analysis of cancer recurrence rates shows that women who metabolize tamoxifen slowly are much more likely to redevelop breast cancer than women with faster metabolism rates. This means that when developing drugs for cancer therapy, it is important to take genetics into account. In some cases, genetic testing is done before a drug is prescribed to ensure maximum benefits from the therapy.

The genetic and environmental interactions in cancer will be discussed in Chapter 12.

BB3.6 How Is Behavior Affected by the Interaction of Genes and the Environment?

We all know that stress can cause people to change their behavior. In certain situations, one person might become depressed, but others may be able to better cope with the stress. Scientists have found that a variation in one gene is associated with some of these behavioral responses. Two alleles, long and short, in a gene called **5-HTT** seem to control

5-HTT a gene that encodes a protein that regulates nerve impulses

autoimmune disease condition in which a person's immune system attacks his or her own tissues

these differences. This gene encodes for a protein that is important in the transmission of nerve impulses. In researching the genetics of *5-HTT*, one study analyzed people between the ages of 21 and 26 who had experienced four or more stressful life events (such as divorce, debt, unemployment, or death of a loved one). Those who had two copies of the short *5-HTT* allele were more likely to become depressed than those carrying two long alleles. This finding adds to the evidence that genetics (the type of *5-HTT* allele) and environmental factors (the stressful event) interact to influence behavior and our reactions to changes in our lives. The role of genes, behavior, and the environment will be discussed in Chapter 13.

BB3.7 How Do Environmental Factors and Genes Interact with the Immune System?

Autoimmune diseases are the third most common form of disease in the United States (after heart disease and cancer). There are 15 to 80 such diseases that affect some 5 to 8% of the population. Autoimmune diseases arise in the body's immune system, which produces antibodies that attack specific cells and tissues in the body. Almost all tissues and organs can be affected by autoimmune attacks, including skin, connective tissue, the digestive system, and the nervous system. Some autoimmune disorders are type I diabetes, multiple sclerosis, and rheumatoid arthritis. As shown in Figure BB3.3, more than 75% of people with autoimmune diseases are women. The reason for this is unknown.

Studies of identical twins have clearly demonstrated that both genetic factors (the presence of specific alleles) and environmental factors (infection) are involved in triggering these diseases. Often patients have no symptoms until they are infected by a specific virus. This infection triggers the start of an autoimmune response or increases its severity. Table BB3.2 lists some autoimmune diseases that are linked to infections.

In addition to viruses, bacteria, fungi, and some parasites can cause an autoimmune disease or increase the severity of an autoimmune response already in progress. The connection between infection and autoimmune disease is still being

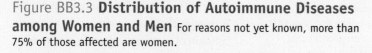

Figure BB3.3 **Distribution of Autoimmune Diseases among Women and Men** For reasons not yet known, more than 75% of those affected are women.

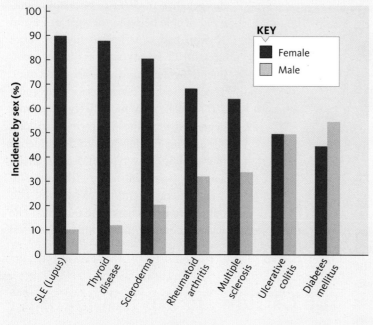

A gene complex on chromosome 6, the HLA genes, controls proteins that are part of the immune system, and specific allele combinations of these genes may determine who will develop an autoimmune disease. The genetics of the immune system and the HLA complex will be discussed in Chapter 14.

BB3.8 How Does the Environment Affect Where and How Often a Genetic Disorder Occurs?

Sometimes both populations and the environment play a part in genetics. One example is sickle-cell anemia. This recessively inherited genetic disease is most common among people who live in West Africa, in the lowland areas around the Mediterranean Sea, and in some regions of the Middle East and India, but it is almost nonexistent in most other populations.

In Chapter 5, Marcia Johnson and her husband both had ancestors who came from West Africa, which put them at higher risk for carrying the sickle-cell allele. But why do people in this region of Africa have a higher frequency of sickle-cell anemia? Geneticists often look to the environment for factors that may be responsible for differences in allele frequencies across populations. In the case of sickle-cell anemia, the relationship between a gene and the environment has been clearly established; in other cases, the link is elusive and still unknown.

As discussed in Chapter 4, sickle-cell anemia is an autosomal recessive disorder caused by a single amino acid substitution in the oxygen-carrying protein, hemoglobin. The gene for this disorder is present in very high frequencies in some populations where malaria, an infectious disease, is present. Scientists discovered that carrying the mutant allele for sickle-cell anemia confers resistance to malaria, making sickle-cell anemia and resistance to malaria interrelated. The reasons for this relationship will be discussed in Chapter 15.

studied. It is sometimes difficult to establish a connection between an infection and an autoimmune response, because often there is a long interval between infection and the abnormal immune response. Oddly enough, recent studies also show the opposite effect. Some infections can actually offer protection from autoimmune disease, which happens when the immune system stops attacking a patient's own tissues and starts to defend the body against the infection. Understanding the relationship between infection and autoimmune disease may lead to the development of effective treatments for autoimmune conditions.

BB3.9 How Is Genomics Used to Study Populations?

Genome sequences from dozens of individuals are now available in public databases. Once multiple genomes became

Table BB3.2 **Autoimmune Diseases and Infections**

Autoimmune Disease	Associated Infection(s)
Multiple sclerosis	Measles virus, Epstein-Barr virus
Rheumatoid arthritis	Hepatitis C virus, Lyme disease
Type 1 diabetes	Mumps and rubella virus

available, scientists began comparing the sequences and discovered a surprising amount of variation in the sequence and arrangement of nucleotides in humans. Once this variation was identified, scientists began to examine the type, amount, location, and effects of these variations, in a study called the Hap-Map project. The simplest and most widespread type of variation they discovered is a single nucleotide change in a genome sequence, called a **single nucleotide polymorphism**, or **SNP** (pronounced "snip") (Figure BB3.4). Over 11 million SNPs have been identified, and scientists are using clusters of neighboring SNPs called **haplotypes** as markers. These haplotypes have been mapped to each human chromosome, and their location at specific chromosome regions is known.

SNP haplotypes are used to scan the genomes of hundreds or thousands of individuals, looking for links between these SNPs and common complex traits and disorders such as cardiovascular disease, diabetes, and mental illness. If a SNP haplotype and a genetic disorder are usually inherited together, this means that genes associated with the disorder must be near the location of the haplotype. The genes in those regions can be isolated and studied to find these disease genes. These **genome-wide association studies (GWAS)** have helped identify and locate genes associated with type 2 diabetes, cancers, neurodegenerative diseases (Alzheimer disease and Parkinson's disease), mental illness, and cardiovascular diseases. These technological advances are helping unravel the number and identities of genes associated with complex diseases and are rapidly changing the study of human genetics.

Preview

In the next section, Chapter 11 examines the interaction of genes and environmental factors in complex traits. Chapter 12 focuses on the cell cycle and how mutations that disrupt the cycle are associated with cancer. Chapter 13 discusses behavior and how the study of behavioral genetics is being

Figure BB3.4 **Single Nucleotide Polymorphisms**

SNPs in four individuals are shown in red. Clusters of SNPs are haplotypes and are used in genome scans as genetic markers to locate genes associated with complex genetic disorders.

	SNP	SNP	SNP	SNP	
Person 1	...AA**C**CTTCGCC.......TTGAGGCATC...				Haplotype 1
Person 2	...AAGCTTCGCC.......T**A**GAGGCATC...				Haplotype 2
Person 3	...AAGCTTC**C**CC.......TTGAGGCATC...				Haplotype 3
Person 4	...AAGCTTC**T**CC.......TTGAGGCA**A**C...				Haplotype 4

revolutionized by new genomic methods. Chapter 14 outlines the components of the immune system, blood groups, organ transplants, and disorders of the immune system. Chapter 15 summarizes the principles of population genetics and the factors that change allele frequencies. Chapter 16 reviews the origins and migrations of human species, using the fossil record as well as molecular and genomic evidence. Chapter 17 provides an overview of genetics in the past, present, and future.

single nucleotide polymorphism (SNP) single nucleotide differences in the genome sequences among individuals in a population

genome-wide association study (GWAS) analysis of genetic variation across an entire genome, searching for associations (linkage) between variations in genome sequence (SNP haplotypes) and a genome region containing genes associated with a genetic disorder

haplotype sets of closely linked (clustered) SNPs that are used as genetic markers to locate genes of interest

11

Inheritance of Complex Traits

REVIEW

How does a baby develop from fertilization to birth? (1.2)

CENTRAL POINTS

- Polygenic traits are controlled by two or more genes.

- Multifactorial traits are polygenic with a significant environmental component.

- Spina bifida is an example of a multifactorial trait.

- There are many other multifactorial traits.

CASE A: Prenatal Pills

Just after her friends Brian and Laura went to see Dr. Franco (see Case A in Chapter 2), Vera Smith found out she was pregnant. She was 33 and single. She knew that she should see a physician right away, but her schedule was hectic. Her friends told her to eat right and not to take any medication, not even an aspirin, because it might hurt her fetus.

When Vera was almost three months pregnant, her mother called from California. She reminded Vera that she should see a physician for prenatal care. Vera called Laura and got Dr. Franco's number. At her appointment, the nurse first asked whether she was taking folic acid. Vera had never heard of folic acid and said proudly, "I am not taking anything; I don't want to hurt the baby." The nurse pointed to a sign that said all women should take folic acid when pregnant to reduce the chance that the baby would be born with a birth defect called spina bifida. Vera wondered what that meant; the nurse gave her a pamphlet about nutrition, folic acid, and pregnancy.

Faye Norman/Science Photo Library/Photo Researchers

Some questions come to mind when reading about Vera and the risk of conditions like spina bifida. Before we can address those questions, let's look at how genes and environmental factors work together to affect phenotypes and the outcomes of pregnancy.

A Young Boy with Spina Bifida

11.1 What Are Complex Traits?

Have you ever wondered why members of a large family or a group of people in a room, on a train, or in a coffee shop all look somewhat different but yet still pretty much the same? This is probably because some traits in humans are determined by the action of multiple genes, whereas many others are controlled by a combination of several genes and environmental factors. Such traits are called **complex traits**.

Up until now, we have focused on simple traits, each controlled by a single gene. Complex traits controlled by two or more genes are **polygenic traits**. Complex traits controlled by two or more genes and environmental factors are **multifactorial traits**.

Eye color (Figure 11.1) and hair color are polygenic traits because they are controlled by more than one gene, often located on different chromosomes. The result is the slight and often variable range of different eye and hair colors we see throughout a population or in a family.

complex traits inherited traits and conditions caused by any combination of genes and environmental factors

polygenic trait a trait controlled by two or more genes

multifactorial trait a trait controlled by two or more genes and environmental factors

Figure 11.1 **Eye Color** This polygenic trait has a continuous distribution of phenotypes.

Frank Cezus/FPG/Getty Images

Frank Cezus/FPG/Getty Images

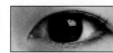
© 2001 PhotoDisc

Ted Beaudin/FPG/Getty Images

Stan Sholik/FPG/Getty Images

Richard Hutchings/PhotoEdit

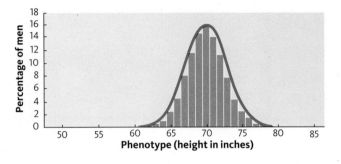

Polygenic traits can usually be measured in some way. When measured, the result is called the trait value.

Figure 11.2 shows the distribution of phenotypes of a typical polygenic trait in a population. This **bell curve** shows that the trait values for most people are clustered around the average, with only a very few individuals clustered at one or the other extreme. Polygenic traits, by definition, involve a number of genes; each gene contributes only a small amount toward the phenotype. Imagine a polygenic trait controlled by 10 genes. There would be many more than 10 possible combinations of phenotypes from these genes, resulting in a bell-shaped distribution of phenotypes in a population.

11.1 Essentials

- **A polygenic trait is one that is controlled by two or more genes.**
- **Complex traits usually have a bell-shaped distribution of phenotypes in a population.**

11.2 What Is a Multifactorial Trait?

A multifactorial trait is controlled by two or more genes *and* the action of environmental factors. Although each gene controlling a multifactorial trait is inherited in Mendelian fashion (see Chapter 4), the interaction of genes with the environment can produce a wide range of phenotypes. Height is a good example. In Figure 11.3 you can see a wide distribution of height among the students. Height is a multifactorial trait because it is determined by more than one gene *and* because environmental factors such as nutrition can contribute to adult height.

What are the characteristics of multifactorial traits? Multifactorial traits have several important characteristics:

- Several genes control the trait.
- The trait is not inherited as dominant or recessive.
- Each gene controlling the trait contributes a small amount to the phenotype.
- Environmental factors interact with the genes to produce the phenotype.
- There are many phenotypic differences in a multifactorial trait within a population; when these distributions are graphed, they form a bell-shaped curve.

Height is clearly a multifactorial trait; it fulfills all of these requirements.

Which diseases are multifactorial traits?

Many common diseases are multifactorial traits (Figure 11.4), and people with these diseases make up the largest part of patients hospitalized with genetic disorders.

Some diseases that show multifactorial influences are cancer (discussed in Chapter 12), spina bifida (discussed shortly), hypertension, and cardiovascular disease.

bell curve a frequency distribution that resembles the outline of a bell

Figure 11.3 **Distribution of Height among Males in a Biology Class**

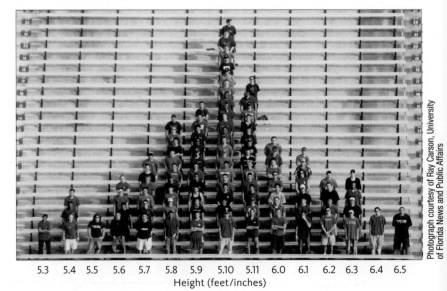

Photograph courtesy of Ray Carson, University of Florida News and Public Affairs

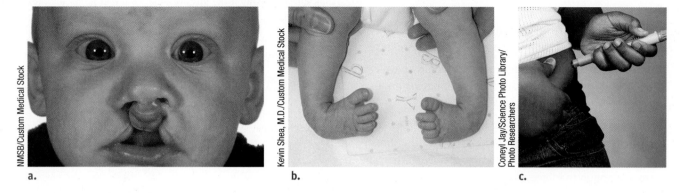

Figure 11.4 **Several Multifactorial Traits** (a) Cleft lip and palate, (b) club foot, and (c) diabetes.

a.

b.

c.

11.2 Essentials

- **When a polygenic trait has a significant environmental component, it is called a multifactorial trait.**
- **The environment can interact with the genes controlling a multifactorial trait in many ways.**

11.3 How Do We Study Multifactorial Traits?

The genetics of multifactorial traits can be studied in several ways: (1) by defining conditions under which a phenotype is expressed, (2) by using family studies to calculate how often the trait reappears, (3) by constructing and studying animal models of these traits, and (4) by using genomics to identify the genes. Here we will briefly consider the first two approaches. The use of animal models will be explored later in this chapter, and genomic methods were discussed in Chapter 10.

Although a bell-shaped distribution of phenotypes is characteristic of multifactorial traits, there are some exceptions. In disorders such as cleft palate or club foot, individuals are either affected or unaffected (see Figure 11.5). In this situation, called the threshold model, the curve represents the

distribution of genotypes in the population, instead of phenotypes. The number of alleles present in a genotype for a specific trait is called **genetic liability**. This liability can be measured by examining the number of relatives who have the condition. However, only those who have a specific combination of alleles *and* are exposed to certain environmental factors will develop the phenotype. In other words, those who have the highest level of genetic predisposition to the disorder will be most affected by the environmental agent.

The risk for these multifactorial disorders should decrease as the degree of genetic relationship to relatives decreases, lowering the level of genetic predisposition. In a family, parents and children have half their genes in common, grandparents and grandchildren have one-fourth of their genes in common, and cousins have one-eighth of their genes in common. As predicted, the risk for some multifactorial disorders declines along with a decline in genetic relatedness.

Because of the relationship between genetic predisposition and genetic relatedness, several factors can be used to

genetic liability in complex traits, the number of genes in an individual's genome that affect phenotype along with environmental influences

Figure 11.5 **The Threshold Model** This model explains the distribution of some multifactorial traits. In a population these traits have a bell-shaped phenotypic distribution. Individuals with these traits must have a combination of a certain number of genes and exposure to certain environmental conditions.

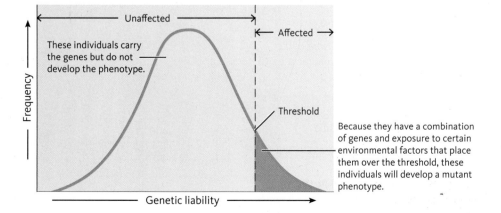

Unaffected

These individuals carry the genes but do not develop the phenotype.

Affected

Threshold

Because they have a combination of genes and exposure to certain environmental factors that place them over the threshold, these individuals will develop a mutant phenotype.

Frequency

Genetic liability

predict the risk that a multifactorial disorder will recur in a family. These factors include

- **consanguinity**: Because of the genes they have in common, parents who are first cousins have a much higher risk of having an affected child than two unrelated parents.

- **number of affected children**: If parents already have two affected children, this increases the risk of having another affected child.

- **severity of the disorder**: Having a severely affected child indicates that the child's genotype is well over the threshold, and that the combined parental genotypes may be extreme enough to increase risk of another affected child.

11.3 Essentials

- **There are several methods for studying multifactorial traits.**
- **The threshold model helps explain the noncontinuous phenotypic distribution of certain multifactorial traits.**

11.4 What Is Spina Bifida?

Spina bifida (SB) is a birth defect involving the nervous system. It appears in the first month of embryonic development when the spinal column forms. SB belongs to a group of conditions called **neural tube defects**. These disorders get their name because they involve problems with the development of the embryo's nervous system (Figure 11.6).

In the embryo, the neural tube gives rise to the brain and the spinal cord (the central nervous system). Other cells form the membranes (**meninges**) that cover and protect these structures. Figure 11.7 shows the major parts of the central nervous system.

There are three basic types of neural tube defects. The most extreme form is called **anencephaly**. In this condition, the neural tube at the head end of the embryo does not close. As a result, major portions of the brain

spina bifida (SB) a neural tube defect caused by the failure of the neural tube to close

neural tube defects a group of disorders that result from defects in the formation or development of the neural tube

meninges the membranes that surround the brain and spinal cord

anencephaly neural tube defect in which major parts of the brain and skull do not form

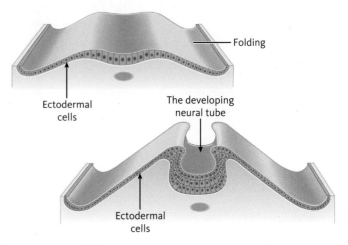

Figure 11.6 **Neural Tube Formation** During embryonic development, a cell layer called the ectoderm folds over the cells that will eventually make up the spinal cord and forms the neural tube.

and skull do not form, and the remaining portions of the brain may not be enclosed in the skull. (See photo in Case B on page 233.) Fetuses with this condition can survive only within the mother, and most are not born alive. Those that are born alive usually die within a few hours or days due to heart and breathing problems. A second, rare form of neural tube defect results from openings in the skull that allow brain tissue to extend beyond the brain case.

When the neural tube at the lower end of the embryo does not close, the result is SB (Figure 11.8). This is the most

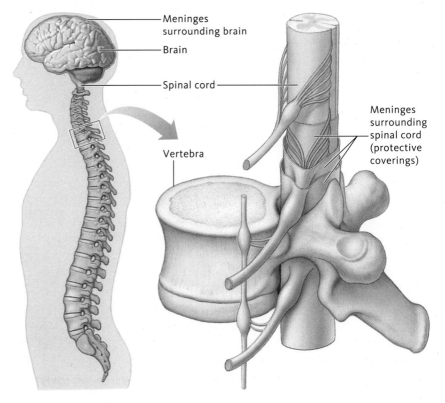

Figure 11.7 **Nervous System** The major parts of the nervous system formed from the neural tube

Figure 11.8 **Three Major Forms of Spina Bifida**

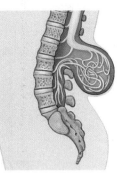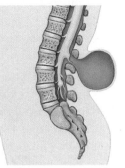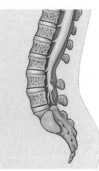

The spinal cord and meninges protrude from the base of the spine.

The spinal cord develops normally, but the meninges protrude.

One or more vertebrae are malformed, but nothing protrudes from the body.

common form of neural tube defect and affects about 70,000 individuals in the U.S. In many cases of SB, the spine can be surgically repaired, but damage to the nervous system is permanent. Children born with SB have varying degrees of paralysis, depending on the location and size of the opening. In addition, most individuals with SB have learning disabilities and may have bowel and bladder problems caused by a lack of nerve control over these functions.

There is no cure for SB. The damaged nerves cannot be repaired; treatment consists of therapy and assistance in the form of crutches, braces, or wheelchairs (Figure 11.9). Complications that arise later in childhood may require further surgery or other therapies. Most individuals with SB live into adulthood.

Spina bifida can be diagnosed by several methods of prenatal testing (see Chapter 8). Vera Smith waited too long to take folic acid, and its preventive effects will not help her fetus. However, maternal blood tests and ultrasound scanning might tell whether or nor her fetus will be affected with a neural tube defect.

11.4 Essentials

- **Spina bifida is one example of a class of genetic disorders called neural tube defects.**
- **Taking folic acid reduces a woman's chances of having a child with spina bifida.**

11.5 How Do We Know That Spina Bifida Is a Multifactorial Trait?

How do we know that spina bifida and other multifactorial disorders actually have a genetic component? Geneticists use several methods to determine whether a trait has a genetic basis, including family studies, twin studies, and adoption studies. If a disorder runs in a family, affects both members of an identical twin set, and is found in children from affected families adopted into unaffected families, then genetics is assumed to play a role in the trait. Using these methods, spina bifida is identified as a multifactorial trait. It tends to cluster in families and when one child is affected, the risk of having another child with SB or another neural tube defect is significantly increased. These methods of identifying SB as a multifactorial trait are indirect; the best evidence comes from the identification of a gene involved in the trait. Recently, researchers have identified a gene associated with neural tube defects in mice and then in humans.

Figure 11.9 This child with spina bifida is functioning well.

© Ellen Senisi/The Image Works

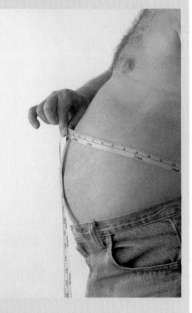

image100/CORBIS

68%
Percentage of adults in the United States who were overweight, according to a 2008 study

folic acid a B vitamin that is a key factor in cell growth and development

What genes are involved in spina bifida?

In mice, the *VANGL1* gene acts early in the development of the neural tube, and mutations in this gene cause conditions similar to SB. In humans, the *VANGL1* gene was mapped to chromosome 1 (Figure 11.10) and shown to also be involved in neural tube formation.

Based on the results from mice, a study of more than 100 patients with neural tube defects showed that all affected individuals had mutations in the *VANGL1* gene. One of these mutations, called V239I, may cause a partial loss of function in the VANGL1 protein. This mutation, along with environmental factors, may produce some forms of SB. More research will be needed to understand how mutations in this gene result in SB and to identify other genes involved in this disorder.

What are the environmental risk factors for spina bifida?

Multifactorial traits depend on both genetic and environmental factors. One significant environmental risk factor for SB has been identified. Diets deficient in **folic acid**, a B vitamin, are a risk factor for SB. Dietary sources of folic acid include whole grains, green leafy vegetables, and fruit. To ensure that all women of childbearing age get enough folic acid, it is recommended that they take 0.4 mg of folic acid every day for at least three months while trying to become pregnant and continue taking it until the 12th week of pregnancy. This preventive step is necessary even though many foods in the United States have been fortified with folic acid since 1998.

For women who have previously had a child with SB or other neural tube defects, the recommended daily dose is 4 mg/day. This dose is ten times higher than for women who have not had a child with a neural tube defect. This dosage has been shown to reduce the risk of SB and related conditions by about 70%. How folic acid interacts with genes that control the formation of the neural tube is still unknown. Other environmental factors, if any, remain to be discovered.

Figure 11.10
Location of the *VANGL1* Gene on Chromosome 1

VANGL1 gene

Chromosome 1

CASE A: QUESTIONS

Faye Norman/Science Photo Library/Photo Researchers

Now that we understand what spina bifida is and how it can be prevented, let's look at some questions involved in Vera's decision whether to take folic acid.

1. Should Vera be skeptical about the use of folic acid because the way it acts has yet to be discovered?

2. Should Vera take the folic acid pills even though she is three months pregnant?

3. Many dietary supplements that were once considered good for us have been shown to be either useless or harmful. Is it possible that this may be the case with folic acid? Should this influence Vera's decision? Why or why not?

4. How much research should Vera do if she decides to take folic acid?

5. Suppose that Vera had taken folic acid but her child is born with SB. What should she do? List four things.

11.5 Essentials

- **Several methods, including twin and adoption studies, help establish that a multifactorial trait has a genetic basis.**
- **One gene associated with spina bifida has already been identified.**
- **Folic acid deficiency is a known environmental risk factor for spina bifida.**

11.6 What Are Some Other Multifactorial Traits?

Fingerprints are a multifactorial trait; they form in the first three months of embryonic development (weeks 6–13). The skin ridges on fingers, palms, toes, and feet are called dermatoglyphic patterns. During this short developmental period, the prenatal environment can influence fingerprint patterns. Some of these environmental influences include nutrition of the mother, rate of finger formation, and overall growth rate of the embryo.

Everyone, including identical twins, has a unique set of fingerprints. Even though identical twins share the same set of genes and occupy the same uterus simultaneously, each is exposed to a slightly different prenatal environment. These subtle environmental factors are enough to create different fingerprint patterns (Figure 11.11).

Is obesity a multifactorial trait?
In the United States, obesity is a rapidly worsening national health problem. More than 68% of adults are overweight, and more than 33% are obese. Family studies confirm that obesity runs in families, and twin and adoption studies show that it has genetic components. The dramatic increase in obesity has occurred in the last 30 to 40 years. It is unlikely that large-scale and rapid changes in our genetic makeup are responsible for this increase. Instead, we must look to changes in environmental factors (such as diet and exercise) interacting with our genotypes that have led to the spread of obesity.

Figure 11.11 **Fingerprints** The fingerprints of identical twins (a) and (c) are slightly different from each other. All three fingerprints, including that of an unrelated individual (b), are different from one another.

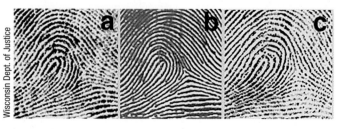

Twin studies have been used to estimate the genetic contribution to obesity. Identical twins, also called **monozygotic (MZ) twins**, are genetically identical because they are formed from a single fertilized egg. If a trait is present in both MZ twins, it is considered genetically based; if not, it is thought to have an environmental origin or component. Obesity occurs in both MZ twins approximately 70% of the time. The numerical calculation of the presence of the same trait in both members of a pair of twins is called **concordance**.

The search for genes involved in obesity includes the use of animal models and genome scans.

What have we learned about obesity from animal models?
Recent breakthroughs in understanding how genes regulate body weight have come from studies in mice. Several mouse genes that control body weight have been identified, isolated, cloned, and analyzed. Mice that are homozygous for a mutant form of the *obese* (*ob*) gene (Figure 11.12, page 232) or the *diabetic* (*db*) gene weigh about three times more than a normal mouse and have about five times as much body fat. The *ob* mice are fat because they do not make the weight-controlling hormone leptin, normally produced in fat cells. The hormone moves through the bloodstream to the brain and signals the brain to stop eating.

Mice with the *db* mutation lack the receptor on brain cells that receive and process the leptin signal. These genes control how energy is used in the body. Using the results from mice, scientists found the same genes in humans. Mutations in the leptin gene are associated with several obesity-related genetic conditions. Some affected individuals have been successfully treated with leptin, but others have not responded, indicating that other, still unidentified genes are involved in weight control.

Obesity is clearly a complex disorder involving the action and interaction of multiple genes and environmental factors.

Scientists are studying large populations and families, searching for additional genes associated with obesity. Some of these studies have shown that important genes for obesity are located on chromosomes 2, 3, 5, 6, 8, 10, 11, 17, and 18 (Figure 11.13, page 232).

Further work may identify additional genes involved in obesity and provide a foundation for studying how these genes interact with environmental factors to cause obesity.

Is intelligence a multifactorial trait?
The idea that intelligence could be measured was first addressed in the late 18th and early 19th centuries. In the beginning, measuring

monozygotic (MZ) twins genetically identical twins derived from the fertilization of a single egg by a single sperm

concordance in twins, a condition in which both twins have or do not have a given trait

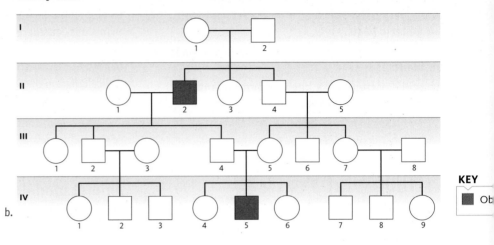

Figure 11.12 Inheritance of Obesity (a) On the left is a mouse homozygous for the *ob* gene. At right is a normal mouse. (b) A mouse pedigree showing inheritance of the obesity trait.

Reprinted with permission of Amgen Inc.

a.

b.

KEY
■ Ob

someone's head size was used to determine how smart he or she was; the bigger the better. But by the beginning of the 20th century, psychological rather than physical methods were used to measure intelligence. Alfred Binet, a French psychologist, developed a graded series of tasks related to basic mental processes, and he tested children for their ability to perform these tasks. Their mental age was calculated by recording how well they performed these tasks compared to other children of the same age.

Wilhelm Stern, another psychologist, divided the mental age derived from these tests by chronological age and multiplied the resulting number by 100. This value became known as the **intelligence quotient (IQ)**. The underlying assumption in using IQ tests to measure intelligence is that intelligence is biological and that it can be measured and the results expressed as a single number. Whether this assumption is correct can be answered only by looking at whether intelligence is a multifactorial trait.

IQ does have certain inherited components. This evidence comes from studying MZ twins raised together and apart. A high concordance in MZ twins raised together (80%) indicates that genetics plays a significant role in determining IQ. But the differences are greater if the MZ twins are raised apart. Clearly, this indicates that both genetic and environmental factors make important contributions to intelligence; however, the relative amount each contributes cannot be measured accurately at this time.

Figure 11.13 Some of the human chromosomes known to carry genes associated with obesity.

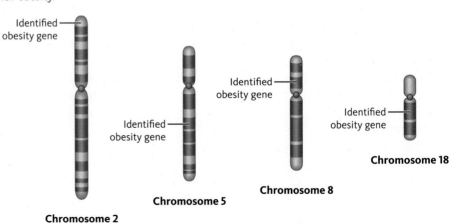

Identified obesity gene

Identified obesity gene

Identified obesity gene

Identified obesity gene

Chromosome 18

Chromosome 8

Chromosome 5

Chromosome 2

intelligence quotient (IQ) a score derived from a standardized test of intelligence

Figure 11.14 **Animal Models** *Drosophila* (fruit flies) are useful animal models to study learning and memory because the biochemical pathways involved in these processes are identical to those in humans.

© Graphic Science/Alamy

How are scientists studying genes that control intelligence?

As in the search for obesity genes, scientists use animal models to help locate genes for intelligence. They are searching for single genes that control aspects of learning, memory, and spatial perception. These animal models include the fruit fly *Drosophila* and the mouse. Some may be surprised to learn that *Drosophila* (Figure 11.14) has many biochemical pathways that are identical to those found in humans. In the nervous system of both fruit flies and humans, some of these pathways play important roles in learning and memory.

Another approach for studying multifactorial traits uses recombinant DNA techniques and information from the Human Genome Project to identify genes that affect specific polygenic traits, such as reading ability and IQ. Using these techniques, scientists have uncovered genes associated with reading disability (developmental dyslexia). Figure 11.15 is a drawing of chromosomes 4, 6, and 15 showing the location of some genes for cognitive ability. Cognitive ability goes beyond

Figure 11.15 Some of the human chromosomes known to carry genes associated with reading ability and cognitive ability.

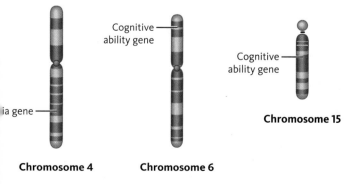

Cognitive ability gene

Cognitive ability gene

...ia gene

Chromosome 15

Chromosome 4   **Chromosome 6**

CASE B:
Donation of a Baby's Organs

When Samantha found out she was pregnant, she was extremely happy. She already had a wonderful, healthy little girl. At her first ultrasound, Samantha was shocked that the physician didn't say everything was all right; in fact, her face showed concern. When Samantha asked what was wrong, Dr. Chang said the head of the baby looked small, and she was going to send the ultrasound out to an expert.

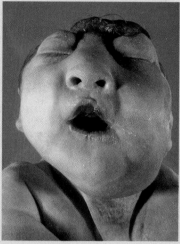

Joseph R. Siebert, Ph.D./Custom Medical Stock

The expert's analysis confirmed that Samantha's baby had anencephaly, a condition in which the brain and head are abnormally small, sometimes absent. A baby with anencephaly is shown above.

"Will my baby live?" Samantha asked.

"It will probably live to be born, but it will not survive very long after birth," Dr. Chang told her. "In California, parents who had a child with a similar condition donated its organs to babies who needed transplants. If you might be interested in doing this, it can be done soon after your baby is born."

Dr. Chang had read that the couple in California placed their baby on a respirator at birth. Then, they donated her heart, lungs, and kidneys to babies waiting for organs.

Samantha thought about her other choices. Should she carry a baby to term just to have it die? Could she do this?

1. **What should Samantha do? List three options.**

2. **What reasons could Samantha give for carrying her baby and donating the organs?**

3. **What reasons could she give to abort the baby?**

4. **Where could Samantha go for help with this decision?**

5. **If Samantha finds out that her baby may be born dead and the organs would not be usable, should this make a difference in her decision?**

IQ and involves verbal and spatial abilities, memory, speed of perception, and reasoning.

More recently, a gene associated with cognitive ability has been identified on chromosome 4, also shown in Figure 11.15. Mutations in this gene are responsible for dyslexia. The accumulated results of research indicates that intelligence is a polygenic and multifactorial trait.

11.6 Essentials

- **Obesity is a multifactorial trait.**
- **Intelligence is another example of a multifactorial trait.**
- **The interaction of environment and heredity is an active area of research in genetics, psychology, and other fields.**

11.7 What Are the Legal and Ethical Issues Related to Multifactorial Traits?

Vera's case brings a number of legal issues to mind. For example, if a patient does not want to take folic acid, does she have the right to refuse? If she did refuse, could her physician force her to take it?

This situation is related to the question of consenting to one's medical treatment. We have discussed *informed consent* in previous chapters. In Chapters 5 and 8, Marcia and Victoria consented to have testing for a genetic condition.

Marcia had herself tested and Victoria allowed her child to be tested. How does the law handle this and the right to refuse both testing and treatment?

All patients can consent to treatment but can also refuse treatment for themselves and sometimes their child. Under U.S. law, any adult can give consent or refuse medical treatment or testing without a problem. Problems often arise when a child is involved.

In the cases in this chapter, both Vera and Samantha are pregnant. A pregnant woman can consent to medical treatment for herself and her fetus, based on the fact that the fetus is inside her body. After her baby is born, Samantha can donate its organs because she is the parent and a child cannot consent to or refuse donation.

Informed consent and consent to medical treatment have a number of exceptions, and one of them is age. When children are younger than 16 (the age of consent varies by state), they cannot consent, and their parents or guardians must sign a consent form. If neither is available, a court must step in. No state gives fetuses or newborns the right to consent.

If a woman is pregnant, no matter what her age, she is considered an adult as far as medical decisions are concerned. Many pregnant women have medical problems (for example, diabetes) and need special treatment. If they refuse, physicians will try to convince them to accept treatment, but they cannot be forced.

Table 11.1 addresses some of the ethical and legal questions regarding consent to treatment:

Table 11.1 **Legal and Ethical Questions about Consent to Treatment**

Question	How Are These Questions Decided?	Related Case or Legal Issue
What might happen if a pregnant woman takes drugs that would harm her fetus?	Usually, not much can be done to force a woman to stop taking drugs, even if those drugs are illegal.	Some judges have tried to put pregnant women in jail to stop their drug intake, but this action has been shown to be unconstitutional.
		In one case, a judge tried to put a pregnant woman in jail when it became known she had abused her previous children. She was released and he was taken off the bench.
Can a woman be forced to take medication that is beneficial to her fetus but not to her?	States do not force women, or any other patient, to have medical treatment because they are considered adults and can, therefore, make their own decisions.	No cases have tried to give a woman medication or vitamins if she refuses. Judges would probably not allow this. Some religions do not allow medical treatment and the right to refuse has been upheld only for adults.
If a newborn or fetus needs special treatment but the mother refuses, what can be done?	Laws allow physicians to go to court and file an injunction for treatment of critically ill newborns or children, overriding the parents' refusal. This law has not been extended to fetuses.	Members of the Jehovah's Witnesses have refused blood transfusions for their children, and in life-and-death situations courts have issued injunctions forcing treatment.

Spotlight on Ethics

The Bell Curve Revisited

Scientists have been trying to define and measure intelligence for quite a long time. Francis Galton (Darwin's first cousin) thought that all inherited traits, including intelligence, could be measured. But, as mentioned in this chapter, it was a Frenchman, Alfred Binet, who developed an intelligence test when the French government paid him to find a way to distinguish normal children and what were then termed inferior children.

Binet's test came to the United States in 1917 and was used to test military recruits in World War I to see what assignments they would be suited for. Soon after, a Binet test was administered to a prisoner on trial for murder in Wyoming. Because the prisoner fared so poorly on the test, the jury acquitted him by reason of his mental condition. Schools around the world began using the Binet test to place children into learning groups. The tested IQ value followed a person throughout his or her life.

Originally, IQ was calculated as the following ratio: 100 times the person's mental age (as determined by the Binet test) divided by his or her chronological age. An average IQ, as shown by the bell curve in Figure 11.16, was considered to be 100.

This type of calculation was originally applied only to children. In 1939 David Wechsler published the first intelligence test explicitly designed for an adult population: the Wechsler Adult Intelligence Scale (WAIS).

Although the inheritance of IQ has been investigated for nearly a century, much controversy remains about its definition, causes, and the methods and accuracy of testing IQ. Many studies and educators asked: "How can we test for intelligence?" and "What should be included in an intelligence test?" In schools, children were being placed in classes based on the results of IQ tests. But these tests turned out to be biased against certain groups because of the ways in which questions were asked. As a result, the IQ test (Figure 11.17) slowly became discredited but was still used in many areas.

In 1994, the publication of *The Bell Curve*, written by psychologist Richard J. Herrnstein and political scientist Charles Murray, changed perception of IQ tests yet again. In their book, the authors tried to analyze the validity and importance of IQ testing. However, educators and scientists are still arguing about the conclusions. Some of their controversial conclusions are:

1. Low measured intelligence and antisocial behavior are linked.

2. Low test scores of African Americans (compared to those of whites and Asians) were said to be caused by genetic factors.

3. The bell curve of IQ test scores was said to prove that some groups were inferior and linked poverty with low intelligence.

This book caused an uproar and caused many people to rethink the use of IQ tests as a measurement of intelligence. Many asked whether this was the intended result for writing the book. If not, it did the exact opposite.

QUESTIONS

1. If we could determine that intelligence (as measured by IQ) was caused by certain genes, what would this mean to our society?

2. Research what happened after *The Bell Curve* was published.

3. The following question is commonly used to show bias in an IQ test for young children: After three drawings are shown, the child is asked to pick out the handbag. Why do you think this could be considered biased?

4. Many schools do not use IQ to place children in special-education classrooms. Do some research to find three reasons why.

5. Research Mensa, a group that claims to have members with very high IQs. Who are some members of this organization?

Figure 11.16 **IQ Scores** The distribution of adult IQ scores follows a typical bell curve. The average score is 100; approximately half the scores are above 100 and half are below.

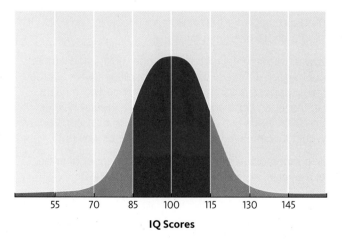

IQ Scores

Figure 11.17 **An IQ Test Booklet**

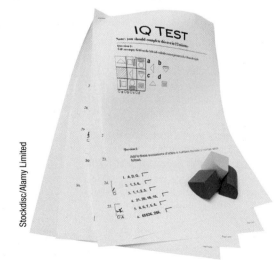

Stockdisc/Alamy Limited

1 A polygenic trait is one that is controlled by two or more genes. (Section 11.1)
The term polygenic means many genes.

2 Polygenic traits usually have a bell-shaped phenotypic distribution curve in populations. (Section 11.1)
This means that the traits or conditions have varying degrees of expression, from slight to severe.

3 Traits controlled by two or more genes with an environmental component are called multifactorial traits. (Section 11.2)
This means that both genetic and environmental factors affect expression of the condition.

4 Multifactorial traits are studied using the threshold model, recurrence risks, animal models, and genomic methods. (Section 11.3)
The risk for a disorder decreases as genetic relatedness decreases.

5 The environment can interact with a multifactorial trait in many ways. (Section 11.4)
Environmental factors can alter the expression of a trait.

6 Spina bifida is one member of a class of genetic conditions called neural tube defects. (Section 11.5)
Spina bifida affects the formation of the spinal cord and brain.

7 Taking folic acid reduces a woman's chance of having a child with spina bifida. (Section 11.5)
Adding folic acid to the diet is one way of altering the environment that can affect the risk and outcome of a multifactorial condition.

8 Obesity is a multifactorial trait. (Section 11.6)
Genes and environmental factors such as diet affect this phenotype.

9 Intelligence is another example of a multifactorial trait. (Section 11.6)
Defining intelligence is the first step in researching its inheritance and the effects of environment on its phenotype.

10 The interaction of environment and heredity is an active area of research in genetics, psychology, and other fields. (Section 11.6)
As more multifactorial traits are identified, investigators work to understand how each affects the body.

Review Questions

1. List two environmental factors that may influence cancer and heart conditions.

2. If we haven't yet identified a gene for a condition, can the condition still be considered to be multifactorial? Why or why not?

3. Name one other human genetic condition that you think might be multifactorial. Then identify the genetic and environmental influences.

4. Make a chart of all the multifactorial conditions listed in the chapter and include their genetic and environmental components. Do some research and add to this information.

5. What can a pregnant woman do to reduce her risk of having a child with spina bifida?

6. Explain what is meant by the term trait value and give an example.

7. Is it harder or easier to find the genes for polygenic traits than for traits controlled by single genes? Why?

8. Research information on club foot. What is done to treat this condition?

9. Explain why fingerprints are considered a multifactorial trait.

10. Height is considered a multifactorial trait. What environmental factors may affect the expression of someone's genotype for height?

11. IQ has a genetic component. Does this necessarily mean that IQ is equivalent to intelligence? Why or why not?

12. What is the value of studying cognitive development rather than IQ as a way to identify genes associated with intelligence?

Application Questions

1. What if a gene whose mutant allele caused a decrease in intelligence was identified? What might be done if a baby was shown to carry this gene at birth?

2. Pregnant women have been jailed to keep them from harming their children. Usually this occurs, albeit rarely, with drug addicts who will not stop taking drugs while they are pregnant. Is this a solution for a woman who will not take folic acid? Why or why not?

3. Skin color is a multifactorial trait. Consider a family in which one parent has very dark skin and the other very light skin. Do some research and discuss what their children will look like and why. If you can, find a photo of such a family.

4. As we know, environmental factors influence multifactorial traits. What might happen if the environmental conditions were altered? For example, suppose a parent learns that her newborn has a mutation in the leptin gene and therefore has a greater risk for obesity. She then feeds him only the right foods and keeps an eye on him. Would this make a big difference? Why or why not?

5. Is it possible that most human traits, even those thought to be polygenic, are really multifactorial? Explain your answer.

6. If educators study SAT scores in a certain group of Americans and find that the group had especially high scores, could the conclusion be that it is hereditary? Why or why not?

7. A man who is 7 feet tall is attracted to women who are short (5 feet tall). He really wants to have tall children. If he marries a woman who is short, does he have a chance to have a tall son or daughter? Explain your answer.

8. Why do the chances of having a child with a complex trait depend to some extent on how many children in the family already have this trait?

9. Although children with spina bifida usually live good lives, a number of wrongful birth lawsuits (see Chapter 3) have been brought on behalf of these children. During a trial of this type, it is necessary for the parents to state that if they had known their child would have this condition, they would have had an abortion. Do you think this might stop some people from suing? Why or why not?

10. In case A, Vera was told it was important for her to take certain pills (folic acid) to protect her pregnancy. When physicians are confronted with a patient who does not want treatment, how should they handle it? Look up one of the cases involving Jehovah's Witnesses and blood transfusions. Is there a difference between a person refusing treatment and a parent refusing treatment for his or her child? Did the courts allow this?

11. Find an IQ test online and take it. What do the results tell you?

12. Should children be separated into classes in school based on IQ tests? Why or why not?

WHAT WOULD YOU DO IF...?

. . . your child were asked to take an IQ test?

. . . based on your child's IQ test, he or she were going to be placed in a class for students with low IQ?

. . . based on your child's IQ test, he or she were going to be placed in a class for students with high IQ?

. . . your prenatal diagnosis showed that your child has spina bifida?

. . . you were given supplements with folic acid for your pregnancy?

Online Resources

Preparing for an exam? Log on at www.cengagebrain.com for study tools to help you assess your understanding. If assigned by your instructor, the Case B and Spotlight on Society activities for this chapter, "Donation of a Baby's Organs" and "The Bell Curve Revisited," will also be available.

In the Media

The Wichita Eagle, April 18, 2011

Surgery before Birth Enhances Life After
Sara Avery

Spina bifida is a leading cause of paralysis in newborns and has no cure. When diagnosed during pregnancy, there were few effective interventions before now. Although not new, fetal surgery has had a great impact on this condition.

The operation, performed in the second or third trimester, requires a risky incision through the mother's uterus to the developing fetus. Physicians then repair the spinal lesion on the fetus's back, which is sometimes an open wound, and other times a bulging cyst.

By intervening before birth, researchers had hoped the brain and spinal column could then develop more normally and reduce problems.

The trial, analyzing 158 pregnant women, bears out those hopes.

To access this article online, go to www.cengagebrain.com.

QUESTIONS

1. Doctors involved in this study feel that the success will make this surgery commonplace. What one obstacle do you see to this hope?

2. Would this surgery work for Vera's baby? Why or why not?

Evansville (Indiana) Courier & Press, October 9, 2007

Taking Aim to Reduce SIDS
Libby Keeling

SIDS is the sudden and unexplained death of an apparently healthy baby aged one month to one year.

Looking at the risk for a specific infant, physicians see that a triple-risk model may be the culprit. They theorize that the risks combine three things: first, an infant's genetic predisposition, such as a brain abnormality; then, an unstable period of growth; and finally, an environmental trigger, such as tobacco smoke.

The program, called "Back to Sleep," urged parents to make sure their babies were sleeping on their backs. For some reason, this has caused a drop in SIDS deaths.

Specialists say it's a multifactorial trait, but police still investigate every death due to SIDS.

To access this article online, go to www.cengagebrain.com.

QUESTIONS

1. Twenty years ago a specialist in childhood death wrote a book saying that SIDS was genetic, and this was the primary thinking for ten years. A woman who had five babies die of SIDS was found not guilty of murder based on his books. Research this case and write a short report on the surprising result.

2. Why do you think babies are kept safe by sleeping on their backs?

12

Cancer and the Cell Cycle

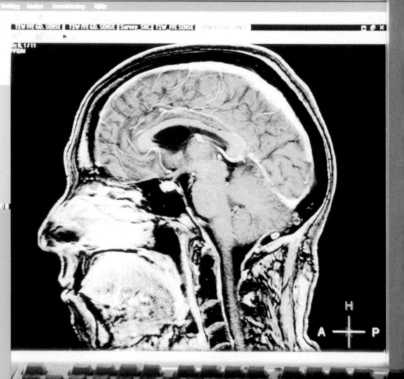

REVIEW

What is the relationship between cancer and the environment? (BB3.4)

Is there a relationship between genetics and cancer therapy? (BB3.5)

CENTRAL POINTS

- Cancer involves uncontrolled cell division and is a disease of the cell cycle.

- Mutations in certain types of genes may lead to cancer.

- Breast cancer is a common form of cancer.

- Chromosomal changes are often associated with cancer.

- Environmental causes of cancer are being studied.

- Some lawsuits have addressed smoking as a cause of cancer.

CASE A: Patient Offered New Cancer Treatment

Harriet Abeline had faced bad news before. She was diagnosed with breast cancer five years ago and underwent a lumpectomy and radiation treatment. Two years later, a routine mammogram revealed another lump. She used similar treatment for this tumor.

Harriet understood medicine; for years she had worked as a recruiter for a major drug company interviewing and hiring scientists and physicians. The company was a reputable and honest business and developed drugs that were used all over the world.

So with each recurrence of cancer, she researched her situation and the available drugs being used for treatment. With each Google search, she read about the experimental treatments that were available for end-stage breast cancer, keeping them in the back of her mind.

Now, as her doctor told her that the cancer had spread to her bones, Harriet was frightened. Dr. Hill understood her fear. She had been working with cancer patients for years and had recently been using some of the newer treatments available for late-stage cancers.

Dr. Yorgos Nikas/Photo Researchers, Inc.

She told Harriet that she could enter an FDA phase 2 clinical trial for a new, genetically engineered drug that would target her specific type of tumor. Harriet knew that phase 2 trials are designed to test for side effects and risks of the treatment. She also knew there was no guarantee that the treatment would cure her cancer.

Some questions come to mind when reading about Harriet's dilemma and the decision she faced. Before we can address these questions, let's look at the biology behind cancer.

© Deco/Alamy

MRIs Are Used to Diagnose Cancer

12.1 What Is Cancer?

Cancer is a complex disease that affects many different cells and tissues in the body. It is characterized by uncontrolled cell division that leads to tumor formation and by the ability of these cells to spread, or *metastasize*, to other sites in the body. These cells are generally referred to as **malignant**. This type of unchecked growth may result in death, making cancer a devastating and feared disease. Figure 12.1a (page 242) shows some of the differences between normal and malignant cells.

Not all tumors are cancerous. Noncancerous growths, called **benign** tumors, increase in size but do not spread (metastasize) to other tissues. Some of these tumors, including cysts, usually cause problems only when they become large enough to interfere with the function of neighboring organs.

Three things identify tumors as cancerous:

1. The tumors begin within a single cell that reproduces by *mitosis*.
2. The cells in these tumors divide continuously.
3. Some of these cells become invasive and move to other sites in the body, a process called **metastasis** (see Figure 12.1b).

Improvements in medical care have reduced deaths from infectious disease and have led to increases in life span, but as people live longer, cancer becomes a major

cancer condition where cells in the body multiply uncontrollably and spread to new sites

malignant cancerous

benign not cancerous

metastasis migration of cancer cells from the primary tumor to other sites in the body

Figure 12.1 (a) Difference Between Normal Cells and Cancer Cells (b) Stages in Metastasis of Cancer Cells

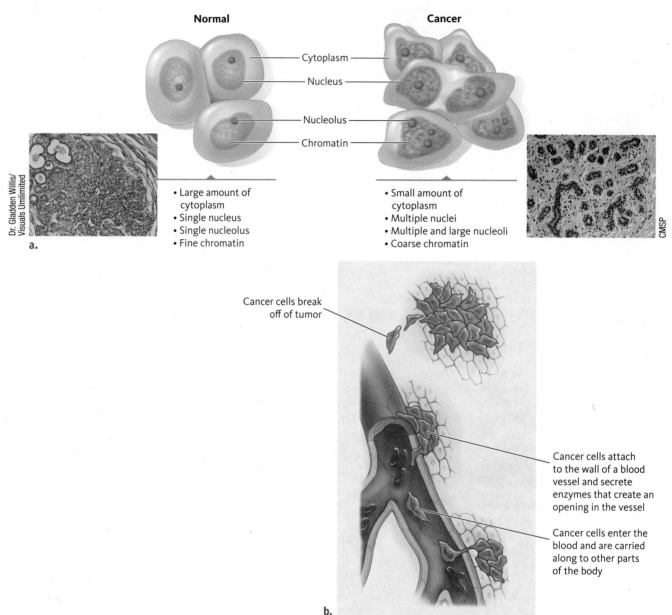

Normal

Cancer

Cytoplasm

Nucleus

Nucleolus

Chromatin

Dr. Gladden Willis/ Visuals Unlimited

a.

- Large amount of cytoplasm
- Single nucleus
- Single nucleolus
- Fine chromatin

- Small amount of cytoplasm
- Multiple nuclei
- Multiple and large nucleoli
- Coarse chromatin

CMSP

Cancer cells break off of tumor

Cancer cells attach to the wall of a blood vessel and secrete enzymes that create an opening in the vessel

Cancer cells enter the blood and are carried along to other parts of the body

b.

cause of illness and death. This is because risks for many types of cancer increase with age, and because more Americans are living longer, they are at increased risk of developing cancer.

Theodore Boveri discovered the link between cancer and genetics early in the 20th century. He proposed that normal cells become malignant because of changes in their chromosomes. Studying the pedigrees of families with cancer reveals that, in some cases, the disease has a genetic component. The pedigree in Figure 12.2 shows a pattern of breast cancer in a family, although clear-cut patterns of inheritance are often difficult to identify.

Mutations can cause a cell to become cancerous and to divide uncontrollably. Cancer, then, is a genetic disorder that acts at the cellular level.

Some factors in our environment, called **carcinogens** (such as ultraviolet light, chemicals, and viruses), and certain behaviors (such as eating an unhealthful diet and smoking) can increase the rate of these mutations, thereby playing a part in cancer risks. These risks are usually expressed as a frequency (such as one in a million).

carcinogens environmental agents associated with cancer

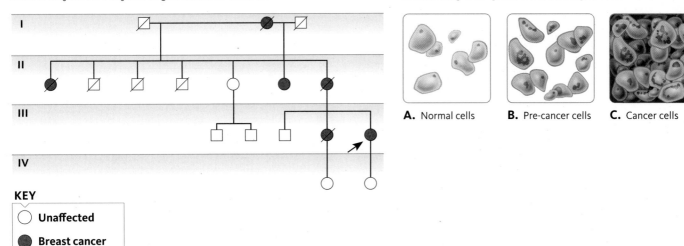

Figure 12.2 In this pedigree, women seem to be at risk for breast cancer. Why doesn't anyone in generation IV have it?

Figure 12.3 HPV can cause changes in cervical cells, accelerating their growth and causing cancer.

A. Normal cells   **B.** Pre-cancer cells   **C.** Cancer cells

KEY

○ Unaffected

● Breast cancer

This frequency describes the possibility that someone might develop cancer after exposure to a toxic substance or an increase in age. Although the numbers do not show an individual's chances of developing cancer, they can be used to educate people as to what may be dangerous to their health.

12.1 Essentials

- **Cancer is defined as uncontrolled cell division.**
- **Cancer cells are able to spread or metastasize to other parts of the body and form new malignant tumors.**

12.2 How Are Genes Involved with Cancer?

Mutations in two general classes of genes can cause normal cells to become cancerous. These classes are **oncogenes** and **tumor suppressor genes**. Normal alleles of these two types of genes work in opposite ways to control cell division and, therefore, cancer.

Normal cells carry genes that turn on cell division, usually in response to signals from outside the cell. These genes are called **proto-oncogenes**. When these genes mutate and cause cells to become cancerous, they are called *oncogenes*. Activation of oncogenes causes cells to begin uncontrollable division, transforming them into cancer cells.

Normal alleles of *tumor suppressor genes* turn off cell division and keep cells from forming tumors. When they are mutated, control of cell division is lost, the cells begin to divide continuously, and cancerous tumors form.

How does a normal proto-oncogene become an oncogene?
As discussed earlier (see Chapter 6), mutations can be caused by errors in DNA replication,

exposure to environmental agents, or lifestyle choices. When a proto-oncogene is exposed to these agents, it may mutate and become an oncogene, causing a normal cell to become cancerous.

Certain viruses can also play a part in precipitating cancer-causing mutations. After such a virus enters a normal cell, it triggers a series of events that transform the cell into a cancerous cell. An example is infection by the **human papillomavirus (HPV)**, a virus associated with some forms of skin cancer, oral cancer, and cervical cancer. In these cancers, interaction among the proteins made by HPV and those made by the cell cause uncontrolled divisions. Infection of cervical cells can result in cancer of the cervix (Figure 12.3). Each year, almost 12,000 women will be diagnosed with this form of cancer.

Recently a vaccine has been developed to immunize young women against some types of HPV. This vaccine is marketed under the name Gardasil; it protects women by activating their immune systems to fight off HPV if they are exposed.

Mistakes made during DNA replication can also cause mutations (see Chapter 6), and some of these may lead to cancer. Earlier chapters described how mutation at a single base

oncogenes a gene that when mutated causes a cell to divide uncontrollably

tumor suppressor genes genes that normally function to suppress cell division but when mutated cause cells to divide uncontrollably

proto-oncogene a normal gene that works to initiate or maintain cell division; mutant alleles, called oncogenes, are associated with cancer formation

human papillomavirus (HPV) virus that is one cause of cervical cancer

Figure 12.4 **The Cell Cycle** During the cell cycle, DNA is duplicated and passed on to the two daughter cells.

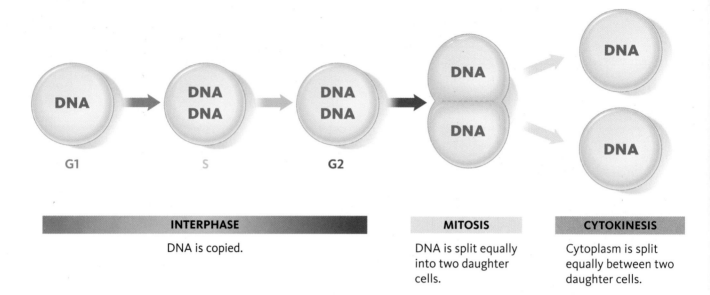

INTERPHASE	MITOSIS	CYTOKINESIS
DNA is copied.	DNA is split equally into two daughter cells.	Cytoplasm is split equally between two daughter cells.

pair can cause a serious condition such as sickle-cell anemia. In cancer, the mutation may be at one or many sites in a gene. A cancer-causing mutation can occur in one or both copies of a gene.

What is the cancer genome? With the discovery of oncogenes, proto-oncogenes, and tumor suppressor genes, along with faster methods of DNA sequencing, scientists have been examining the genomes of certain cancers. The results indicate that some cancers contain many different mutant genes. One study found over 90 mutations in cells of one breast cancer tumor. In addition, different cancers contain different sets of mutations. For example, colon cancers have sets of mutant genes that are very different from those carried by breast cancers.

A catalog of cancer mutations, The Cancer Genome Atlas (TCGA), is being compiled by the National Cancer Institute and the National Human Genome Institute. This database will allow scientists to identify and study the many different mutations that contribute to cancer. Looking ahead, genomic research may allow us to provide personalized therapy tailored to the set of mutations carried by a specific cancer genome.

12.2 Essentials

- **Cancer appears to run in some families.**
- **Proto-oncogenes and tumor suppressor genes control cell division.**
- **When these genes are mutated, cancerous tumors can form.**
- **Understanding the cancer genome will make treatment more effective.**

12.3 What Is the Cell Cycle?

Cells in the body alternate between two basic states: division and nondivision. The time between divisions varies from minutes to months or even years. During the time between divisions, the DNA in a cell is copied, and during division, each daughter cell receives a complete copy of the DNA (Figure 12.4). The sequence of events from one division to another is called the **cell cycle** (Figure 12.5). The cell cycle consists of three parts: **interphase**, or a resting phase; **mitosis**, the separation of the chromosomes; and **cytokinesis**, division of the cytoplasm to form two cells.

The first part of the cell cycle, called *interphase*, is the time between divisions. It has three stages: **G1**, **S**, and **G2**. Cell division occurs during the other parts of the cell cycle: *mitosis* (division of the chromosomes) and *cytokinesis* (division of the cytoplasm).

cell cycle series of events within a cell that eventually result in cell division; also called the cell division cycle

interphase the stage of the cell cycle between divisions

mitosis a form of cell division

cytokinesis division of the cytoplasm in cell division

G1 the first stage of interphase

S part of the cell cycle in which DNA is replicated

G2 the last stage of interphase

Figure 12.5 **The Cell Cycle: Interphase and Mitosis** Mitosis is followed by cytokinesis. Interphase has three parts: G1, S, and G2.

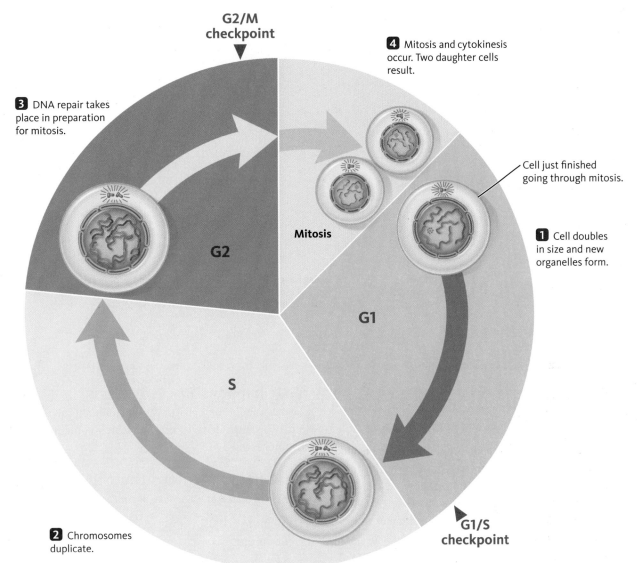

G2/M checkpoint

3 DNA repair takes place in preparation for mitosis.

4 Mitosis and cytokinesis occur. Two daughter cells result.

Cell just finished going through mitosis.

Mitosis

G2

1 Cell doubles in size and new organelles form.

G1

S

G1/S checkpoint

2 Chromosomes duplicate.

Certain checkpoints in the cell cycle maintain control of cell division, ensuring that it is regulated. Cancerous cells are not regulated at these checkpoints; the result is uncontrolled cell division and the formation of malignant tumors. Since this fact was discovered, studies of the cell cycle have become an important part of cancer research. If we can find ways to restore control of the cell cycle, we might be able to stop cancerous cells from dividing.

Follow the stages shown in Figure 12.5 as we discuss the cell cycle, beginning with a cell that has just finished division (1). After division, the two daughter cells are about half the size of the parent cell. Before they can divide again, they must undergo a period of growth. Most of this growth takes place during G1, the first of the three stages of interphase (*G1, S,* and *G2*).

1. G1 begins immediately after division. New organelles (including membranes and ribosomes) and cytoplasm are

formed at this time. By the end of G1, the cell has doubled in size.

2. During the S phase, a duplicate copy of each chromosome is made (in humans, this means that all 46 chromosomes are copied).

3. During the G2 phase, the cell prepares for mitosis by repairing any breaks or damage to its DNA. By the end of G2, the cell is ready to divide.

4. During the M phase, the cell divides by mitosis. Following mitosis, the cytoplasm divides and two cells are formed. This phase is called cytokinesis.

What are the stages of mitosis? When a human cell completes G2, it enters mitosis, the second major part of the cell cycle. During mitosis, a complete set of chromosomes is distributed to each daughter cell, and in cytokinesis, the cytoplasm is divided between the new cells. Although mitosis

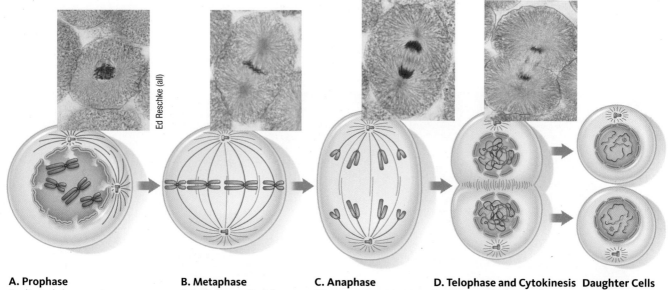

A. Prophase
Replicated chromosomes become visible.

B. Metaphase
Spindle fibers attach to chromosomes and align across cell.

C. Anaphase
Chromosomes move to opposite poles of the cell.

D. Telophase and Cytokinesis
Two new nuclei form and cytoplasmic division occurs.

Daughter Cells
Two new cells are formed.

is a continuous process, for the sake of discussion, it is divided into four stages: prophase, metaphase, anaphase, and telophase. Follow along with Figure 12.6 as we discuss these stages.

1. **Prophase**: At the beginning of prophase, the replicated chromosomes (remember, they were copied during the S stage of interphase) condense and become visible. Each of the 46 duplicated chromosomes consists of two sister chromatids joined by a centromere. Near the end of prophase, the nuclear envelope breaks down, and a network of specialized fibers called spindle fibers form in the cytoplasm and stretch across the cell.

2. **Metaphase:** Metaphase begins when the chromosomes, with spindle fibers attached, move to the middle of the cell.

3. **Anaphase**: During anaphase, the centromeres divide, converting each sister chromatid into a chromosome. After this, the newly formed chromosomes migrate toward opposite ends of the cell. At the end of anaphase, there is a complete set of 46 chromosomes at each end of the cell.

4. **Telophase:** In the final stage of mitosis, the chromosomes begin to unwind, the spindle fibers break down, cytokinesis begins, and the nuclear envelope begins to re-form.

G1/S checkpoint a regulatory point in the cell cycle

G2/M checkpoint a regulatory point in the cell cycle

signal transduction transfer of signals from outside the cell to the nucleus that result in new patterns of gene expression

What happens in cytokinesis?
In this stage, the plasma membrane constricts and gradually divides the cell into two daughter cells, each of which contains a set of 46 chromosomes.

What helps regulate the cell cycle?
The cell cycle is regulated at a number of points, including:

1. a point in G1 just before cells enter S, known as the **G1/S checkpoint**

2. the transition between G2 and mitosis, called the **G2/M checkpoint**

The tumor suppressor genes and proto-oncogenes discussed earlier control these checkpoints. Proteins made by the tumor suppressor genes can turn off or decrease the rate of cell division. Proto-oncogenes encode proteins that turn on or increase the rate of cell division. Proteins encoded by these two classes of genes normally act at either the G1/S or G2/M control points to regulate the movement through the cell cycle.

In normal cells, signals from outside the cell can activate either tumor suppressor genes to turn off cell division or proto-oncogenes to turn on cell division. This process, which regulates cell division, is accomplished through **signal transduction** (Figure 12.7).

These signals originate outside the cell and can be proteins, hormones, or nerve impulses. Signals can also come from the environment outside the body and may include

Figure 12.7 **Signal Transduction** In this process, signals from outside the cell are received and processed by receptors in the plasma membrane. The activated receptors cause a chain of events in cytoplasmic proteins that produce changes in gene expression.

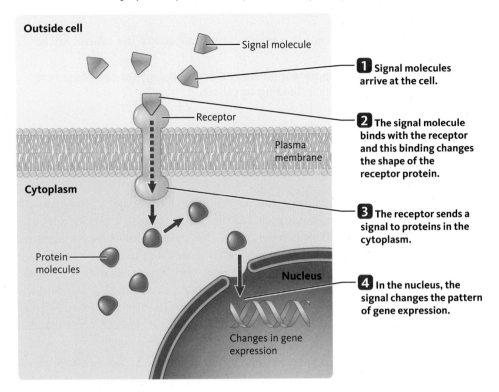

Outside cell

Signal molecule

1 Signal molecules arrive at the cell.

Receptor

Plasma membrane

2 The signal molecule binds with the receptor and this binding changes the shape of the receptor protein.

Cytoplasm

3 The receptor sends a signal to proteins in the cytoplasm.

Protein molecules

Nucleus

4 In the nucleus, the signal changes the pattern of gene expression.

Changes in gene expression

steroids, pollutants, and other molecules. It is easy to see that any change in these signals might cause the cells to change their pattern of division.

Signal transduction begins when a signal molecule binds to a receptor in the plasma membrane. This binding sets off a series of interactions inside the cell, although the signal molecule may remain outside the cell. The binding of the signal changes the shape of the receptor and allows it to transmit a signal to other proteins in the cytoplasm. These proteins in turn alter other proteins and a chain reaction occurs, generating a response in the nucleus, leading to an increase or decrease in cell division.

Generally, this process occurs when one protein is changed, or *transduced*, from one form to another and affects another protein. For example, a signal may bind to and change the shape of a receptor; this may, in turn, cause a protein within the cell to become chemically modified, allowing it to interact with and activate a third protein, causing a domino effect within the cell.

How is cell cycle regulation related to cancer? Cancer is related to a loss of cell cycle control. This often is caused by mutations in the genes themselves or changes in the mechanisms of cell regulation or in the cell cycle control machinery.

Let's look at an example of how a mutation in a proto-oncogene called *RAS* might lead to cancer. When functioning normally, the *RAS* proto-oncogene produces the RAS protein, which attaches to the inside of the plasma membrane. When activated by a signal, the RAS protein changes shape and becomes switched on, transmitting a signal to another protein in the pathway. Once the signal has been transmitted, the RAS protein changes shape again, switches off, and becomes inactive.

Mutations in the *RAS* gene can cause the protein to become stuck in the "on" position, constantly transmitting its signal, calling for cell division. This causes the cell to lose control of the cell cycle and to divide in an uncontrolled fashion, becoming

RAS a proto-oncogene involved in signal transduction

cancerous. *RAS* gene mutations are found in 20 to 30% of all cancers, including colon cancer, pancreatic cancer, and stomach cancer.

12.3 Essentials

- **The cell cycle is important in understanding cancer and the control of cell division.**
- **Mutations in tumor suppressor genes and proto-oncogenes can cause loss of cell division regulation, leading to cancer.**

12.4 How Does a Woman Develop Breast Cancer?

In the United States, breast cancer is the most common form of cancer in women. An estimated 178,000 new cases are expected in the United States in any given year, and each year, more than 40,000 women die from breast cancer. Although environmental factors may be involved, geneticists struggled for years with the question of whether there is a genetic link to breast cancer. After more than 20 years of work, the answer is clearly yes. Mutations in two different genes, *BRCA1* and *BRCA2*, can predispose women to breast cancer and ovarian cancer. These are classified as high-risk cancer genes, because over 80% of women carrying a mutant allele of either of these genes will develop breast cancer.

In addition to these two genes, scientists studying the cancer genome have found that mutations in two other types of genes, called moderate- and low-risk genes, are also involved in breast cancer. Mutations in these genes play a part in triggering cancer in those who do not carry high-risk genes and may also play a role in cancer development in those who carry mutations of *BRCA1* and *BRCA2*. This may help explain why women who carry normal alleles of *BRCA1* and *BRCA2* still develop breast cancer.

The story of how breast cancer was linked to specific genes began in the 1970s, when Mary-Claire King (Figure 12.8) and her colleagues analyzed the pedigrees of 1500 families with breast cancer. They found that approximately 15% of the families had multiple cases of breast cancer. In the 1980s, Dr. King decided to test the blood of as many families as possible, trying to locate a genetic marker for breast cancer. Finding such a marker was a long shot because most cases of breast cancer, like other cancers, occur at random and would not carry the same genetic marker.

Finally, in 1990, using PCR and other recombinant DNA techniques, her team found a link between breast cancer and a genetic marker on chromosome 17 (Figure 12.9). As discussed in Chapter 9, a *genetic marker* is a small segment of DNA on a chromosome that can be used to find the location of a gene. Many of the original families in Dr. King's study were tested for this marker, and a correlation was found between the marker and breast cancer. This meant that a gene for a predisposition to breast cancer was located somewhere on chromosome 17, near the genetic marker discovered by Dr. King. Soon thereafter, scientists formed a large working group and identified a gene now called *BRCA1* that is associated with this predisposition.

All women who carry the *BRCA1* mutation inherit an increased risk of developing breast cancer, but not all women who carry the mutant allele will actually get breast cancer. If a woman carries one mutant copy of the *BRCA1* gene, she will develop breast cancer if a mutation occurs in the normal allele in a breast cell. Approximately 82% of women who inherit one mutant *BRCA1* allele will develop a mutation in the other *BRCA1* allele and get breast cancer. Women with a *BRCA1* mutation are also at higher risk (a 44% chance) for ovarian cancer.

Figure 12.8 Cancer Researcher Mary-Claire King

Newscom

Figure 12.9 Chromosome 17 Showing the Location of the *BRCA1* Gene

— BRCA1

Chromosome 17

BRCA1 the first gene discovered that can, if mutated, increase a woman's risk of breast cancer

BRCA2 the second gene discovered that can, if mutated, increase a woman's risk of breast cancer

Figure 12.10

Chromosome 13 Showing the Location of the *BRCA2* Gene

BRCA2

Chromosome 13

A second breast cancer predisposition gene, *BRCA2*, was discovered in 1995. As with *BRCA1*, mutations in *BRCA2* increase the risk of developing breast cancer. *BRCA2* is found on chromosome 13 (Figure 12.10).

Even though these mutations are rare (1 in 500 in the general population), in some populations the frequency of women carrying a *BRCA* mutation is much higher. One such population is women of Eastern European Jewish ancestry (**Ashkenazi Jews**); their combined frequency of carrying *BRCA1* and *BRCA2* is 1 in 40.

Many questions about the protein products and the functions of *BRCA1* and *BRCA2* remain unanswered, as do questions about how mutations in these genes lead to breast cancer. In the years since the discovery of these genes, they have been shown to play a part in transcription (see Chapter 5) and other cellular functions, including DNA repair. The details of exactly how mutations in these genes start the cell down the path to cancer are still unclear.

Can men get breast cancer?

Although most people think of breast cancer as a women's disease, men have breasts and can also develop breast cancer. Men can inherit the mutant *BRCA1* or *BRCA2* alleles from their parents. Because they have small amounts of breast tissue, breast cancer isn't usually thought of as a disease in males. As a result, men are frequently diagnosed at later stages than women because they mistakenly believe that they cannot get breast cancer. Because such cases are detected later, they are often more difficult to treat.

In the United States, about 1% of breast cancer cases occur in males, with an estimated 2000 new cases reported each year. In parts of Africa, the rates are significantly higher. In Egypt, males account for 6% of all cases, and in Zambia, male breast cancer represents 15% of all cases. Risk factors for male breast cancer include age, family history of breast cancer (usually in female family members),

and occupational exposure to heat, gasoline, or estrogen-containing creams in the soap and perfume industry. Males of Eastern European Jewish ancestry, also called Ashkenazi Jews, and black males have higher rates of breast cancer than other populations. Researchers are working to discover the reasons for increased rates of breast cancer in certain populations.

12.4 Essentials

- **Researchers have discovered two mutant genes, *BRCA1* and *BRCA2*, that increase the risk of developing breast cancer.**
- **Men, as well as women, can develop breast cancer.**

12.5 How Is Breast Cancer Detected and Treated?

In breast cancer, and other forms of cancer, physicians see a tremendous benefit to early detection. If cancer is found early enough, it can often be treated successfully. For this reason, women are encouraged to examine their own breasts and then have mammograms beginning at age 40. Mammograms are digital or radial x-rays that can detect thickening or tumors within the breast tissue (Figure 12.11).

Ashkenazi Jews an ethnic and religious group of Eastern European ancestry studied for the breast cancer gene

Figure 12.11 **Mammogram** (a) A normal breast; (b) a breast containing a tumor.

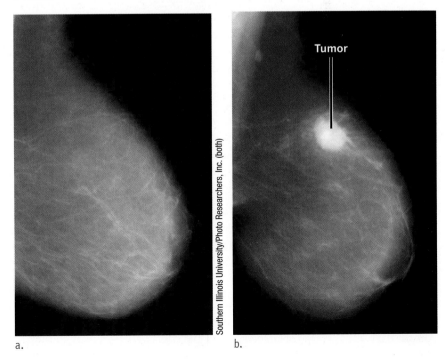

Southern Illinois University/Photo Researchers, Inc. (both)

Tumor

a.

b.

If a growth or irregularity is detected, a biopsy is taken. During a biopsy, tissue is removed from the area and examined by a pathologist to see if cancer cells are present.

If the diagnosis is positive for cancer cells, the physician formulates a method of treatment and discusses it with the patient, just as Harriet's physician did. In Harriet's case, she had already tried some of the treatments discussed here before she looked into whether or not to become part of a clinical trial.

If breast cancer is confined to one area of the breast, the lump can be removed and the area treated with radiation to keep the cancer from recurring. This surgical procedure is called a lumpectomy. If the cancer has spread or is identified as a fast-growing type, a mastectomy may be performed and the entire breast removed. After either procedure, radiation, chemotherapy, or both may be recommended. Both types of treatment are intended to kill cancer cells and, therefore, reduce or eliminate the chance of recurrence.

With radiation treatment, a patient is exposed to concentrated radiation at the specific location where the cancer was removed, killing cells only at that site. With chemotherapy, chemical agents known to be toxic to cancer cells are given intravenously over a prolonged period of time, affecting the cancer cells as well as normal cells in many parts of the body. Because both radiation and chemotherapy will affect healthy cells as well as the cancer cells, there are side effects that may be unpleasant. One of these symptoms is hair loss.

A drug widely used to treat women who have breast cancer or are at risk for breast cancer is tamoxifen. As mentioned in "Biology Basics 3: Genes, Populations, and the Environment," this drug affects individuals in different ways, depending on the genetic makeup of their tumors. Tamoxifen was discovered in 1962 when scientists were examining thousands of plants to see if any of them had medicinal qualities to use with cancer patients. The bark of the Pacific yew (*Taxus brevifolia*) (Figure 12.12) was found to have antiestrogenic qualities. Because many breast cancer tumors respond to estrogen by growing, any drug that blocks estrogen action would shrink the tumor. The yew tree is a very slow-growing plant, and harvesting the tree to extract tamoxifen endangered its survival. Today, these trees, grown on farms, provide all the tamoxifen needed to produce the necessary quantities of the most-used cancer drug, and synthetic versions are also available.

In addition to treating women with breast cancer, tamoxifen is also given to women at high risk of getting cancer (those who test positive for mutations in the *BRCA* genes or

Figure 12.12 The anti-cancer drug tamoxifen is extracted from the bark of the yew tree.

Tom & Pat Leeson/Photo Researchers, Inc.

those with a strong family history). Studies have shown that taking this drug *prior* to the development of any cancer can significantly lower a patient's risk.

Another approach to treating cancer involves the use of adult stem cells, like those found in bone marrow. Bone marrow transplants are a fairly common treatment for chronic myelogenous leukemia (CML), a form of cancer, and are very successful. The stem cells from healthy bone marrow are transferred into the patient so they will produce new, healthy white blood cells.

How are cancer treatments tested? When new treatments for cancer are developed, scientists must perform a specific series of tests to find out if they are effective, cause side effects, or are dangerous to patients. The rules for these tests are written and monitored by the U.S. Food and Drug Administration (FDA). The first step after development is to test the drugs on animals. Mice are commonly used, but larger mammals are also tested.

Since World War II, it has been considered unethical to test treatments directly on people. Physicians in Germany were prosecuted for the experiments they did on prisoners during the war, and from this we learned an important lesson: people must consent to any testing done on them—in other words, they must be true volunteers. These tests are called **clinical trials** and are used to test the effectiveness and side

clinical trial use of human volunteers to test the effectiveness and potential side effects of drugs and treatments

Table 12.1 **Clinical Trials for New Therapies**

FDA Trial	Why Is the Trial Done?	Number of People Tested
Phase 1 trial	To find out if the drug or treatment is toxic (poisonous) in any way	Less than 100
Phase 2 trial	To find out how effective the drug is, whether its use has risks and side effects	Hundreds of participants
Phase 3 trial	To find conclusive evidence of the effectiveness of the drug or treatment and its safety	Thousands of participants

effects of a treatment. In Case A, Harriet was given the opportunity to participate in a clinical trial.

When developing a new drug or treatment, companies must follow the guidelines set forth by the FDA. These guidelines, as seen in Table 12.1, have certain phases and come after the FDA approves the animal studies.

Would you volunteer to be in a clinical trial for a treatment that has not been proven to help you, and might actually harm you? This is the question people with cancer, like Harriet, face.

If we put ourselves in the position of patients like Harriet, we can see some problems. Patients with cancer want to get well, and if modern medicine cannot help them, they may be willing to try anything. Some experts say that this makes it difficult for them to freely consent to participation.

The internet has made it easier for both patients and physicians to find where clinical trials are being conducted, and what diseases are being treated. The National Institutes of Health maintains a comprehensive Web site for clinical trials across the United States. In Section 12.7, we will discuss some of the legal aspects of consent as it applies to clinical trials.

12.5 Essentials
- **Treatments for cancer include surgery, radiation, and chemotherapy.**
- **New drugs are being developed all the time.**
- **Drugs for cancer and other diseases are tested using clinical trials.**

12.6 What Are the Other Genetic Causes of Cancer?

Changes in the number and structure of chromosomes are often seen in cancer cells. For example, as discussed in Chapter 3, Down syndrome is caused by the presence of an extra copy of chromosome 21. In addition to most of the symptoms discussed in Chapter 3, people with Down syndrome are also 18 to 20 times more likely to develop leukemia (a cancer of the white blood cells) than those in the general population. This was a clue to scientists that they might look on chromosome 21 for a gene associated with leukemia. That connection has not yet been discovered.

The relationship between chromosome rearrangements, or *translocations*, and cancer was first discovered in leukemia. Leukemia involves uncontrolled division of white blood cells. Specific chromosome changes in these cells are easy to identify and are used to diagnose this type of cancer.

In one form of leukemia, **chronic myelogenous leukemia (CML)**, a translocation occurs between parts of chromosomes 9 and 22, and they exchange parts. This unusual chromosome rearrangement is called the **Philadelphia chromosome** (see Figure 12.13), after the city in which it was discovered. Other cancers including **acute myeloblastic leukemia**, **Burkitt's lymphoma**, and **multiple myeloma** are also associated with translocations. The finding that certain forms of cancer are consistently associated with specific types of chromosomal abnormalities is significant in the study of these cancers. This discovery can lead to detection, treatment, and cures.

Figure 12.14 shows a number of genes associated with cancer and the chromosomes on which they are located.

12.6 Essentials
- **Chromosomal abnormalities can also cause cancer.**

12.7 What Is the Relationship between the Environment and Cancer?

Epidemiologists, who study the frequency of disease in different populations, have found that certain diseases occur at much higher frequencies in certain geographic areas. A large number of cancer cases found in a restricted geographic area is called a **cancer cluster**. Epidemiologists examine the environment surrounding the cancer cluster to look for a cause-and-effect link. The 1986 book *A Civil Action* by Jonathan

CASE A: QUESTIONS

Now that we understand that some cases of breast cancer are caused by a genetic predisposition, let's look at some of the issues raised in Harriet's case.

1. What should Harriet do? Should she be tested to see if she carries mutant alleles of *BRCA1* and *BRCA2*?

2. If Dr. Hill told Harriet that no other treatment could help her, might that make a difference in her decision to participate in the trial? Why or why not?

3. Many trials need end-stage patients to see how a drug might affect individuals who are very sick or dying. Can these patients really understand what the trials are about and give consent?

4. What should Dr. Hill tell Harriet about the proposed treatment before she agrees? List five items.

5. Is there anything that a researcher or physician shouldn't tell a patient?

6. One serious problem in clinical trials is that researchers often cannot persuade people to participate. Why do you think that is?

Figure 12.13 **The Philadelphia Chromosome** This unusual chromosome arrangement is associated with chronic myelogenous leukemia. The chromosomes involved in this translocation are circled in red.

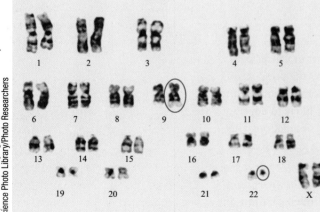

Dept. of Clinical Cytogenetics, Addenbrookes Hospital/Science Photo Library/Photo Researchers

chronic myelogenous leukemia (CML) a form of leukemia

Philadelphia chromosome a translocated chromosome created from parts of chromosome 9 and 22 that is associated with CML

acute myeloblastic leukemia a condition of the blood where white blood cells proliferate, also called blood cancer

Burkitt's lymphoma a cancer of the lymphatic system thought to be associated with a chromosomal translocation

multiple myeloma cancer of the bone marrow cells thought to be associated with a chromosomal translocation

cancer cluster population or group where cancer is widespread

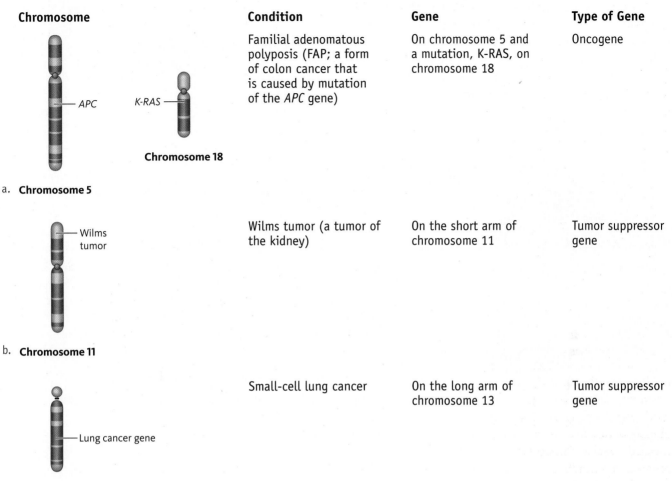

Figure 12.14 **Cancer Genes on Other Chromosomes** (a) chromosome 5 and 18, (b) chromosome 11, (c) chromosome 13.

Chromosome	Condition	Gene	Type of Gene
a. Chromosome 5 (APC) / **Chromosome 18** (K-RAS)	Familial adenomatous polyposis (FAP; a form of colon cancer that is caused by mutation of the *APC* gene)	On chromosome 5 and a mutation, K-RAS, on chromosome 18	Oncogene
b. Chromosome 11 (Wilms tumor)	Wilms tumor (a tumor of the kidney)	On the short arm of chromosome 11	Tumor suppressor gene
c. Chromosome 13 (Lung cancer gene)	Small-cell lung cancer	On the long arm of chromosome 13	Tumor suppressor gene

Harr described a cancer cluster in Woburn, Massachusetts. The environmental trigger that caused the cancer turned out to be industrial solvents that entered the town's water supply.

Another way epidemiologists study the relationship between disease and environmental factors is to examine populations to determine which types of cancer they develop. Because many forms of cancer in the United States are related to our physical surroundings, personal behavior, or both, it is estimated that at least 50% of all cancer can be attributed to some type of environmental factor.

What environmental factors are associated with cancer? Smoking is the number one environmental factor implicated in cancer—it can affect the smoker directly or others indirectly by exposure to secondhand smoke. Smoking is related to cancers of the oral cavity, larynx, esophagus, and lungs and accounts for 30% of all cancer deaths.

As shown in Figure 12.15 (page 254), smokers can take up to twenty years to develop cancers. Most of these cancers have very low survival rates; lung cancer (Figure 12.16, page 254), for example, has a five-year survival rate of 13%, meaning that on average, only 13% of those diagnosed with lung cancer will live for five years. Cancer risks associated with tobacco are not limited to smoking; people who use snuff or chewing tobacco are 50 times as likely to develop cancer of the mouth compared to non-users.

An estimated 1 million new cases of skin cancer are reported in the United States every year; almost all cases are related to ultraviolet (UV) light exposure from the sun or tanning lamps. The most serious form of skin cancer is called melanoma. A melanoma on the skin is shown in Figure 12.17 (page 254). Skin cancer cases are increasing rapidly in most populations, presumably as a result of an increase in outdoor recreation and people moving to regions of the United States with more sun exposure. Epidemiological surveys show that

1 in 500
Number of women who will inherit the mutant breast cancer gene

Nick Hawkes/Alamy Limited

82%
Percentage of women with the mutant *BRCA1* allele who will develop breast cancer

Figure 12.15 Lung Cancer
This graph shows how the number of cigarettes smoked by men is related to a steep increase in lung cancer deaths 20 years later.

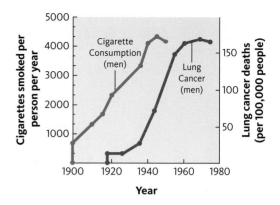

lightly pigmented people are at much higher risk for skin cancer than heavily pigmented individuals. This supports the idea that genetic characteristics can affect the susceptibility of individuals or populations to environmental agents.

Ozone depletion of the atmosphere in certain geographic regions also contributes to increased levels of UV radiation exposure, which in turn is associated with increases in skin cancer frequency. This is because the ozone layer around the planet blocks a great deal of UV radiation. More than 80% of lifetime skin damage occurs by age 18. In spite of this, many Americans think suntans are healthy looking, and only 25% consistently use sunscreen lotions or oils.

12.7 Essentials
- **Smoking is only one lifestyle factor that may affect cancer risks.**
- **Environmental agents often cause normal cells to become cancerous.**

12.8 What Are the Legal and Ethical Issues Associated with Cancer?

Certain kinds of lawsuits, called **tort** cases, address the issue of whether someone has been legally wronged. These cases affect the individuals involved, but they are actually intended to change society. Some of the most famous of these lawsuits have involved tobacco companies. In these suits, people with cancer, and sometimes the families of those who have died of cancer, sued tobacco companies for negligence in causing the cancer.

The primary evidence presented in these cases was that the tobacco companies knew, when they sold the cigarettes, that tobacco is harmful and possibly fatal, and that it is addictive.

Figure 12.16 A Lung of a Smoker Who Died of Lung Cancer

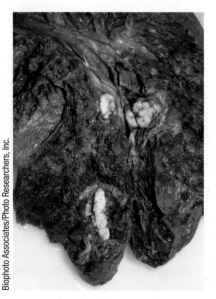

Biophoto Associates/Photo Researchers, Inc.

Figure 12.17 Melanoma Is a Deadly Form of Skin Cancer

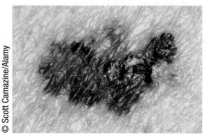

© Scott Camazine/Alamy

At first, these lawsuits were not successful, but later, memos found in the files of the companies showed that they did know that tobacco is both harmful and addictive, and that they did not warn smokers. Today, because of these lawsuits, warnings appear on all packages of tobacco products, and that may stop many people from smoking.

An interesting ethical problem arises with smokers as well. One of the ethical principles that most societies hold dear is

CASE B:
Government Experiments on Children

Even though it happened in 1969, Robin Matchison still had nightmares about it. Early one morning, her mother told her to get up and say good-bye to her brother—he was going away for a long time. Robin was only 10 years old, but she remembered it clearly. A big car was parked in the driveway, and her 5-year-old brother, Jack, was being carried into the car.

"Where is he going?" she asked her mother. "To a place where they can take care of him," her mother said. "But we can take care of him, can't we, Mommy?" Robin asked.

Six years later, Robin's mother told her that Jack had been taken to a home for children with mental retardation in Massachusetts. They went to visit Jack, who at that time was quite ill. The doctors told her mother that Jack had leukemia, which is often associated with Down syndrome (the cause of Jack's retardation).

Recently, Robin read an article in the newspaper about experiments conducted by the U.S. government laboratories. Some of these involved exposing people to radiation when they were not aware of it. One of the places it happened was the home in Massachusetts where Jack had lived until he died of leukemia in 1977. Robin didn't know much about leukemia, but she went to the library and learned that one of the causes of leukemia is exposure to radiation, especially in children. She wondered if this was what caused her brother to die.

It seemed awfully unfair that the U.S. government had used children like her brother as guinea pigs. A friend at work told Robin she should see a lawyer about it.

A school for developmentally disabled children.

1. **Should Robin see a lawyer? Why or why not?**

2. **Who might the lawyer sue as part of this lawsuit?**

3. **How does this case fit into today's informed consent rules (read Section 12.8)?**

4. **Give three arguments why experimentation on people with mental retardation would be wrong.**

5. **Give three arguments why such experimentation might be considered all right.**

6. **What problems do you think may arise because this occurred so many years ago?**

7. **What might the government have done in 1969 to make this type of experimentation more acceptable?**

8. **It is very difficult to get people to volunteer for clinical trials. Does that excuse this experiment? Why or why not?**

autonomy. This means that we have the personal right to make decisions for ourselves, no matter what the outcome. In most states, laws have been passed saying that people cannot make legal or health decisions until they are 18 years of age, or in some cases, until they are 21. Should there be restrictions on a person's choice to smoke? When laws were passed to outlaw selling cigarettes to minors, this restricted their access to tobacco products. Later, recognizing the dangers of secondhand smoke, lawsuits were filed to stop smoking inside buildings, and today most states have laws prohibiting smoking inside buildings. Table 12.2 addresses some of the ethical and legal questions related to smoking.

What is the major legal question in clinical trials?
Laws require that physicians and hospitals conducting clinical trials inform participants of the following:

a. information about the procedure or research

b. alternatives to the trial

c. risks of participating in the trial

d. patient's rights as far as the trial is concerned

In addition, an indication that the patient understands and consents must be included (signature).

This may not be as simple as it sounds. Study participants need to understand what is happening to them, but the science can be confusing. If there are significant risks, how can the doctors convey them and not frighten patients? Finally, how can you know if someone really understands?

Often participants in trials are paid for their time, which can muddy the issue. Are the participants doing this for the right reason? Did the payment convince them to go against

tort type of lawsuit based on an act that injures someone

autonomy a legal and ethical term meaning the right to make decisions about yourself

Table 12.2 Legal and Ethical Questions in Smoking

Question	How Are These Questions Decided?	Related Case or Legal Issue
If a person wants to smoke, can a law say he or she can't?	Most states have a minimum legal age for purchasing cigarettes.	For people who are old enough to decide for themselves, autonomy is upheld.
Can laws control where a person can smoke?	Most states have laws about smoking in either public buildings or all indoor space, with the exception of private homes.	These laws protect not only the smoker, but nonsmokers inhaling secondhand smoke. Employees of bars who were exposed to a great deal of secondhand smoke first introduced this legal issue by suing their employers when they became ill.
Could laws be passed that would control other environmental causes of cancer?	Both the federal government and state legislatures have regulated the control of dangerous chemicals such as pesticides and pollution.	Even with these laws in place, lawsuits have been brought to demand payment for treatment of patients who became ill from these exposures.

what they know is dangerous? It is very difficult to know someone's true motives. In addition, volunteers are often scarce. Studies advertise for participants (see Case A photo), but they still get few volunteers. The information provided to a prospective participant about all the possibilities may be frightening, causing volunteers to back out. What is a researcher to do?

The basic principles of clinical trials are based on three essential human rights outlined in the Belmont report, compiled after World War II. These three rights are respect for persons, autonomy, and justice. Table 12.3 looks at some of the questions raised by clinical trials and how they relate to the Belmont report.

Because of the large number of clinical trials being conducted in the United States, the FDA cannot closely supervise all of them. To ensure that all ethical guidelines are followed, local hospitals and drug companies have **Institutional**

Review Boards (IRB) that oversee each clinical trial. These boards are made up of physicians, attorneys, patients, clergy, and volunteers from the community. They discuss problems and report regularly to the FDA, which resolves problems and refines the guidelines accordingly.

12.8 Essentials

- **Successful lawsuits have been brought against cigarette companies.**
- **Clinical trials have their own ethical and legal problems.**

Institutional Review Boards (IRB) groups of physicians, patients, and clergy that oversee clinical trials

Table 12.3 Ethical and Legal Questions about Informed Consent

Question	How Are These Questions Decided?	Ethical Issue
What if a participant does not understand the terms of the clinical trial because of a language problem?	The FDA demands that the informed consent document be translated.	Respect for persons
Is payment for these services a type of coercion?	This is up to the FDA and the local review board to decide.	Justice
What if you need children for testing?	Parents decide what is best for their children.	Autonomy
How are you sure someone really understands what is going to happen to them?	A few ways to clear this up are to have a friend or relative accompany the person, train a colleague to explain the science simply, or show a video.	Respect and Autonomy
What if the participant is terminally ill?	This is determined on a case-by-case basis.	Respect for persons

Spotlight on Ethics

Henrietta Lacks and Her Immortal Cells

A cell line is made from a single cell or a group of cells that is kept alive *in vitro* (in the lab) and used for scientific study. The *HeLa* cell line is a famous cell line used in cancer research. Most cell lines divide only a limited number of times and then die, but the HeLa cell line is an exception. It was derived from cervical cancer cells taken from Henrietta Lacks (Figure 12.18), who died of cancer in 1951. The cells were removed during a biopsy taken from her cervix as part of her cancer diagnosis.

Figure 12.18
Henrietta Lacks, whose cancerous cervical cells were used to create a cell line.

New York Public Library/Photo Researchers, Inc.

HeLa cells are termed *immortal* because they can divide an unlimited number of times in a laboratory as long as they are given the right nutrients. Medical students and scientists all over the world use the HeLa cell line to study these cervical cancer cells. These cells are shown in Figure 12.19.

The cells in this cell line were originally grown by George Otto Gey without Lacks's knowledge or permission and later sold to medical schools. This gave Dr. Gey a nice income even though he had never patented the cell line.

Recall from Chapter 10 the case of *Moore v. Regents of the University of California,* in which the court said that the doctors did not have to inform Moore of their use of his cells after they were removed. The same was true for the HeLa cell line and Henrietta Lacks.

But even if no law forces physicians or scientists to inform patients of the use of parts removed, are they somehow ethically obligated to do so? In the transcript of the Moore case,

Henrietta's case was discussed. It is difficult to know whether this might have influenced the California Supreme Court decision.

Recently, these cells were used to find a connection between certain strains of the human papillomavirus, discussed earlier, and human cervical cancer. A vaccine has been developed for this form of HPV and is being marketed to 9- to 14-year-olds and their parents.

QUESTIONS

1. The company that marketed the vaccine for the human papillomavirus used an interesting advertising strategy. It ran television ads that appeared to be educational. They were telling women that a virus causes cancer. Soon after these ads ran, an ad for the vaccine appeared, with girls saying they wanted to be "one less" woman with cervical cancer. Make a list of the pros and cons of advertising a product this way.

2. Next, the company began to lobby state legislatures to make vaccination of 9- to 14-year-old girls mandatory for entering school. Is this ethically right? Why or why not?

3. Make a checklist comparing and contrasting the *Moore* case with Henrietta's case.

4. After all these years, would there be any way to determine what Henrietta Lacks would have wanted to be done with her cervical cancer cells? How?

5. A best-selling book, *The Immortal Life of Henrietta Lacks* by Rebecca Skloot, chronicles this case. Read the book and summarize it in a PowerPoint presentation.

Figure 12.19 **HeLa Cells**

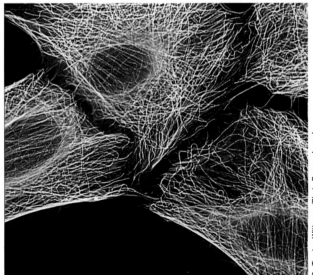

Dr. Torsten Wittmann/Photo Researchers, Inc.

The Essential 10

1 Cancer is defined as uncontrolled cell division. (Section 12.1)

Uncontrolled division can occur in any cell.

2 Cancer appears to run in some families. (Section 12.2)

Pedigrees of families often show a number of members with different forms of cancer.

3 Proto-oncogenes, oncogenes, and tumor suppressor genes control cell division. (Section 12.2)

Depending on which type of gene is mutated, uncontrolled cell division may result.

4 When these genes are mutated, cancerous tumors form. (Section 12.2)

Some of these genes turn on cell division, while others turn it off.

5 The cell cycle is important to understanding cancer and the control of cell division. (Section 12.3)

The normal cell cycle gives us clues as to where uncontrolled growth can occur.

6 Researchers have discovered two mutant genes, *BRCA1* and *BRCA2*, that increase a woman's risk of developing breast cancer. (Section 12.4)

Testing for these two genes is becoming popular and marketed directly to the public via the internet and other media.

7 Chromosomal abnormalities can also cause cancer. (Section 12.5)

An extra copy of chromosome 21 has been shown to increase the risk of certain forms of leukemia.

8 Early detection of breast cancer can increase the long-term recovery rate. (Section 12.6)

Mammograms are recommended for women after the age of 40.

9 Smoking is only one lifestyle factor that may affect cancer risks. (Section 12.7)

Warnings are now present on cigarette packages to inform consumers of this risk.

10 Successful lawsuits have been brought against cigarette companies. (Section 12.8)

Evidence shows that the cigarette companies knew that their product caused cancer.

Review Questions

1. Why is cancer so difficult to treat?

2. What is the difference between benign and cancerous tumors?

3. Not so long ago, it was thought that cancer did not have a genetic component. Now cancer's genetic causes are discussed in a human genetics book. What has changed?

4. What are the differences between tumor suppressor genes and oncogenes? List three.

5. How much should we do to keep carcinogens out of our environment? What do you do?

6. List the parts of the cell cycle.

7. What is signal transduction and how might it relate to cancer?

8. How did scientists find the markers for breast cancer?

9. What is the Philadelphia chromosome and why was it called this?

10. If we suspect which environmental agents might cause cancer, why can't we just get rid of them?

11. Why is tamoxifen used for cancer treatment and prevention?

12. List the three treatments most commonly used for cancer.

13. What is a clinical trial?

Application Questions

1. Give three reasons why people are so afraid of cancer. Apply the science that you have learned.

2. Recently, a vaccine has been developed to protect women from cervical cancer. Research this vaccine and how it is marketed.

3. It is easy to understand how older people's cells can become cancerous, but many children, even babies, get cancer. Do some research on this topic to see why it occurs, and write a short report.

4. A large number of cancer cases in a certain geographic area is called a cancer cluster. Create a chart of cancer clusters around the country and include where they are located and what type of cancer the patients had.

5. Plastics in our environment have been discussed as a possible cause for cancers. List two or three chemicals used in the manufacturing of plastics that are considered dangerous.

6. Now that you know about signal transduction, go back and read over the example of the *RAS* gene. How might you use this knowledge to prevent cancer from occurring?

7. Do some research on environmental causes of cancer. Make a list of how you might change what you do to cut your cancer risk; then evaluate each and see if you could actually change your habits.

8. Research and write a short biography of Mary-Claire King.

9. Large settlements have come from lawsuits against cigarette manufacturers. Compare cigarette smoking lawsuits to the suits overweight patrons file against fast-food restaurants. Why did cigarette lawsuits work?

10. When clinical trials of cancer patients are started, many want to work with patients who are dying. Give three reasons why these patients would be desirable for researchers and three reasons they would not be.

11. In 1951, a well-known scientist from the University of Illinois, Dr. Andrew Ivy, held a press conference to announce a breakthrough in cancer treatment. It was a drug called Krebiozen that was extracted from the blood serum of horses. Create a timeline of this drug and the result of its invention.

12. In Chapter 8, screening of all newborns was discussed. Now that there is a test for the *BRCA1* and *BRCA2* genes, should it be mandatory for all newborns? Why or why not?

13. Consider the following hypothetical case based on the HeLa cell line (Spotlight on Ethics):

Heidi Hagawa remembered her mother, Frieda, very well. Heidi was 20 when her mother was diagnosed with cancer and underwent many tests at the hospital, and Heidi stayed by her side. Heidi's parents were divorced, and Heidi was the only family member her mother had.

Heidi remembered when the physicians told them that Frieda's blood cells made special molecules called *monoclonal antibodies*. These cells might be used to fight off cancer and some viral infections in other patients. They asked for samples of her blood, which would be used to help her and others. Frieda signed a consent form stating that it was all right to use her blood to treat others with cancer. She died eight months later.

Heidi didn't really think about the blood samples after her mother died, but a few years later she heard about a man named Moore in California who had his spleen removed. His spleen cells produced lymphokines, which could be used to treat others. He was suing his doctors for part of the millions of dollars they had made from selling his cells (along with their DNA) to a biotechnology company to make a cell line. A cell line keeps reproducing cells and their products for many, many years, and these products can then be sold.

Heidi wondered if her mother's case was the same situation. She called Irv Kutler, an attorney friend of hers, to ask about it. Irv said he thought they had a case. The doctors had sold Frieda's cell line, and a great deal of money and stock had changed hands. The rights to Frieda's cell line were sold to a company called Montansic, and even though Frieda did sign a consent form, she was never told what they would do with the cells. Irv said they should sue.

Could they win? Why or why not?

Online Resources

Preparing for an exam? Log on at www.cengagebrain.com for study tools to help you assess your understanding. If assigned by your instructor, the Case A and Spotlight on Ethics activities for this chapter, "Patient Offered New Cancer Treatment" and "Henrietta Lacks and Her Immortal Cells," will also be available.

WHAT WOULD YOU DO IF...?

. . . you were offered a space in an experimental study to cure cancer?

. . . you were advising your mother about whether she should join such a study?

. . . you were advising young people not to smoke?

. . . you were voting on a bill to take cigarettes off the market?

. . . you were working on marketing strategies to keep people out of the sun and had to reach people who must work outside (construction workers, letter carriers, etc.)?

Learn by Writing

In Chapter 1 we suggested that you start a blog with members of your class or others who are interested in our topics. Now is the time to revisit your blog and consider some of the questions in Chapters 9 and 10. E-mail others you think might be interested and invite them to contribute.

Here are some ideas to address in your blog:

- Should there be changes in the way patients with cancer are treated, now that we know there is a genetic component?

- Is it important that patients like Harriet know about the clinical trials being done on cancer?

- Do some research on clinical trials being done in your area and share the Web sites in your blog.

- Should the law actually get involved in any of the legal and ethical questions raised in Chapters 9–12?

- Discuss any other questions or comments you want to make or address with your fellow students.

In the Media

Top News: New Zealand, April 18, 2011

Milk's Cells Can Detect Breast Cancer

A study conducted by U.S. researchers at the University of Massachusetts, Amherst, has revealed that cells present in breast milk can help in detecting breast cancer. For the study, the researchers collected samples from 250 mothers who were about to undergo biopsies to check for breast cancer.

After collecting samples, they tested the DNA of the milk to see whether genes had been altered in any way, which might suggest early signs of breast cancer. The study found that women whose biopsies suggested that they had breast cancer also showed signs of the disease in milk cells.

To access this article online, go to www.cengagebrain.com.

QUESTIONS

1. Will testing breast milk be a good way to screen women for breast cancer? Why or why not?

2. Do you think the milk cells that showed signs of the disease might somehow affect babies who are nursing? Explain.

Times of India, September 8, 2007

Blood Test to Spot Cancer Early

British scientists tracked more than 11,000 women over a 30-year time frame. The central premise of the study was to see which of the women developed breast cancer and then to develop a blood test that might spot the cancer before symptoms appear.

The women all gave blood samples to be tested for biomarkers that are proteins created by cancer cells. Many of these biomarkers have already been linked to certain cancers.

After studying the women's blood samples, markers were found in those who developed cancer.

The test has been used to detect signs of breast cancer in some women, and it is hoped that this will help in early screening and better prevention.

To access this article online, go to www.cengagebrain.com.

QUESTIONS

1. Would you take the blood test described in the article? Why or why not?

2. Sometimes researchers take samples and test them but do not tell study participants about the results. Do you think the scientists should have told the women of their possible diagnosis?

13

Behavior

REVIEW

How are autosomal dominant traits inherited? (4.2)

How is the sex of a child determined? (1.1)

CENTRAL POINTS

- Behavior is a reaction to stimuli from our environment.

- Animals and humans have many similar behaviors.

- Brain chemicals play an important part in human behavior.

- Behavior can result from the action of single genes or multiple genes.

- Twin studies are an important part of behavioral genetics.

- Courts are unclear on how to address the issue of genetics and behavior.

CASE A: Twins Found Strangely Alike

Jim Lewis didn't learn he had a twin until he was 5, and it never sank in. Jim Springer learned he had a twin when he was a little older, but thought his twin was dead. When the two men were 39, researchers at the University of Minnesota Twin Study Group, who had been studying twins reared apart, reunited them.

The "two Jims" are famous because they have so many similarities. It amazed the scientists that they even weighed exactly the same (180 lb. [81.6 kg]), had the same first name, and were both 6 ft. (1.8 m) tall.

In addition to their similar physical traits:

- Both had a dog named Toy.
- Each had been married twice.
- Each one's first wife was named Linda.
- Each one's second wife was named Betty.

- Both had a son named James Allan.
- Both smoked Salem cigarettes.
- Both drank Miller Lite beer.
- Both had worked as sheriffs, part time.
- Both had recurring migraine headaches.
- Both bit their nails.
- Many of their mannerisms and speech patterns were alike.

Jim Springer and Jim Lewis.

Enrico Ferorelli Photography

Some questions come to mind when reading about the two Jims. Before we can address those questions, let's look at the genetics of behavior.

Students Taking an IQ Test

13.1 How Is Behavior Defined?

In its simplest form, behavior can be defined as a reaction to **stimuli** in our environment. This behavior was shown by studying simpler organisms such as amoeba and worms; these behavioral studies led to work with larger animals. We still look at animals as having simpler forms of behavior, which makes them easier to study, but human behavior has many similarities to animal behavior. Human behavioral responses can be more variable than those of animals, and because people react differently to similar situations, their behavior can be more difficult to study and define.

Psychologists often separate behaviors (either human or animal) into two categories: innate and learned. Simply put, **innate behaviors** are those we are born with (such as reflexes) and **learned behaviors** are those we acquire through experience.

As an example, when birds are born, they are usually helpless, and their parents feed them. However, as they grow older, they must learn to feed themselves by watching their parents search for food. In many ways, humans are no different. We are born helpless and, over time, learn behaviors that will help us become independent from our parents.

There are more specific forms of behavior that do not fit into these two categories and whose origins are not as clear. One of these is addictive behavior, such as alcoholism. Alcoholism can be defined as the development of characteristic behaviors associated with excessive consumption of alcohol. But is this definition clear enough to be useful as a phenotype in genetic analysis of alcoholism? Is there too much room for

stimuli environmental agents that generate a response from an organism

innate behavior behavior that is present at birth

learned behavior behavior that is acquired by experience or through teaching after birth

interpreting what is alcoholism or what is too much drinking? As we will see, whether a behavioral phenotype is defined narrowly or broadly can affect the outcome of the genetic analysis and even the model of inheritance for the trait. More importantly, how can we determine whether we are born with a genetic predisposition to become addicted to alcohol or whether this is a learned behavior with no genetic basis?

Behavioral genetics is the study of the influence of genes on behavior. One of the big questions in behavioral genetics is determining the roles of genetics and environmental factors in producing a specific behavior. How we react to certain situations may be influenced by genetics, things in our environment, and social factors such as our parents, siblings, or educational experiences. The interactions among all these factors make the study of our behavior somewhat complicated but always interesting.

13.1 Essentials

- **Behavior is an innate or learned response to stimuli.**
- **A person behaves in a specific way based on his or her heredity and environment.**

13.2 How Does the Body Work to Cause Behavior?

When we think of behavior as simple, we are referring to the body's response to stimuli in the environment. Because the nervous system perceives and directs response to stimuli, and stores the experience as memory, the nervous system (Figure 13.1) is the focus of behavioral genetics.

The **central nervous system (CNS)** includes the brain and spinal cord; the **peripheral nervous system (PNS)** includes the nerves that radiate from the CNS and the sense organs. To react to a stimulus in our environment, we must receive the information from our sense organs, process it in the brain, and react. A simple reflex reaction illustrates this process. If you touch a hot stove, sensors in your skin send signals to your peripheral nerves and central nervous system. These signals generate a reaction (quick movement of your hand) that is immediate and done without conscious thought. You may not even feel pain until you move your hand away. This reflex action is one of many that protect us from dangers in the environment.

behavioral genetics a form of genetics that studies its effect on human behavior

central nervous system the portion of the nervous system consisting of the brain and spinal cord

peripheral nervous system portion of the nervous system outside the brain and spinal cord

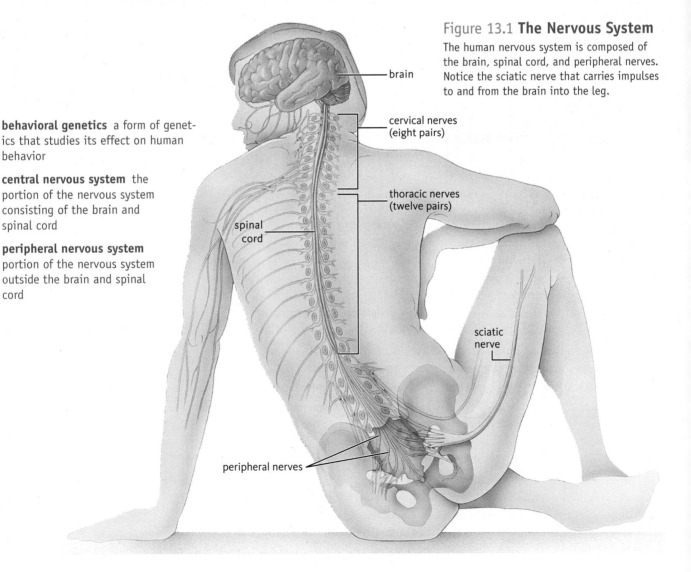

Figure 13.1 **The Nervous System**
The human nervous system is composed of the brain, spinal cord, and peripheral nerves. Notice the sciatic nerve that carries impulses to and from the brain into the leg.

brain

cervical nerves (eight pairs)

thoracic nerves (twelve pairs)

spinal cord

sciatic nerve

peripheral nerves

Table 13.1 Selected Neurotransmitters and Some Processes They Affect

Neurotransmitter	Brain Activities
Acetylcholine	Muscle stimulation, heart rate, learning, sleep
Dopamine	Learning, memory, mood, positive reinforcement
GABA	Inhibitory neurotransmitters, motor control, anxiety
Serotonin	Sleep, mood, appetite, pain, body temperature

Table 13.2 Selected Recreational Drugs and the Neurotransmitters They Mimic

Drug	Neurotransmitters They Mimic
Cocaine, morphine, heroin	Endorphins
Mescaline	Dopamine
Nicotine	Endorphins
Amphetamines	Norepinephrine

As simple as this reflex reaction seems, most aspects of human behavior are much more complex and can involve many of your senses and your memory, as well as innate and learned responses.

Chemical signals called **neurotransmitters** are an important component of human behavior (Table 13.1). The action of these signals can change our moods and our actions; the signals are released during the transfer of a nerve impulse from one cell to another in the nervous system. This process is called **neurotransmission**. The speed or frequency of neurotransmission can alter how we react in a certain situation.

As shown in Figure 13.2, nerve cells in the brain are separated by a small gap called the **synapse**. When a nerve impulse travels to the end of one cell (nerve cell A), it must cross the synapse to nerve cell B to continue its journey. The nerve impulse causes the release of stored neurotransmitter molecules to expedite movement of the nerve impulse across the synapse. The neurotransmitter molecules bind to receptors on cell B, initiating a nerve impulse that will travel to the other end of this cell. Changes in the timing or duration of the action of neurotransmitters may change a person's behavior. In addition to mediating behavior, neurotransmitters can affect our mood, our memory, and our sense of well-being.

Endorphins are one of the several types of neurotransmitters. When endorphins are released, often during exercise, stress, or excitement, we feel a rush of exhilaration. Many drugs that change our mood or reactions, such as cocaine, bind to the endorphin receptors on nerve cells and mimic the naturally occurring neurotransmitters, creating a similar feeling (Table 13.2). This may be why these drugs are popular and addictive.

Because nerve cells produce neurotransmitters, information for synthesis of these chemicals is carried in our DNA. As discussed in Chapter 5, proteins are synthesized in cells. In the case of neurotransmitters, proteins are involved in making these molecules—and feelings and behaviors result. In addition, the receptors that bind neurotransmitters are proteins. Therefore, it is clear that at least some of our behavior has a genetic basis.

13.2 Essentials

- **The release of neurotransmitters in the brain causes us to act and react.**
- **The primary actor in our behavior is the CNS.**

neurotransmitters chemicals secreted across synapses in cells of the brain

neurotransmission the process of transmitting a nerve impulse across a synapse between two cells in the nervous system, mediated by neurotransmitters

synapse the cleft between nerve cells in the brain

endorphin neurotransmitter associated with happy feelings

Figure 13.2 A Synapse Is the Gap Between Two Nerve Cells
When a nerve impulse arrives at the end of cell A, stored neurotransmitters are released, cross the synapse, and bind to receptors on nerve cell B. This event is necessary for the nerve impulse to move along cell B.

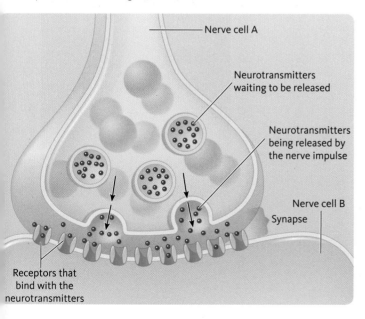

Nerve cell A

Neurotransmitters waiting to be released

Neurotransmitters being released by the nerve impulse

Nerve cell B

Synapse

Receptors that bind with the neurotransmitters

13.3 How Can Changes in Genes Cause Changes in Behavior?

A variety of genetic changes including single-gene defects, chromosome aberrations, and multifactorial traits can alter our behavior.

An example of a single-gene defect that changes behavior is Huntington disease, an adult-onset neurodegenerative disorder inherited as an autosomal dominant trait that affects about 1 in 10,000 individuals. (Review Alan's case in Chapter 4.) The symptoms of HD include behavioral changes such as involuntary movements, progressive personality changes, and dementia. The HD gene codes for a large protein called huntingtin, which is necessary for the survival of certain brain cells. When a person carries the mutant HD allele, his or her cells produce an altered version of huntingtin. This altered protein is toxic and kills cells in specific regions of the brain and nervous system. As these cells die, behavioral changes appear. The photo in Figure 13.3 shows the difference between a normal brain and one with HD.

Changes in behavior can also be caused by mutations associated with chromosomal abnormalities. People with a mutation in the *FMR-1* gene have specific behavioral changes. This condition, called **fragile X syndrome (FXS)**, is the most common form of mental retardation in males and is a significant cause of mental retardation in females. The condition is associated with hyperactive or impulsive behavior and is the best understood cause of autism. The fragile site, located at the tip of the X chromosome (Figure 13.4), is where the *FMR-1* gene is located. The mutation changes the DNA at the location of the gene and causes the constriction at the chromosome tip, which is fragile and easily broken. People with FXS fail to make a protein called FMRP. The lack of this protein alters normal patterns of synaptic transmission by knocking out key receptors, causing the symptoms associated with this inherited disorder.

A multifactorial condition (multifactorial traits are discussed in Chapter 11) that affects behavior is **schizophrenia**. Schizophrenia is a collection of mental disorders that affects about 1 in 100 individuals. This behavioral disorder can have many symptoms, including hallucinations, delusions, disordered thinking, and changes in behavior. Genes associated with schizophrenia have been found on the X chromosome and several autosomes. There is also an environmental component to schizophrenia. Many scientists are studying the genetic and environmental links to schizophrenia.

fragile X syndrome genetic disorder caused by a mutation in the *FMR-1* gene, which maps to the tip of the long arm of the X chromosome

schizophrenia set of mental disorders characterized by hallucinations and loss of touch with reality

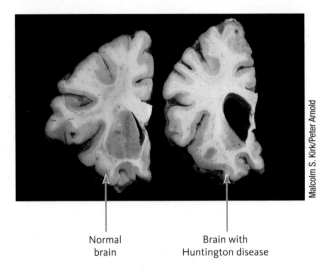

Figure 13.3 Section of a normal brain (*left*) and a brain from someone with Huntington disease (*right*). The brain from the HD patient shows extensive cell loss from a region of the brain called the striatum.

Normal brain

Brain with Huntington disease

Malcolm S. Kirk/Peter Arnold

Figure 13.4 **Fragile X Syndrome** This scanning electron micrograph of an X chromosome shows the constriction at the tip caused by a mutation in the *FMR-1* gene at that location. This is symptomatic of fragile X syndrome.

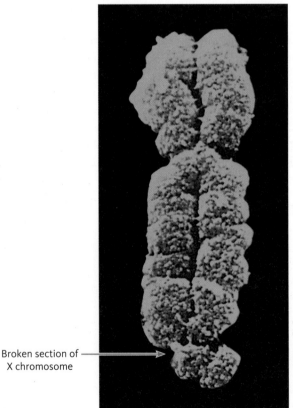

Broken section of X chromosome

Christine J. Harrison, Ph.D.

- **Behavioral disorders can be caused by a single-gene disorder, by chromosome aberrations, or can be a multifactorial trait.**
- **Huntington disease is an example of a genetic condition that causes changes in behavior.**

13.4 How Do We Study the Genetics of Behavior?

Behavioral geneticists search for genetic influences on human behaviors, an area that is also studied by **psychologists** and **psychiatrists**. Medical geneticists study the genetic basis for mental illnesses that can be treated by physicians. However, the differences between these fields are blurring as we learn more and more about genetic and environmental causes of behavior disorders.

Take, for example, the study of depression. At first, depression was treated by psychologists or psychiatrics and was considered to be an environmentally based behavioral disorder. Now, with information from pedigree analysis, molecular genetics, and neurobiology, it is clear that depression is a medical condition with genetic and environmental components. Physicians as well as specialists in psychology now treat depression as a physical illness.

How do we know that behavior has a genetic component?

One way scientists study the genetic basis of behavior is by twin studies. The Jim twins are a special case because they are identical twins adopted by different families and raised in separate environments. As we know, identical twins have identical genomes, and, therefore, each has a set of genes exactly the same as that of their twin. If a specific phenotype or behavior is the same in both twins, it most likely has a significant genetic component. If the two twins are raised in separate environments and have different behavioral traits, those traits have a significant environmental component.

For example, if one identical twin is alcoholic, there is a 55% chance that his or her twin will also be alcoholic regardless of his or her environment. This statistic was determined by studying groups of identical twins raised apart. This relationship is called **concordance**. Concordance is a way of calculating how often a trait occurs in both members of a pair of twins. Because identical twins are genetically identical, it follows that any genetic mutation carried by one would also be carried by the other. If this is true, the concordance rate for such a trait would be 1.0. For example, in cystic fibrosis (genotype *cc*), if one identical twin has the condition, the other will also have it (genotype *cc*) and the concordance will be 1.0.

Different degrees of concordance occur when a phenotype is present in one twin but not the other. One example is schizophrenia. The concordance in identical twins has been

Table 13.3 Concordance for Selected Traits

Trait	Concordance in Identical Twins	Concordance in Fraternal Twins
Blood type	1.0	0.66
Eye color	0.99	0.28
Schizophrenia	0.69	0.28
Diabetes	0.65	0.19
Alcoholism (males)	0.41	0.17
Attention deficit disorder/ hyperactivity	0.58	0.31
IQ score	0.69	0.88

found to be 0.69. This means that, according to studies of schizophrenia in identical twins, 69% of the time both twins had schizophrenia, and in 31% of the cases only one twin had the condition.

Thus, the concordance for a given trait helps establish whether or not it has a genetic component. Concordance is also studied in *fraternal twins*. Fraternal twins do not have identical DNA, but like all brothers and sisters, they share half their genes. Table 13.3 shows the concordance of schizophrenia and a few other traits.

If a specific behavioral trait is to be studied in a family, a pedigree is constructed, and the trait is traced through as many generations as possible. In alcoholism, many members of a family may have the condition, which will show up clearly in a family's pedigree. As already discussed, however, there are often problems in defining behavioral phenotypes. In the case of alcoholism, is it someone who drinks every day? Or is it someone who can't stop drinking? At what age can a diagnosis of alcoholism be made? Figure 13.5 (page 268) shows pedigrees of some behavioral traits.

Can we use animal models to study human behavior?

Animal studies are another way scientists study behavior. Because mice reproduce quickly and their behaviors are observable, many behavioral geneticists study them. If a gene is suspected to be involved in a behavior, researchers can use recombinant DNA techniques to insert a human gene, mutate a mouse gene, or change its pattern of expression and study the effects on the behavior of these genetically modified mice and their offspring.

psychologists non-physicians who specialize in treating psychological and psychiatric disorders

psychiatrists physicians who specialize in treating psychiatric disorders

concordance pairs or groups of individuals identical in phenotype; in twin studies, a measurement of how often both members of a twin pair show or fail to show a trait of interest

Figure 13.5 **Pedigrees of Some Behavioral Traits**

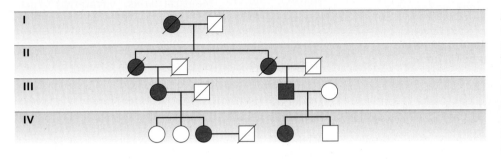

A pedigree from a family whose members suffer from Huntington disease (similar to Alan's family in Chapter 4). HD is inherited as an autosomal dominant trait, and patients show behavioral changes including depression, personality changes, and dementia. The gene for HD is located on chromosome 4.

KEY

■ Huntington disease

A pedigree from a family whose members suffer from schizophrenia (a mental illness). Affected individuals often hear voices, can be violent, and often have visual hallucinations. Genes on chromosomes 2, 10, and 15 are associated with this condition.

KEY

■ Schizophrenia

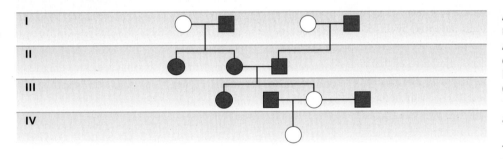

A pedigree from a family whose members suffer from alcoholism. Addictions such as this are difficult to categorize, and they have many different phenotypes. Genes on chromosome 1, 4, and 11 have been associated with alcoholism and other addictions. However, a single gene for this condition has not been identified.

KEY

■ Alcoholism

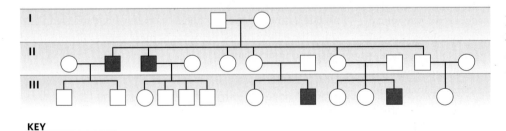

A pedigree from a family in which some members identify themselves as homosexual. Genes on chromosomes 7, 8, and 10 are associated with male homosexuality.

KEY

■ Homosexuality

In 1999, scientists working at Princeton University published the results of one such study. They inserted a human gene that codes for a protein known to be associated with memory into a mouse. The transgenic animals performed better in mazes and other tests used to measure intelligence in mice.

How can all of the methods just discussed be used to study one behavioral trait?
The genetic causes of many behaviors, such as sexual orientation, are studied using combined methods that include twin studies, chromosomal analysis, and pedigree analysis and genome-wide association studies. Sexual orientation is a complex trait, affected by several genes and environmental factors including sociocultural and experiential influences. In one twin study, 56 sets of male identical twins and 54 sets of male fraternal twins were studied. Table 13.4 shows the results.

This study and others like it have shown that there is a strong genetic component to homosexual behavior.

Other researchers have used linkage and pedigree studies to identify an area on the X chromosome associated with homosexual behavior (Figure 13.6). To do this, they first identified 38 families that included two homosexual brothers. After finding a number of molecular markers on the X chromosome of the brothers, they then searched for these markers on the X chromosomes of family members. The location of a gene or genes associated with these markers is shown in Figure 13.6. Subsequent work in other laboratories did not support the conclusions of this study but did provide a foundation for further work on genetics and homosexuality.

Figure 13.6 A region on the X chromosome, which in one study was thought to contain a gene for male homosexual behavior.

Possible gene for male homosexual behavior

X chromosome

Genome scans of 456 individuals from 146 families with at least two homosexual brothers showed that markers on chromosomes 7, 8, and 10 have a strong association with male homosexual behavior. Other evidence suggests that homosexual behavior associated with the region on chromosome 10 may be associated with epigenetic modifications in the form of imprinting (see Chapter 6 for a discussion of epigenetics). Further genome scans using markers from these regions may help identify specific genes associated with this behavior.

13.4 Essentials
- **Behavioral genetics can be studied in both animals and humans.**
- **Because of the identical genomes in identical twins, scientists study twins reared together and apart to assess the role of genetics and environment in behavior.**
- **Recombinant DNA techniques have been used to transfer human behavior genes to mice and observe changes in behavior.**
- **Scientists are still studying the roles of heredity and the environment in influencing behavior.**

Table 13.4 **Concordance for Homosexuality**

Concordance in Identical Twins	Concordance in Fraternal Twins
0.52	0.22

13.5 Can a Single-Gene Defect Cause Aggressive Behavior?

In 1993, scientists began studying a large European family that showed aggressive and violent behavior (Figure 13.7). Many of the men in this family have been jailed for committing violent offenses. Notice that the phenotype is expressed only in men. The gene for this condition was mapped to the short arm of the X chromosome (Figure 13.8) and is, therefore, an X-linked trait (see Chapter 4).

This gene encodes an enzyme called **monoamine oxidase type A (MAOA)** that breaks down neurotransmitters in synapses. Once an impulse has moved on to the next nerve cell, the synapse must get ready to receive the next impulse, and does so by clearing away neurotransmitters in the synapse. When mutated, this gene causes a condition called *MAOA deficiency*. Failure to rapidly break down neurotransmitters such as **serotonin** disrupts the normal transmission of nerve impulses. This changes normal functions in the nervous system and can cause abnormal behavior.

To study the role of these neurotransmitters in behavior, researchers knocked out the genes that encode one type of serotonin receptors on brain cells in mice. As a result, these mice did not make any of that receptor, altering the ability of nerve cells to bind and process nerve impulses. In tests, an unfamiliar mouse was placed in a cage with mice without these receptors, and the knockout mice exhibited highly aggressive behavior, as shown in Figure 13.9.

monoamine oxidase type A an enzyme that breaks down neurotransmitters such as serotonin

serotonin a neurotransmitter

CASE A: QUESTIONS

Now that we know more about genetic control of behavior, let's ask some questions about the case of the two Jims.

Jim Springer and Jim Lewis.

1. Why were these two men an ideal set of twins in which to study behavior?

2. Do you think the evidence shows that the behaviors listed in the case are 100% genetic?

3. Do you know any identical twins? Are they alike or different? List four things about them that are alike and four that are different.

4. Can you see any problems that might arise from studying twins raised apart? List three.

Figure 13.8 The Location of the MAOA Gene on the X Chromosome

Gene for *MAOA* deficiency

KEY

■ Aggressive behavior

X chromosome

Figure 13.7 MAOA Deficiency A pedigree with family members who carry a mutant allele of the MAOA gene and who exhibit violent or aggressive behavior. Notice that all the affected individuals are males.

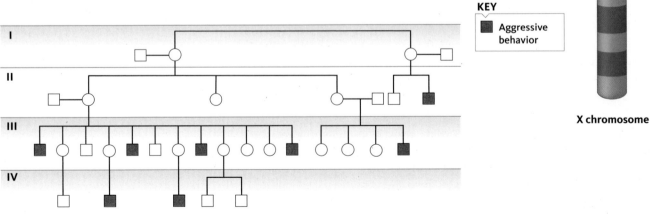

Figure 13.9 **Aggressive Behavior** Knocking out the genes for one type of serotonin receptor led to highly aggressive behavior in mice.

13.5 Essentials

- **MAOA deficiency is another genetic condition that is studied in behavioral genetics.**

13.6 Can We Treat Complex Behavioral Disorders?

As we have seen, humans have many complex behaviors, and no two are the same. In any given situation, a number of people may have different behavioral responses to the same stimulus. This complexity makes human behavior disorders difficult to analyze and treat. If genetics plays a part in these behaviors, then knowledge of the molecular basis of the behavior opens the possibility that drugs can be developed to treat these conditions. Over many years, we have found that conditions such as depression, bipolar disorder, and schizophrenia are due to an imbalance in brain chemicals. Since then, drugs have been developed to try to correct this imbalance. Is this type of treatment the future of human behaviors?

Drugs that affect the synthesis, uptake, and turnover of neurotransmitters such as dopamine, serotonin, and endorphins are now used successfully to treat Parkinson's disease, chronic depression, and schizophrenia. In some cases of bipolar disorder, affected individuals have low levels of lithium in the brain, and lithium treatment helps bring brain and nervous system levels back to normal. In the past, such conditions were treated only with therapy or even institutionalization. Today, drug therapy allows people with these illnesses to lead normal lives without the stigma of mental illness.

CASE B: Important Conference on Hold

Dr. David Wasserman envisioned his planned scientific meeting as a gathering of an academically diverse group of professionals to discuss cutting-edge research about genetics and behavior. He had organized other symposia, and, as in the past, he carefully planned whom to invite as speakers and guests, and even what food to serve.

Not so many years before, such an academic meeting would not have attracted the attention of the national press, but, in this case, the planned meeting made headlines. Dr. Wasserman was really surprised when the first reporter called.

He gave the reporter all the information requested, including the title of the symposium, "Genetic Factors in Crime: Findings, Uses, and Implications." Of course, Dr. Wasserman and his colleagues knew that the topic might raise some eyebrows because of the past role of eugenics and genetics associated with behavior, but they didn't think of it as a sensational topic. After all, they were scientists and approached the topic in a scientific way.

Then the problems began. Protests against the meeting came from many directions. Critics argued that there should be no public funding for behavioral genetic research and that the conference would reinforce racist assumptions about crime. The controversy caused the meeting to be canceled.

1. **What should Dr. Wasserman and his colleagues have done in response to the criticism?**

2. **How could they have had the conference without causing any problems?**

3. **Years ago the public would never have known about such academic conferences; why has this changed?**

4. **Some people see a racial component to these kinds of studies. Research this topic and give your opinion.**

5. **The National Institutes of Health revoked the funding for this meeting; do you think it should have done so? Why or why not? How should controversial issues be discussed and how should people be educated about the facts surrounding these topics?**

6. **Research what actually happened with this conference. Was it ever held?**

However, because these are complex, multifactorial conditions controlled by several to many genes, identifying these genes continues to be slow and often laborious. When major genes controlling these conditions are identified, it will lead to better treatments and understanding of how these conditions work.

In Chapter 6, we discussed epigenetic alterations and their effects on gene expression. Scientists have been studying epigenetic changes in several groups, including animal models, children who have suffered abuse, and individuals with mental illnesses. For example, in animal models, research has shown that maternal behavior permanently affects the adult behavior of the offspring. This work and other studies have shown that exposure to certain environmental factors early in life or even prenatally has profound effects on behavior later in life. These effects are mediated by epigenetic changes in the pattern of gene expression in the brain and nervous system. As we learn more about the genes associated with development, behavior, and signal reception and their mechanisms of action, we will learn more about interactions between genomes and the environment. This information will help guide the development of new generations of drugs that can go beyond treating symptoms to the treatment of the genetic and molecular causes of these disorders.

13.6 Essentials

- **With the discovery of more genes, the genetic control of complex behavior will lead to more treatments.**

13.7 What Are the Legal and Ethical Issues Associated with Behavior?

Scientists are not sure about the root causes of many aspects of human behavior. Because the science is in an early stage of development, courts are finding it difficult to deal with this topic. The most important question in *Frye* (see "Spotlight on Law: *Frye v. U.S.*" in Chapter 1) is whether a specific scientific theory is generally accepted among members of the scientific community. Behavioral genetics is a science still in its infancy; therefore, experts do not agree on many findings and, as a result, individual courts have difficulty deciding how these findings should be used to determine guilt or innocence.

In our society we often ask the question, why does someone commit a crime? In legal circles, this reason is referred to as *motive* and is often a component of a criminal trial. If we could determine that certain genes or groups of genes cause criminal behaviors, motive would no longer be relevant.

In the 1970s, some scientists studied individuals with XYY syndrome (also called Jacobs syndrome; see Chapter 3).

Figure 13.10 **Richard Speck** A mass murderer, he claimed an extra Y chromosome as part of his defense. Karyotypic analysis showed this was not true, and he dropped this defense.

© Bettmann/CORBIS

In the United Kingdom, a study of prisoners incarcerated for violent crimes showed a higher percentage of men with XYY syndrome than men in the general population. Initially, this finding sparked similar studies all over the globe, including research in the United States. This work was done to see if there is a relationship between criminal behavior and an XYY chromosome constitution. Generally, XYY males are taller, and some are aggressive, but this does not mean they are criminals. Most XYY males do not show antisocial behavior and are not violent criminals. Studies in the United States have been halted, so information on this possible relationship is still incomplete.

In 1966, Richard Speck (Figure 13.10) was tried and convicted of killing eight student nurses in Chicago, Illinois. He was identified by the testimony of the one surviving witness to the crime. During his sentencing, his attorney prepared a defense using the XYY karyotype as an argument against the death penalty. It was not used at the hearing because tests showed that Speck did not have the XYY karyotype. However, since that time, many unanswered questions about genetics and behavior remain in the minds of both prosecutors and defense attorneys. For example, can a genetic cause of criminality be established and used in court (see "Spotlight on Law: *Mobley v. Georgia*" later in this chapter)?

Another question often discussed is whether a person can be rehabilitated after committing a crime. The idea of rehabilitation is the basis for setting up our penal system. But to change a person's behavior, it is necessary to control that behavior. If criminal or antisocial behavior is genetically controlled, how can we expect a person with a genetic behavioral disorder to control or change his or her behavior?

In addition, the questions raised in *Frye* were set up to separate **junk science** from "real" science. To date, most judges

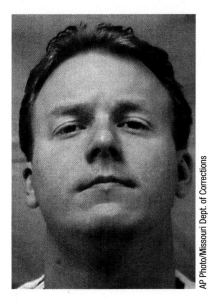

Figure 13.11
Christopher Simmons Was Simmons, who was 17 when he committed murder, unable to understand the consequences of his actions?

AP Photo/Missouri Dept. of Corrections

and juries have considered a genetic predisposition to crime to be in the junk science category.

Roper v. Simmons addressed some of these issues. In 1993, Christopher Simmons (Figure 13.11) and two other boys were convicted of pushing a woman off a railroad trestle into a river, killing her. At the hearing to decide whether Simmons would receive a death sentence, his attorney argued that because he was only 17 when the crime was committed, he was too young to realize the consequences of his actions. To prove the point, the attorney presented an **amicus curiae** ("friend of the court") brief written by scientists who studied the brains of adolescents. MRI brain scans from thousands of teens showed significant differences from those of adults. The scans showed that an area of the brain that controls impulsive behavior (the frontal lobe) was underdeveloped in teens when compared with adults (Figure 13.12). In addition, the emotional responses of teenagers were different than those of adults.

In this case, the court decided that the scientific information was not strong enough to stop an execution. The case was taken to the U.S. Supreme Court and argued in October 2004. The Court determined that the execution of offenders who were 18 years of age or younger when their crimes were committed was unconstitutional. Therefore, Simmons was not executed and is serving life in prison. In the written opinion Justice Kennedy wrote, "Retribution is not proportional if the law's most severe penalty is imposed on one whose culpability or blameworthiness is diminished, to a substantial degree, by reason of youth and immaturity."

One of the first behavioral issues to be addressed in criminal trials was insanity. If a person could be proven to be insane, the courts asked, would this make him or her not responsible for the crime? For the most part, a medical diagnosis of schizophrenia can allow a person to be found not guilty by reason of insanity. Often people who suffer from schizophrenia have both auditory and visual hallucinations that may tell them to commit crimes. Because of this, and the fact that they cannot distinguish between right and wrong, they are often not held responsible for their actions. In court cases throughout the United States, judges have pondered the definition of insanity, and their answers vary.

In general, courts have held that only if a defendant cannot tell right from wrong and has a definite diagnosis of a medical condition, such as schizophrenia, he or she can be found not guilty by reason of insanity. Schizophrenia has been shown to have a genetic basis. Does this mean a person with this condition cannot control his or her behavior?

Table 13.5 addresses some of the ethical and legal questions related to genetics and behavior.

13.7 Essentials

- **The response to the question of genetic control of behavior has not been definitively answered in courts of law.**

junk science legal term meaning science that has yet to be generally accepted by members of the scientific community

amicus curiae a legal term translated as "friend of the court," applied to briefs submitted to a judge providing new information before a decision is rendered

Figure 13.12 **Adolescent and Adult Brains**
These drawings of an adolescent brain (a) and an adult brain (b) show large differences in areas that control impulsive behavior. In adults, brain tissue has expanded into the circled region, indicating maturation of this part of the brain.

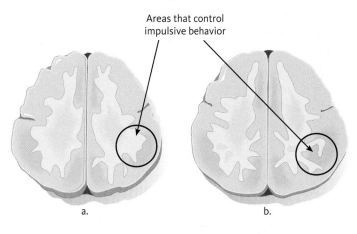

Areas that control impulsive behavior

a.   b.

Table 13.5 **Ethical and Legal Question in Behavioral Genetics**

Question	How Are These Questions Decided?	Related Case or Legal Issue
Is a person who has been diagnosed as insane responsible for his or her actions?	Courts make their decisions based on legislation passed on a state-by-state basis. Some states allow pleas of "not guilty by reason of insanity"; others use "guilty but insane."	When John Hinckley tried to kill President Reagan (*U.S. v. Hinckley*), he argued that because he had schizophrenia, he was not guilty by reason of insanity. His attorney showed brain scans that indicated he had a smaller than normal brain. He was found not guilty by reason of insanity.
If a person has a genetic condition that makes him or her aggressive, such as MAOA deficiency, can he or she be considered guilty of a crime?	In general, courts today in the United States are not likely to accept genetic defenses.	See "Spotlight on Law: *Mobley v. Georgia*" in this chapter.
Can a person argue that he or she is not responsible for a crime committed under the influence of drugs or alcohol?	The courts have held that voluntarily taking drugs and alcohol cannot mitigate a person's responsibility for a crime.	
If a person has a genetic condition that causes aggression, could this condition be used in his or her defense?	This issue has not been addressed in U.S. courts; however, some scientists believe that if genes control certain behaviors, people may be unable to control them.	The question of what should be done with people who have this type of mutated gene may become important. See "Spotlight on Law: *Mobley v. Georgia*" in this chapter.
If a behavior is obviously "insane," can that be used in a case?	Many people believe that women who kill their children should be considered insane.	Andrea Yates drowned her five children in the family's bathtub in 2001. In 2003, Deanna Laney beat her two young sons to death and injured a third with stones, and Lisa Ann Diaz drowned her two daughters in a bathtub. Dena Schlosser fatally severed her 10-month-old daughter's arms with a kitchen knife in 2004. All four of these women were found not guilty by reason of insanity. Yates initially was convicted of capital murder, but that verdict was overturned on appeal.

Spotlight on Law

Mobley v. Georgia
455 S.E.2d 61(1995)

On February 17, 1991, John Collins was shot in the back of the head during a robbery of a Domino's Pizza where he worked in Oakwood, Georgia. On March 13, after robbing a dry cleaner, Stephen Mobley was arrested and confessed to the murder and armed robbery at Domino's.

During this confession he boasted about how John Collins fell on his knees and begged for mercy. After committing the crime, Mobley had himself tattooed with a Domino's logo and plastered his jail cell with Domino's boxes. His criminal history included rape, robbery, assault, and burglary. The prosecution said, "Mobley is evil, a cold-blooded, heartless killer."

Daniel Summer, Mobley's court-appointed attorney, tried to enter a guilty plea and arrange a deal for life in prison, but the deal was rejected. The prosecutors wanted the death penalty for Mobley. During questioning of the family, Summer met Mobley's aunt, Joyce Ann Childers. She told Summer that "volcanic, aggressive, physical abuse and violent behavior" prevalent throughout the family tree." Summer then remembered an article he had read in the *Chicago Tribune* in which scientists at Harvard and the National Institutes of Health, well as overseas, were conducting research on genetic ties

lence. The pedigree of Mobley's family based on his aunt's testimony is shown in Figure 13.13. Compare it to the pedigree of the family with MAOA deficiency (Figure 13.14).

ere are obvious differences between the two families. The most glaring one is that women in Mobley's family are identified with the trait, but women are not affected in the family with MAOA deficiency. Could members of Mobley's family have MAOA deficiency?

fore the sentencing hearing, Summer contacted Dr. Xandra akefield at Harvard Medical School and Dr. Bahjat A. rat at Emory University. When Mobley's story was revealed, ese researchers began to see an emerging pattern of a tory of violence through several generations in his family. th offered their services free of charge; Mobley needed

specialized blood and urine tests to determine whether he had a mutation of the MAOA gene.

One might think Mobley came from a bad family environment, but the opposite is actually true. Even though many of the people in Mobley's family tree were violent, many were amazingly successful. His father, for example, even though he refused to help Mobley's defense, is a self-made millionaire. He tried sending his son to private school, then to psychiatrists, and finally to jail, but stated, "He never developed a value system or a conscience." In the end he washed his hands of all responsibility for his son.

Mobley could not afford the cost of the MAOA tests (about $1000), and Summer asked the court to cover the expense. This allowed the trial court, and eventually the Supreme Court of Georgia, to weigh the question of validity of genetic causes of criminality.

RESULTS

The Georgia trial court could find no reason to allow for testing Mobley; the decision stated, "The theory of a genetic connection is not at a level of scientific acceptance that would justify its admission." In other words, this did not meet the *Frye* standards of scientific evidence.

QUESTIONS

1. Would you have allowed the testing if you were a member of the Georgia Supreme Court?

2. If a gene or group of genes were found to cause violent behavior, how could defense attorneys use this information?

3. How might the prosecution argue for *not* including genetic information in Mobley's case?

4. Is a person who is drunk, taking drugs, or mentally ill responsible for any crimes he or she commits? Why or why not?

5. How does this case relate to a genetic cause of crime?

gure 13.13 **A Pedigree of the Mobley Family** Many individuals in this digree exhibit violent or aggressive behaviors, and many are incarcerated. The arrow dicates Stephen Mobley.

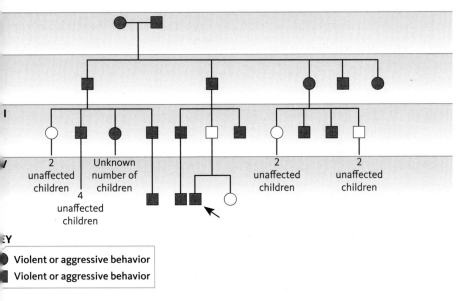

2 unaffected children

Unknown number of children

4 unaffected children

2 unaffected children

2 unaffected children

EY

● Violent or aggressive behavior
■ Violent or aggressive behavior

gure 13.14 **A Pedigree of a Family with MAOA Deficiency (see age 270)** What similarities and differences do you see when you compare this pedigree th that of the Mobley family in Figure 13.13?

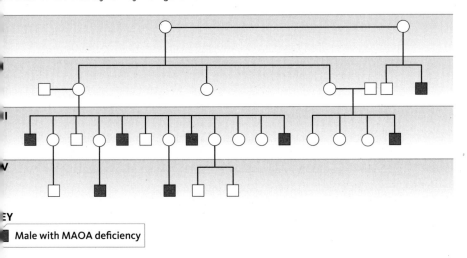

EY

■ Male with MAOA deficiency

The Essential 10

1 Behavior is a response to stimuli. (Section 13.1)

This response can be a trivial or a major behavioral change.

2 A person behaves in a certain way based on his or her heredity and environment. (Section 13.1)

Twins are often used to study the influence of heredity.

3 The action of neurotransmitters in the brain causes us to act and react in certain ways. (Section 13.2)

These chemicals cause changes in the nervous system and produce changes in our behavior.

4 Huntington disease is a genetic condition that causes changes in behavior. (Section 13.3)

The mutation in Huntington disease affects the structure and function of the brain and nervous system.

5 Behavioral genetics is studied in animal models as well as humans. (Section 13.4)

The study of behavior in animal model systems can teach us something about human behavior.

6 Scientists are still evaluating the roles of heredity and environment in behavior. (Section 13.4)

Human behavior can be difficult to study because people can react differently to the same situation.

7 Because identical twins have identical genomes, many scientists study twins reared apart to investigate the role of genetics and environment in controlling behavior. (Section 13.4)

In this situation, people with identical genomes are exposed to differing environments.

8 Recombinant DNA techniques have been used to generate and study behavioral changes in mice. (Section 13.4)

Changing specific genes in mice can cause them to act aggressively.

9 MAOA deficiency is a human genetic condition that is being studied in behavioral genetics. (Section 13.5)

Men with this condition show extreme aggression and may commit crimes.

10 The answer to the question of genetic control of behavior has not been decided in scientific circles or in courts of law. (Sections 13.6, 13.7)

If our behavior is caused largely by genetics, are we responsible for what we do?

Review Questions

1. How do neurotransmitters work to cause our feelings and behaviors?

2. Research examples that show how humans are similar to animals in their behaviors and write a short report.

3. Is the study of animal models of behavior a good way to identify human behaviors? Why or why not?

4. What are the main parts of the central nervous system?

5. What is concordance?

6. Behavior seems to have a simple definition. What is it?

7. How do we compare the study of animal behavior to human behavior? Why are some animals better for this comparison than others?

8. List three brain chemicals and their effect on behavior.

9. Schizophrenia is a fairly well-known human condition. What external (environmental) factors affect it?

10. What is the difference between a psychiatrist and a psychologist?

11. List three parts of the peripheral nervous system.

12. Epigenetic modifications and environmental influences seem to play a part in mental illness. Explain how this can happen.

Application Questions

1. Find an example of another set of twins who participated in the Minnesota twin study. How do they compare to the two Jims?

2. If you were going to adopt a child who was an identical twin, do you think it would be better to adopt both or just one? Why?

3. In this chapter we discuss how twins are studied to determine if they have similar behaviors. Design a way to study behaviors in people who are not twins. What trait (such as anger or kindness) would you study? How would you set up the experiment? What would you use as a control in this experiment?

4. If sexual orientation were shown to be genetically caused, would this change the way people view homosexuals? How?

5. If MAOA deficiency were shown to cause a predisposition to homicidal behavior, should we test all newborns for this trait? Why or why not?

6. Look at the pedigree for alcoholism in Figure 13.5. What do you think might occur if we could identify the genes responsible for this pattern of inheritance?

7. How does the fact that some drugs mimic naturally occurring brain chemicals explain why people get addicted to them?

8. Research some legal cases where schizophrenia was used as a defense. Does this mean that because schizophrenia has a genetic component, these defendants have a true defense?

9. In the law and ethics section of this chapter there is a quote from the Supreme Court decision in *Roper v. Simmons*. How does it apply to genetics?

10. Create a drawing of the central nervous system and show a simple reflex using arrows.

11. If a drug mimics a normally occurring brain chemical and the result is an unpleasant feeling, do you think addicts would use this drug? Why or why not?

12. One problem that has yet to be solved is how to repair a damaged part of the CNS. Quite a bit of work has been done using stem cells to regenerate spinal cords. Find one study and summarize it.

13. A young local artist recently began doing work with leather. His work is very creative and he can create clothing, book covers, and other items. His parents say his works look almost exactly like his grandfather's, a leather craftsman. He had never met his grandfather. Could artistic and musical talent be inherited? Compile a list of some artists that have a family history of creative work.

14. Read the book *Touched With Fire: Manic-Depressive Illness and the Artistic Temperament* by Kay Jamison and write a short report. Does this book establish a relationship between creativity and mental illness?

15. What would you have included in the amicus curiae brief for *Simmons*? List three things.

16. Epigenetics and other environmental influences seem to play a part in mental illness. Research and summarize a study that dealt with child abuse and epigenetics.

WHAT WOULD YOU DO IF...?

. . . you were a juror for a trial in which a defense attorney said her client was not responsible for the crime because he had an extra Y chromosome?

. . . you had been asked to speak at Dr. Wasserman's conference?

. . . you were on a jury being asked to force a man with schizophrenia to take his medication?

. . . your identical twin daughters were asked to be in a twin study?

. . . you were a reporter asked to write a story about genetics and criminal behavior?

Online Resources

Preparing for an exam? Log on at www.cengagebrain.com for study tools to help you assess your understanding. If assigned by your instructor, the Case B and Spotlight on Law activities for this chapter, "Important Conference on Hold" and "*Mobley v. Georgia*," will also be available.

In the Media

Baltimore Sun, August 14, 2007

Insanity Defense Muddles Case
Justin Fenton and Andrea F. Siegel

Should a person with schizophrenia be found guilty and placed in prison if he or she has committed a crime? This is the question raised in the case of Zachary Thomas Neiman, who was accused of killing his mother with two shotgun blasts as she sat on her sofa.

He has refused to take his medication that would make him competent to stand trial. If he continues to refuse treatment he won't be released from custody. His treatment for schizophrenia allows him to function in society. Without medication, he has severe symptoms.

If he were found not guilty by reason of insanity, he would still have to spend time in a state institution where psychiatrists, psychologists, therapists, nurses, and other experts would continually watch, test, and reevaluate him. If he continued to take his medication he might be allowed parole.

To access this article online, go to www.cengagebrain.com.

QUESTIONS

1. If medications can treat mental illness, can the courts force defendants to take them?

2. Is a person insane if he or she is taking medications that allow "normal" behavior?

Daily News Analysis, April 30, 2011

Genetics Influence Alcohol Dependence, Brain Activity

Researchers have uncovered a new link between genetic variations associated with alcoholism, impulsive behavior, and a region of the brain involved in craving and anxiety.

The results of this study suggest that variations in the *GABRA2* gene on chromosome 5 contribute to the risk of alcoholism by influencing impulsive behaviors, at least in part through a portion of the cerebral cortex known as the insula.

Individuals under stress who have this variant tend to act impulsively, a behavior that may lead to the development of alcohol problems. It is unsure whether this gene in its mutated form can cause alcoholism.

To access this article online, go to www.cengagebrain.com.

QUESTIONS

1. Does this study give you an idea of what can cause alcoholism? Why or why not?

2. Could these studies be harmful to the participants involved in them?

3. From whom would you keep the information about participation?

14

Immunogenetics

REVIEW

Can transgenic organisms be used to study human diseases? (7.5)

What are some other uses for DNA profiles? (9.4)

CENTRAL POINTS

- Genetics plays a central role in the development of the immune system.
- Immune system compatibility is an important consideration in organ transplantation.
- Human blood types are inherited.
- Disorders of the immune system can cause serious problems.
- Allergies are related to the immune system.
- Carriers of immune system disorders can be identified.

CASE A: His Sister Wants to Donate a Kidney

Kidney disease has devastated the Sanchez family. Mr. Sanchez had failing kidneys; his 15-year-old son, Julio, was already on dialysis, but his older daughter, Maria, was healthy. When her biology class studied heredity, Maria did a family pedigree using kidney disease as the trait to be studied. She discovered that her great-grandmother had died of kidney disease in Mexico, as had her grandfather.

Maria desperately wanted to help Julio, who was on a list for a kidney transplant, but his tissue type was rare. She knew from reading her biology book that siblings are often the best match for transplants. She went to see Dr. Tulley, her family doctor, and asked him to test her to see if her tissue type matched Julio's. He was hesitant, but she said, "I am your patient, and I want you to do this—and don't tell my parents." He agreed.

Transplantation of a human organ.

The test results showed that she was a close match. So Maria asked her father for permission to donate a kidney to Julio, but he became angry. "No," he said, "I want you to stay healthy . . . we'll wait."

Maria still wanted to donate her kidney to Julio, but she was only 17 (not yet legal age) and didn't know what to do.

Some questions come to mind when reading about Maria's family. Before we can address those questions, let's look at the biology of immunity and transplantation.

Circulation in a Human Hand

14.1 What Does the Immune System Do?

The human immune system protects the body from infection caused by bacteria, viruses, and other foreign invaders. Our body's first line of defense against infection is the skin, which acts as a barrier and blocks these invaders when they try to enter the body. The second line of defense is the inflammatory response, which activates the immune system in response to bacteria, fungi, and viruses. The third line of defense is called the adaptive immune system. It has two components: white blood cells (*lymphocytes*) called **T cells** (Figure 14.1a, page 282) and other white blood cells, the **B cells** (Figure 14.1b). Each of these cell types mount specific targeted responses that work to attack and destroy infectious agents and the molecules they secrete. In this chapter we will focus on the role of the immune system in infection, blood types, and transplants. We will also discuss disorders of the immune system.

Most agents that infect the body carry protein molecules called **antigens**. These molecules are detected by the immune system and trigger a response that usually involves several stages. Follow these stages in Figure 14.2 (page 282).

1. Detection of the antigen activates a type of T cell called a **T4 helper cell**, which in turn activates B cells.

2. The activated B cells divide and produce and secrete proteins called **antibodies** that bind to and inactivate the antigen on the surface of the invader.

T cells one of two types of immune system cells (B cells are the other)

B cells antibody-secreting cells of the immune system

antigens molecules that enter the body and trigger an immune response

T4 helper cells specialized T cells that serve as the on switch for the immune system

antibodies proteins synthesized by B cells that inactivate antigens

Figure 14.1 **Adaptive Immune System** (a) A T cell with a large nucleus surrounded by cytoplasm containing little endoplasmic reticulum (ER). (b) An activated B cell with large amounts of endoplasmic reticulum active in the synthesis of antibodies that are released into the blood.

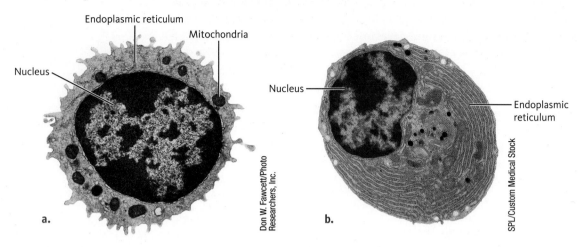

3. The infective agents marked by the antibodies are destroyed by other white blood cells.

Antigens may enter the body on cell surfaces via a blood transfusion, a cut (that carries in bacteria), or a transplanted organ. They may also be part of a disease-causing agent such as a virus, bacteria, or fungus.

B cells are activated when a T4 helper cell delivers the antigen to a receptor on the surface of the B cell. Once the

antigen is bound to the receptor, signals are transmitted to the nucleus of the B cell. In response to these signals, the B cell begins to divide and synthesize antibody molecules. Once secreted by the B cell, these antibodies bind to the antigen, wherever it is in the body. Binding of the antibody to the antigen marks it for destruction by other cells of the immune system.

Some of the activated B cells form **memory cells** that allow the body to mount a quick and massive response if the

Figure 14.2 **Stages in the Immune Response**

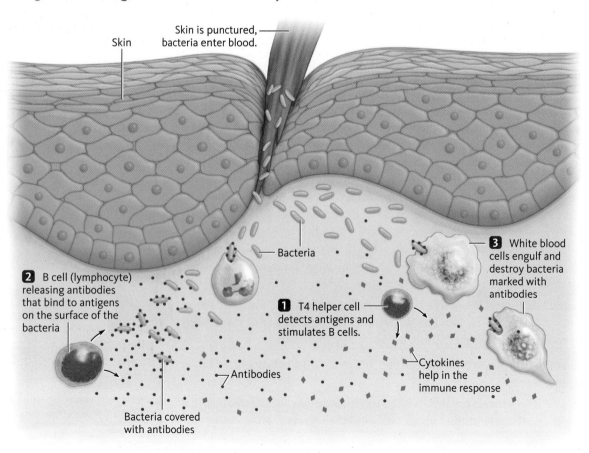

body is exposed to the same antigen a second time. The immune system reacts quickly, preventing a second infection.

Memory cells are the basis of **vaccination** against infectious diseases. A **vaccine** contains an inactivated or weakened disease-causing agent. When injected, the antigens are not powerful enough to cause an infection, but they do stimulate the immune system to produce antibodies and memory cells against that antigen. Later, if the person is exposed to the disease-causing agent (the virus, bacteria, or fungus that was the source of the antigen), he or she will be immune to infection by that agent.

The immune system is controlled by several groups of genes. The study of these genes and how they function is called **immunogenetics**. These genes encode proteins on the cell surface as well as the antibodies that directly attack foreign antigens. The receptors on the surface of our T and B cells are also encoded in our DNA. Understanding how these genes and their respective proteins work has helped us cure infectious diseases, prevent infection, and make organ transplants possible. Because these genes control the immune response, mutations in these genes can also cause diseases of the immune system, including **autoimmune disorders**, and *allergies*.

Autoimmune disorders, where the body fails to recognize its own cells and attacks and destroys them, include juvenile diabetes, arthritis, multiple sclerosis (MS), and inflammatory bowel disease (see Table 14.1).

14.1 Essentials

- **The immune system protects the body from infections.**
- **Two parts of the immune system are the inflammatory response and the adaptive immune response.**

14.2 How Does the Immune System Affect the Transplantation of Organs or Tissues?

When a new organ is placed in a recipient's body, the cells of the transplanted organ carry antigens that are different from those on the cells of the recipient. These antigens, which serve as molecular identification tags, are encoded by a cluster of 140 genes on chromosome 6 known as the **major histocompatibility complex (MHC)** (Figure 14.3). The MHC is found in most vertebrates, and these genes play important roles in the immune system and organ transplants. A set of nine genes within the MHC are known as human leukocyte antigen (HLA) genes. Each of the HLA genes has many different alleles, and the combinations of these alleles are nearly endless. The set of HLA alleles present on each copy of chromosome 6 is called a **haplotype**. Because there are

Figure 14.3 A Map of Chromosome 6 Showing the Position of the HLA Gene Complex

MHC genes

Chromosome 6

memory cells cells produced by activated B cells that protect an individual from future infections by the same antigen

vaccination stimulation of the immune system and memory cell production by administration of an attenuated antigen

vaccine preparation containing a weakened or noninfective antigen used to confer immunity against that antigen

immunogenetics the study of the genes that play a part in the immune system

autoimmune disorders conditions in which the body's immune system attacks tissues and organs in the body

major histocompatibility complex (MHC) a gene set that encodes proteins that act as molecular identification tags

haplotype combination of alleles carried close together on a chromosome

Table 14.1 Some Autoimmune Diseases

Disorder	Body Part or System Affected	Frequency in the United States
Type I diabetes	Pancreas and insulin production	1 in 800 people
Rheumatoid arthritis	Skeletal/joints and connective tissue	1 in 108 people
Multiple sclerosis	Central nervous system	1 in 700 people
Crohn's disease	Digestive system	1 in 4500 people
Psoriasis	Skin	1 in 49 people

so many allele combinations, it is difficult to find two people with exactly the same HLA haplotypes . . . except, of course, identical twins.

A transplanted organ may be attacked by the immune system if the antigens it carries do not match those encoded by the recipient's HLA genes. This mismatch and the immune response it generates causes **rejection** of the transplant. Even when the HLA haplotypes are closely matched, drugs are used to suppress any rejection response. The major genes in the **HLA complex** have been identified, and we can test to see if the HLA haplotypes of a potential donor match those of someone in need of an organ transplant.

How do cell surfaces play a role in the immune system? The plasma membrane contains many different protein molecules in and on the surface of the cell (Figure 14.4). Many of these proteins play important roles in the transport of molecules into and out of the cell. In addition, HLA proteins on the cell surface serve as molecular identity tags, identifying the cell to the immune system. If the antigens on a cell do not match those recognized by the immune system, then an immune response will attack and destroy those cells. Consequently, the HLA genes and the proteins they encode play an important part in transplantation and other immune responses.

What determines whether organ transplants are successful? Successful organ transplants and skin grafts depend on matches between the HLA haplotypes of the donor and the recipient. Because so many allele combinations are possible in the HLA complex, two

rejection an immune reaction to a transplanted organ that does not match the recipient

HLA complex a group of genes on chromosome 6 that determine tissue type for transplants

Figure 14.4 **Plasma Membrane Proteins** Cells of the body carry surface proteins encoded by the HLA genes that must be matched with those on a donor organ.

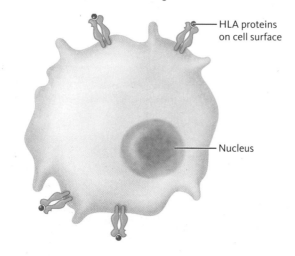

HLA proteins on cell surface

Nucleus

Figure 14.5 Richard Herrick (left) and Ronald Herrick (right) were the first successful recipient and donor in a kidney transplant procedure. The medical team was not sure it would work even though the brothers were identical twins.

AP Photo

individuals rarely have a perfect HLA match. That's why finding a compatible donor for those who need organ transplants often takes a long time.

Richard Herrick received the first successful organ transplant in 1954 (Figure 14.5). He entered Brigham and Women's Hospital in Boston suffering from kidney failure. His brother, Ronald, an identical twin, volunteered to donate a kidney to Richard. As in Maria's case, Ronald just wanted to help his brother. The night before the surgery, Richard wrote a note to his brother that read: "Get out of here and go home." Ronald jotted a quick reply: "I am here and I am going to stay." The transplant was successful, and the surgeon, Dr. Joseph Murray (Figure 14.6), was awarded a Nobel Prize for this medical breakthrough. Ronald Herrick went on to become a teacher

Figure 14.6 Dr. Murray (*right*) won the Nobel Prize for the first successful kidney transplant, with Ronald Herrick the kidney donor. Ronald's brother, Richard, died eight years after the transplant.

AP Photo/Eric Miller

while Richard married and had two children. Unfortunately, eight years after the transplant, Richard died of an infection in the transplanted kidney.

Before an organ transplant is performed, the HLA haplotypes of the donor and the potential recipient are analyzed. If there is at least a 75% match, the transplant will usually be successful. After the surgery, the recipient must take **immunosuppressive drugs** to suppress his or her immune system and therefore reduce the possibility that the transplanted organ will be rejected.

Rejection begins when cells of the recipient's immune system, called killer T cells, attack and destroy the transplanted organ. Once the rejection process begins, the cells of the transplanted organ are killed rapidly, and the organ must be removed. The recipient will then need another organ or he or she will die. Closely matching the HLA haplotypes of the donor and the recipient is necessary to ensure successful transplants.

Overall, there is a 25% chance that siblings will have matching HLA haplotypes. In the case discussed earlier, Maria has already been tested, and her kidney would be an ideal match for Julio. If Maria's father does not allow her to donate a kidney to Julio, he will have to go on a waiting list with over 74,000 other individuals who need a kidney transplant. Each year, there are only about 17,000 kidney transplants performed, and hundreds of people on the waiting list die before receiving a transplant. It is estimated that several thousand lives would be saved each year if enough donor organs of all types were available.

Can animal organs be transplanted into humans?
One way to increase the supply of organs for transplantation is to use animal donors. Animal–human transplants (called **xenotransplants**) have been attempted many times, but with little success. Some important problems related to rejection currently prevent the use of animal organs. The first problem occurs when an animal organ is transplanted into a human. Proteins on the surface of the pig cells (antigens) are so different from those on the human cells that they trigger an immediate and massive immune response, known as a **hyperacute rejection**. This reaction usually destroys the transplanted organ within hours.

To overcome this rejection, several research groups have used recombinant DNA techniques (see Chapter 7) to modify the antigens on the organs of donor pigs to make them more

immunosuppressive drugs drugs given to transplant recipients to suppress their immune system to prevent rejection of a transplanted organ

xenotransplantation the process of transplanting organs between species

hyperacute rejection a rapid and massive reaction to a transplanted organ from another species

Figure 14.7 **Transgenic Pig** Genetically modified mini-pigs are a potential source of organs for transplant because of their size and physiological similarity to humans.

Sasha Radosavljevich/Shutterstock.com

compatible with human HLA antigens. To do this, they isolated and cloned human genes of the HLA complex. These genes were then injected into fertilized pig eggs. The resulting transgenic pigs (Figure 14.7) carry human antigens on all their cells. Organs from these transgenic pigs appear as human organs to a recipient's immune system, preventing a hyperacute rejection. Transplants from genetically modified pigs to monkey hosts have been successful, but the ultimate step will be an organ transplant from a transgenic pig to a human. Some scientists say it could happen in less than 10 years.

Some clinical trials have used pig donors to transplant cells to treat people who suffer from Parkinson's disease. One of the first recipients was Jim Finn (Figure 14.8), who received cells from a pig's brain and had a good recovery. Whether

Figure 14.8 Jim Finn, shown here before treatment, suffered from Parkinson's disease and couldn't walk, talk, or use his hands. As part of a clinical trial, he had fetal pig neural cells injected into his brain. Six months later, he could sit, stand, and walk independently. Today he is an advocate for using animal cells for treatment.

CBS News Archive

these procedures can be used on a large scale will depend on the outcome of more clinical trials.

Even if the hyperacute rejection can be suppressed, transplanted pig organs may cause other problems. For example, even if genetically modified, pig organs may still trigger a stronger-than-normal immune reaction by the human recipient. Another potential problem is that the cells of the pig organs may carry viruses that are potentially dangerous to humans.

A more radical way to ensure that donor organs from pigs will be compatible with human recipients is to transplant bone marrow from a donor pig to the human recipient. The recipient would then have a pig–human immune system, called a **chimeric immune system**. This chimeric immune system would recognize the pig organ as self, would not trigger a rejection response, and would still retain normal immunity to fight infectious diseases. As far-fetched as it may sound, animal experiments using this approach have been successful. A similar method of bone marrow transplants from donor to recipient is already in use in human-to-human heart transplants to increase chances of success, so pig–human transplants would be a logical next step.

One of the most controversial xenotransplants ever done was the transplant of a baboon heart into a newborn baby (Figure 14.9). Baby Faye was born in 1984 in Barstow, California. Her type of heart defect is always fatal, so her parents told physicians they could try the experimental procedure. Twenty

chimeric immune system an immune system created in the recipient when bone marrow is transplanted between species

xenografts organ transplanted between species (same as *xenotransplants*)

Figure 14.9 **Xenotransplant** Baby Faye received the first baboon heart transplant in 1984. She did not survive.

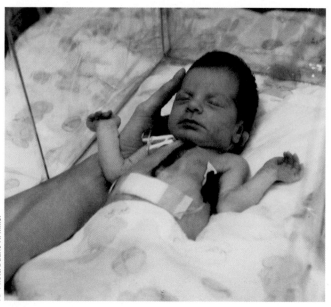

CASE A: QUESTIONS

Now that we understand more about the basics of organ transplantation, let's look at some of the issues raised in Maria Sanchez's case.

Transplantation of a human organ.

1. List four arguments Maria could use to persuade her father to allow her to donate her kidney.

2. List four arguments Maria's father could use to persuade her not to donate her kidney.

3. Does Maria need a lawyer? Why or why not?

4. In the state where they live, the legal age for consent is 18. How does this affect Maria's case?

5. How does family heredity play a part in Maria's argument?

days after the transplant, she died when her body rejected the new heart.

As recently as ten years ago, the possibility of animal–human organ transplants seemed remote, more suited to science fiction than to medical reality. But today more than 200 people in the United States have already received **xenografts** of animal cells or tissues.

The advances described here make it likely that xenotransplants of major organs will be attempted in the next few years.

14.2 Essentials

- **The antigens on the cells of a donor organ determine whether it will be a match for the recipient.**
- **By testing HLA haplotypes, matches between recipients and potential organ donors can be made.**
- **Xenotransplantation may become widespread in the near future.**

14.3 What Are Blood Types?

Blood types are determined by antigens on a blood cell's surface. Humans have more than 30 different blood types, each of which is defined by the presence of specific antigens on the surface of blood cells. These antigens are similar to those found on the cells of transplanted organs. Antigens identify the cells of the donor and the recipient in a transfusion. As with organ transplantation, the immune system of recipients will reject a blood transfusion if the antigens are mismatched. We'll focus on two examples of blood types: the ABO system that is important in blood transfusions, and the Rh factor, which plays a role in a disease called **hemolytic disease of newborns (HDN)**.

What determines the ABO blood types?

ABO blood types are determined by a gene, *I*, which encodes cell surface proteins, or antigens, that identify blood types as A, B, AB, or O. This gene has three alleles: I^A, I^B, and I^O, often written as *A*, *B*, and *O*. The *A* and *B* alleles each encode a slightly different version of the antigen and the *O* allele produces no antigen. Those with type A blood carry the A antigen on their red blood cells and make antibodies against the B antigen. Those with type B blood carry the B antigen on their red blood cells and make antibodies against the A antigen. Therefore, those with type A blood will not accept type B blood and vice versa. Type O cells carry no antigen on their surface and, therefore, do not stimulate an antibody response when used in a transfusion (Figure 14.10).

How are ABO blood types inherited?

As we stated, the *I* gene has three alleles that contribute to blood types A, B, and O. People with type O blood have the genotype

Figure 14.10 **Blood Bank** Blood is collected and can be kept either frozen or refrigerated for use in transfusions.

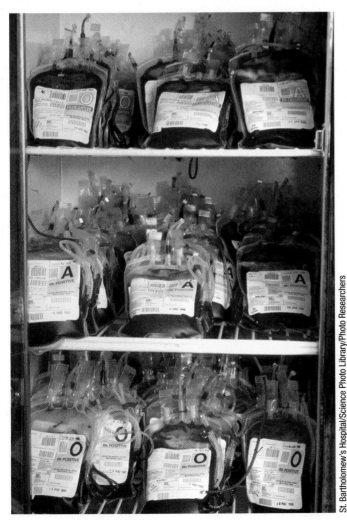

St. Bartholomew's Hospital/Science Photo Library/Photo Researchers

$I^O I^O$ (or *OO*). The I^O allele is recessive to both the I^A and I^B allele. Therefore, someone with genotype $I^A I^O$ has type A blood, whereas someone with genotype $I^B I^O$ would have type B blood. However, when the genotype is $I^A I^B$, neither allele is dominant to the other. Instead, these alleles are **codominant**, and therefore those with the genotype $I^A I^B$ will have an AB blood type.

If you have type A blood, your genotype could be $I^A I^A$ or $I^A I^O$. If you have type B blood, your genotype could be $I^B I^B$ or $I^B I^O$. If you have type AB blood or type O blood, your genotype can be only $I^A I^B$ or $I^O I^O$, respectively.

How do blood transfusions work?

To understand how transfusions work, first look at Table 14.2. It shows that people with type AB blood have both the A and B antigens on their red blood cells and do not make antibodies against either the A or the B antigen. The table also shows that those with type O blood have neither the A nor the B antigen on their red blood cells but do make antibodies against the A and B antigens.

As a result, people with type A blood can donate to others who are type A, type B individuals can donate to type B individuals, and so forth. However, the situation is a little more complicated than that. Because people with an AB blood type do not make antibodies against the A or B antigen, they can receive a transfusion using blood of any type. Such individuals are known as **universal recipients**. Those with the O blood type have neither antigen on their blood cells, and as a result they can donate blood to anyone and no reaction will occur. Type O individuals are known as **universal donors**.

Can there be problems with transfusions?

For blood transfusions to be successful, the ABO antigens of the donor and recipient must match. If there is a mismatch, the antibodies in the recipient's immune system will bind to antigens on the donor cells, causing them to form clumps (Figure 14.11, page 288). The clumped blood cells block circulation

hemolytic disease of newborns (HDN) blood condition that occurs when antibodies from an Rh⁻ mother destroy blood cells in an Rh⁺ fetus

ABO blood types the major blood types in humans

I isoagglutinin, a gene whose three alleles (*A*, *B*, and *O*) determine blood type

I^A an allele that determines the A blood type

I^B an allele that determines the B blood type

I^O an allele that, when homozygous, determines the O blood type

codominant a condition where two alleles are both fully expressed in the heterozygote

universal recipients individuals with type AB blood; they can accept donations from all blood types

universal donors individuals with type O blood; they can donate to all other blood types

Figure 14.11 **Red Blood Cells Clumping Together after a Transfusion Reaction**

Custom Medical Stock

Red blood cells clumping

in small blood vessels, reduce oxygen delivery, and often have fatal results. As the clumped blood cells break down, they release large amounts of hemoglobin into the blood, which forms deposits in the kidneys and can cause kidney failure.

How is the Rh factor inherited?

The **Rh blood group** was named for the rhesus monkey in which it was discovered. To simplify the genetics, we can say it consists of **Rh positive (Rh⁺)** individuals, who carry the Rh antigen on their blood cells, and **Rh negative (Rh⁻)** individuals, who do not carry this antigen. In this blood type, the *Rh⁺* allele is dominant to the *Rh⁻* allele.

As a result, if you are Rh⁺, you might carry the allele combination *Rh⁺Rh⁺* or *Rh⁺Rh⁻*. But if you are Rh⁻, you can carry only the *Rh⁻Rh⁻* genotype.

Rh blood group a human blood type

Rh positive (Rh⁺) blood type with the Rh antigen on blood cells

Rh negative (Rh⁻) blood type lacking the Rh antigen

How does the Rh factor cause problems in newborns?

During pregnancy (Figure 14.12a), a small number of fetal cells may cross the placenta and enter the mother's bloodstream (see Chapter 2). These blood cells can also enter her bloodstream during childbirth. If she is Rh⁻ and her fetus is Rh⁺, fetal blood cells that cross the placenta will stimulate the mother's immune system to make antibodies against the Rh antigen on the fetal cells. If this is the mother's first pregnancy, this situation will usually not harm either the fetus or the mother.

However, exposure to the Rh⁺ antigen creates memory cells with the ability to make antibodies against the antigen. If the mother becomes pregnant again, and the fetus is also Rh⁺, antibodies from her blood will cross the placenta in the late stages of pregnancy and destroy red blood cells of the fetus (Figure 14.12b). The result is a serious disease called *hemolytic disease of newborns*.

To prevent HDN, Rh⁻ women are given an Rh-antibody preparation (called RhoGAM) during their first pregnancy if the child they are carrying is Rh⁺. These antibodies move through the mother's circulatory system and destroy any fetal cells that may be present. To be effective, the RhoGAM must be given before the mother's immune system has had a chance to make antibodies against the Rh antigen.

14.3 Essentials

- **A, B, AB, and O are the major human blood types.**
- **The Rh blood type can cause problems in newborns when the mother's type is Rh⁻ and the fetus is Rh⁺.**

14.4 How Are HIV Infection and AIDS Related to the Immune System?

Acquired immunodeficiency syndrome (AIDS) is a clinical disease that develops after a person is infected with the human immunodeficiency virus (HIV). Once in the body,

Table 14.2 **Summary of A, B, and O Blood Types**

Blood Type	Antigens on Plasma Membranes of Red Blood Cells	Antibodies in Blood	Safe to Transfuse to	Safe to Transfuse from	Possible Genotypes
A	A	Anti-B	A, AB	A, O	AA or AO
B	B	Anti-A	B, AB	B, O	BB or BO
AB	A + B	None	AB	A, B, AB, O	AB
O	None	Anti-A + anti-B	A, B, AB, O	O	OO

Figure 14.12 **Rh Factor** (a) Blood cells from an Rh⁺ fetus can cross the placenta. (b) If the mother is Rh⁻, she can develop antibodies that cause hemolytic disease of the newborn in subsequent pregnancies with an Rh⁺ fetus.

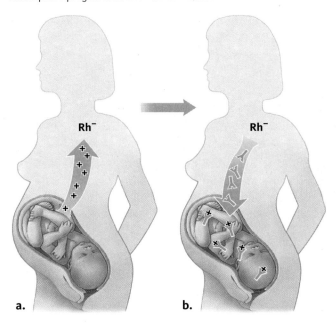

a.   b.

Figure 14.13 **HIV and Its Life Cycle**

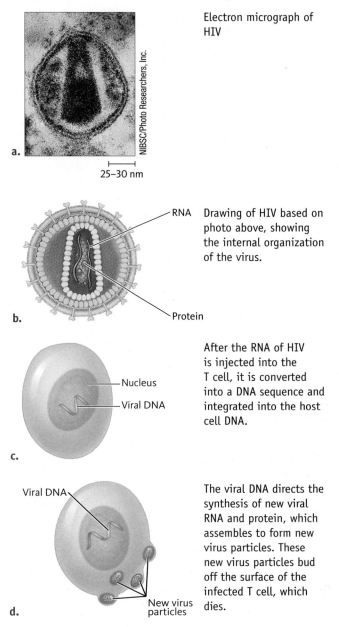

a.

Electron micrograph of HIV

25–30 nm

b.

RNA — Drawing of HIV based on photo above, showing the internal organization of the virus.

Protein

c.

Nucleus
Viral DNA

After the RNA of HIV is injected into the T cell, it is converted into a DNA sequence and integrated into the host cell DNA.

d.

Viral DNA

New virus particles

The viral DNA directs the synthesis of new viral RNA and protein, which assembles to form new virus particles. These new virus particles bud off the surface of the infected T cell, which dies.

HIV infects and kills *T4 helper cells*. As we discussed earlier, these cells are extremely important for the onset of an immune reaction. Without them, the body cannot recognize foreign antigens on infective agents such as bacteria and viruses. Recall that at the beginning of an immune response, the T4 helper cell recognizes the antigen and then activates the B cells, which in turn form antibodies. In other words, T4 helper cells serve as the "on" switch for the immune response.

After HIV invades a T4 cell, it copies its genetic information, which is then inserted into a chromosome in the infected cell. This viral genetic information can remain inactive for months or years. Later, when the infected T4 cell is called upon to participate in an immune response, the viral genes become active. New viral particles are formed in the cell and bud off the surface of the T cell, rupturing and killing it. These new viruses then spread through the body and infect other T4 cells. Over the course of an HIV infection, the number of T4 helper cells gradually decreases, and the body loses its ability to fight infection. Without an active immune system, infected people become ill from many diseases that they would otherwise fight off.

Eventually, by killing the T4 helper cells, HIV infection disables the immune system, resulting in AIDS. In turn, AIDS causes premature death from one or more infectious diseases that overwhelm the body and its compromised immune system. In Figure 14.13a, an electron micrograph is placed above a drawing of the virus (14.13b). Figures 14.13c and d show steps in the reproduction of the virus.

HIV is transmitted from infected to uninfected individuals through body fluids, including blood, semen, vaginal secretions, and breast milk. The virus cannot live for more than one to two hours outside the body and cannot be transmitted by food, water, or casual contact.

Are some people naturally resistant to HIV infection? About fifteen years after the first AIDS case was reported in the United States, researchers discovered a small number of people who engaged in high-risk behavior, such as unprotected sex with HIV-positive partners, but did not become infected with HIV. Shortly thereafter, researchers discovered that these individuals carried two copies of a mutant allele of a gene called *CC-CKR5*.

CC-CKR5 gene that encodes a cell surface protein that signals the presence of an antigen

When the *CC-CKR5* gene functions normally, it makes a protein on the cell surface that signals the immune system when an infection is present. HIV is able to bind to the normal CC-CKR5 protein and use it to infect T4 helper cells. If the *CC-CKR5* gene is mutated by a small deletion (with 32 base pairs missing), the resulting protein is shorter and HIV cannot use this protein to infect and kill T4 helper cells. As a result, individuals who are homozygous for the *CC-CKR5* mutation are resistant to HIV infection.

After this discovery, researchers asked several questions: Which populations carry this mutant allele, and how widespread is it? Can knowledge about how HIV uses the CC-CKR5 protein to infect T4 cells help in the design of anti-HIV drugs?

Which populations carry the mutant *CC-CKR5* allele?
Population studies show that this mutant allele is present only in Europeans and those of European ancestry. The highest frequency of the mutant allele is found in northern Europe, and the lowest frequency is found in Greece and Sardinia.

Scientists speculate that this mutation is more frequent in certain populations because at some time in the past it may have offered resistance to another unknown but deadly infectious disease. Because those who carried the mutant allele survived this unknown disease, they were able to pass this allele to their offspring, while those who did not carry the mutation died. Because of this, populations differ in the frequency of the *CC-CKR5* allele, which also confers resistance to HIV infection.

Researchers have also discovered that other populations carry mutant alleles of different genes that confer resistance to HIV infection. As these are identified, they are being studied to find new ways of treating HIV and AIDS.

Are some people naturally susceptible to HIV infection?
Recently, scientists have discovered why HIV infection might be highest in sub-Saharan Africa (Figure 14.14). An allele of the *DARC* gene, which encodes a cell surface antigen, protects carriers from malaria (a debilitating and often fatal infectious disease) but increases their vulnerability to HIV infection by 40%. In other words, people who carry the anti-malaria allele of *DARC* have a greater chance of contracting HIV if exposed to this virus. In addition to increasing risk of infection, this allele interferes with the immune system's ability to fight HIV infection in its early stages. This study was done over a twenty-five-year period and involved DNA samples from thousands of participants. About 60% of African Americans carry this allele and are more susceptible to HIV infection than the general population.

Figure 14.14 In Africa, large numbers of AIDS patients are cared for in makeshift hospitals, but there are not enough physicians and nurses to care for all infected people.

Karen Kasmauski/CORBIS

How do today's HIV drugs work?
Many of the drugs (Figure 14.15) now used to treat HIV infection work by preventing the virus from replicating once it is inside the T4 helper cells. Other drugs block HIV at different stages of the infection and reproduction cycle. Combinations of these drugs are successful in slowing or stopping the progress of HIV infection and development of AIDS. However, these drugs have serious side effects, and, in addition, drug-resistant strains of HIV have developed.

By studying the way HIV enters cells using the protein encoded by the *CC-CKR5* gene and other proteins on the cell surface, researchers are developing a new generation of drugs that will prevent entry of the virus into its target cells. One of these drugs, **enfuvirtide**, has been approved by the U.S. Food and Drug Administration for clinical trials. Other drugs are under development and will soon be ready for trials.

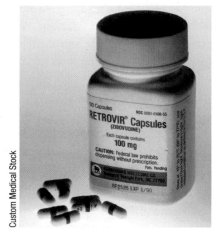

Figure 14.15 AZT was one of the first drugs used to treat patients with HIV/AIDS. Today AZT is used in combination with a number of antiviral agents. This drug cocktail keeps HIV infection under control, making it a chronic rather than fatal condition.

Custom Medical Stock

enfuvirtide a drug in clinical trials that prevents entry of HIV into a cell

14.5 How Are Allergies Related to the Immune System?

Allergies result when the immune system overreacts to antigens that do not cause an immune reaction in most people. These antigens, called **allergens**, are carried by dust, pollen, and certain foods and medicines. One of the most serious food sensitivities is the allergy to peanuts, which is a growing health concern in the United States.

Allergic reactions to peanuts, bee stings, or other proteins may provoke a very serious reaction called **anaphylactic shock**. In this reaction, the bronchial tubes constrict, restricting airflow in the lungs and making breathing difficult. Heart **arrhythmias** and cardiac shock can develop and cause death within one to two minutes.

Parents of children with these severe allergies often carry an injectable medicine (epinephrine contained in an EpiPen®) that can be used as soon as an anaphylactic reaction begins. The pen injects the drug quickly and can be used by the patients themselves when they feel the reaction coming on (Figure 14.16). This drug counteracts the immune response as soon as it is injected and is a life-saving treatment.

About 80% of all anaphylactic shock cases are caused by allergies to peanuts. Individuals with peanut sensitivities must avoid eating peanuts and any products that include peanuts. Even peanut dust in the air can be dangerous to a person with this allergy. The card shown below explains the dangers of peanut allergies to those who work in a restaurant.

The number of children and adults who are allergic to peanuts appears to be increasing. In a 2008 survey, 1.4% of U.S. children had peanut allergies, up from 0.4% in 1997. Now as many as 1 in 50 children may have peanut or tree nut allergies.

BY THE NUMBERS

25%
Chance that a sibling will be a match for an organ transplant

50 million

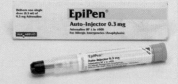

Number of people in the U.S. population with allergies

1500
Number of deaths from anaphylactic shock in the United States each year

3 million
Number of people allergic to peanuts in the United States

WARNING!
FOOD ALLERGY:
PEANUTS

I have a severe food allergy to peanuts, any foods cooked with peanuts, and anything that has touched peanuts. Please help me dine safely by making sure all utensils, cookware, and surfaces are very clean and have not been previously used. Please review the ingredients on the back of this card and read food labels carefully. If I accidentally ingest even a small amount of peanuts, please dial 911 as I will require immediate emergency medical care.

Figure 14.16 To prevent anaphylaxis when an allergic reaction starts, a patient may have to administer an injection of epinephrine.

allergies an abnormal immune response to allergens in the environment

allergens antigens in the environment that cause allergic reactions

anaphylactic shock a potentially fatal allergic reaction involving major organ shutdown

arrhythmias irregular heartbeats

CASE B:
Prisoners Used as Guinea Pigs

The request had come from the governor's office. It couldn't be ignored. But Warden Wilson had lots of doubts about it. He felt that he had an obligation to his prisoners and that they had the same rights as people in the general population.

A few years ago the state legislature passed a law requiring that inmates with HIV (which causes AIDS) be isolated in a special wing of the prison. Warden Wilson didn't like this because he was against these forms of prisoner segregation, but he could understand the safety concerns. Since then, the spread of AIDS had slowed in the prison population, and the prisoners in the special wing got better medical care—he saw to that.

Now he had received a letter explaining that a new study would use inmates to test a vaccine for HIV. The vaccine had been successfully tested in animals and approved by the FDA for testing on humans. The drug company developing the vaccine needed a population to test the side effects of the vaccine on humans.

The letter went on to explain that prisons were the ideal place for testing on humans because the population could be carefully monitored, and each person had already been tested and isolated because of his or her HIV status. Each inmate would be paid $1000 a month (a huge salary for inmates) and receive better food and medical care than the other inmates.

Half the participants would be HIV positive and half would be negative. Each inmate would be given an attenuated virus (a weakened, noninfective form of the virus). Even though the virus is supposed to be noninfective, inmates who began the test as HIV negative might contract HIV from the vaccine. This had not happened with the animal tests, but the scientists mentioned it as one of the possible side effects.

Warden Wilson couldn't sleep; he was worried about whether he should allow this study to be done in his prison.

In the past, prisoners were used as subjects in clinical trials.

1. **What should Warden Wilson do?**

2. **Give three arguments Warden Wilson should use to persuade the governor's office not to allow the testing in his prison.**

3. **Give three arguments the scientists might use to persuade the governor's office to allow this kind of testing in the prison.**

4. **How do you think the prisoners would respond?**

5. **What reasons might the prisoners have for wanting to participate in the study?**

6. **Why do you think the researchers picked prisons for this experimentation? Explain your reasoning.**

7. **Do you think prisoners have the right to refuse such treatment? Why or why not?**

What is causing the increase in peanut allergies? The answer is still unclear, but environmental factors appear to play a major role. For example, peanut allergies are extremely rare in China, but the children of people who immigrate to the United States from China have about the same frequency of peanut allergies as the children of native-born Americans. This suggests that there is involvement of some environmental factors.

Some researchers feel that because peanuts are now a major part of the diet in the United States, the exposure of newborns and young children (age 1–2) to peanuts is more common. This exposure may occur through breast milk after the mother has eaten peanut butter or other peanut-containing foods. The immune system of newborns is immature and develops over the first few years of life. As a result, exposure to some antigens is likely to cause food allergies because the mother sensitizes her fetus by eating or using certain products. This is especially true if the woman herself is allergic to a specific product.

However, until more is known about how all allergies develop, some physicians recommend that mothers avoid eating peanuts and other common allergens. This is stressed if peanut allergies run the family of the pregnant woman, whether she has allergies or not. In addition, peanut products should not be eaten while nursing, and children should not be exposed to peanuts or other nuts for the first three years of life.

14.5 Essentials
- **Some allergies are becoming a serious problem in our society.**

14.6 What Are the Legal and Ethical Issues Associated with Organ Donation?

The ability to transplant organs has created a number of legal and ethical issues. As the Sanchez family discovered, finding a matching donor isn't easy. Siblings are the best donors, but they cannot always consent.

Most laws require an individual to be 18 years of age to consent to medical treatment. Those younger than age 18, or those unable to consent for some other reason, are not allowed to donate an organ. In addition, any person who has been identified as incompetent by a court of law, or patients who are comatose or unconscious, cannot consent to donation or any other medical treatment. However, their next of kin can give consent for them. One problem with this type of permission occurs when a family cannot agree on the decision.

Because the number of people on transplant lists has increased over the years, more donors are needed. As a result, many states have passed laws that make organ donation easier for those who want to do it. Most states have organ donor stickers or cards that are issued with driver's licenses. These cards can be used to inform family members and health care workers of the donor's wishes. Directions for organ donation can also be spelled out in a living will written by the donor or discussed with friends and family. But many people still do not donate or make their wishes known. Some states have been thinking of reverting to **assumed consent**. This would mean that everyone would become an automatic organ donor. Only those who carry a card that states their desire not to be a donor would be exempt.

Another option would be to legalize providing compensation for organ donors. The National Organ Transplant Act of 1984 prohibits the sale of organs or payment of compensation to donors. A lawsuit now before the 9th U.S. Circuit Court of Appeals asks that bone marrow donors be exempt from the ban on receiving payment for donations. At present, women can sell their eggs, men can receive compensation for sperm donation, and blood donors can sell their blood. However, as the law stands, donors cannot sell bone marrow, which is naturally renewable in much the same way as gametes and blood. New methods make marrow donation almost as easy as blood donation. However, marrow donors risk up to five years in prison for receiving compensation if they provide these cells to the sick, including those with genetic disorders or leukemia. The lack of bone marrow donors is a major public health problem. More than 3000 people in the United States die each year while waiting for a marrow donor. Government attorneys are defending the ban against selling bone marrow, arguing that the law protects the poor from exploitation, prevents the development of a black market, and guards against potential donors extorting money from those who need a transplant to remain alive.

Table 14.3 addresses some of the ethical and legal questions associated with organ donation.

14.6 Essentials

- **Those who want to donate organs should make their wishes known to their family.**

assumed consent legal term meaning that consent for treatment is presumed

Table 14.3 Legal and Ethical Issues in Transplantation

Question	How Are These Questions Decided?	Related Case or Legal Issue
What if family members cannot agree about donating the organs of a comatose relative?	Courts can decide whether the comatose patient wanted to donate. They use organ donor cards, living wills, and conversations with friends.	Often these cases need to be sent to a judge to decide what the actual desires of the patient were.
What happens when a wealthy or famous person needs an organ but other people are ahead of him or her on the transplant list?	The waiting list is carefully set up so that those whose condition is the most serious get organs first.	The United Network for Organ Sharing (UNOS), run by the U.S. government, sets up the organ donor list, which does not take wealth or celebrity into account.
Some people are afraid they will not be treated as aggressively in an emergency room if they carry an organ donor card. How is this falsehood dealt with?	Public education, including television ads, has been slowly changing the mistaken idea that organ donors receive less aggressive treatment.	These beliefs may come from television programs as well as fiction books such as *Coma* by Robin Cook. There have been no lawsuits based on this urban legend.
There are still not enough organs; how can the supply be increased?	Many labs have been working on using recombinant DNA technology to create transgenic animals that will be a source of organs.	Courts and legislatures are trying to simplify the method of organ donation by passing laws that make this decision automatic.
Can bone marrow or organs be sold?	Courts must decide if bone marrow can be sold.	Is the sale of bone marrow different from sale of nonrenewable organs like kidneys?

Spotlight on Society

Drug Development and Public Outcry

As discussed in Chapter 12, drugs developed by scientists or companies must be approved by the FDA through a time-consuming and costly process. Approval involves animal testing and then four stages of human testing (clinical trials) before a drug can be put on the market. This process takes years and costs millions of dollars.

But what if the condition is life threatening and affects many people? In the 1980s, this was the situation when drugs were being developed to treat people with HIV/AIDS. Thousands of people were dying of AIDS, and a treatment or cure did not seem to be in the works for many reasons. The social stigma associated with AIDS (most people who had it were gay men, poor, or drug users), the cost, and the time commitment slowed everything down.

The leaders of an organization called ACT UP (AIDS Coalition to Unleash Power) began using social activism to generate change, bring more drugs to the market, and increase awareness about AIDS. This small organization of gay men and women organized a grassroots movement to get AIDS into the national spotlight and force the U.S. government to realize the health risks involved. To get the attention of the public, ACT UP staged a number of events and rallies, including political rallies, large demonstrations in Washington, D.C.,

protests at the International AIDS Conference, and sit-ins a hospitals.

In June 1989, one of the most successful protests was held in front of Sloan-Kettering Hospital in New York City. Prote tors, dressed as health care workers and patients, sat in fr of the hospital for four days while acting out scenarios abc people dying of AIDS (Figure 14.17). The purpose of the p test was to demand that more people with AIDS be include in clinical trials of AZT (zidovudine or retrovir), a drug now routinely used as a treatment for AIDS.

As a result of the pressure brought by ACT UP and other groups, the FDA was forced to make HIV drugs available fo treatment under an already existing rule called the "compa sionate use rule," which allowed new drugs to be moved through trials more quickly if patients were terminally ill. was the first step.

Finally, in 1997, the FDA formally introduced the Fast Track Designation and Priority Review. It was aimed at speeding the process of development and approval for drugs that are identified as important for treatment of serious diseases ar that address unmet medical needs.

To qualify for Fast Track Designation, a drug must meet an unmet need for a serious condition with no existing therap But did HIV qualify?

Figure 14.17 One of the demonstrations organized by ACT-UP the 1980s.

AP Photo/Tim Clary

The use of the word "serious" was considered carefully. The level of seriousness is based on how a disease affects a person's day-to-day living and what might happen if the condition were left untreated. Cancer, Alzheimer disease, and heart failure were considered serious because if left untreated, they can be fatal. The same is true of HIV, and it was included in these conditions; therefore, drugs for its treatment were "fast tracked." ACT UP had been successful in its quest to speed up the approval of drugs for AIDS.

QUESTIONS

1. The FDA created its rules for drug development to protect the public from exposure to untested drugs. But imagine that a member of your family had a fatal condition whose treatment was very close to being approved by the FDA but not close enough. What would you try?

2. ACT UP was criticized for putting AIDS in the face of the public. But it worked. Do you think this was worth upsetting some people? Why or why not?

3. Our government was founded on activism; the ACT UP protests were just one example. Research other similar protests and write a short report.

4. Make a short list of some the drugs used to treat HIV/AIDS that have been fast-tracked.

IMMUNOGENETICS   **295**

The Essential 10

1 The immune system protects the body from infections and foreign invaders. (Section 14.1)

Many of these infectious agents enter the body through openings in the skin.

2 Two of the most important parts of the immune system are the inflammatory response and the adaptive immune response. (Section 14.1)

Cells of the immune system, called lymphocytes, protect us from infectious agents.

3 The antigens on the cells of a donor organ determine whether it will be a match for the recipient. (Section 14.2)

If they are different from the recipient's, the body will reject a transplanted organ.

4 By testing HLA haplotypes, a suitable match for an organ transplant can be determined. (Section 14.2)

The HLA genes encode antigens on cell surfaces.

5 Xenotransplantation may become widespread in the near future. (Section 14.2)

The transplantation of animal organs to humans is being investigated.

6 A, B, AB, and O are some of the major human blood types. (Section 14.3

Blood transfusions are usually based on these blood types.

7 The Rh blood type sometimes causes problems in newborns when the mother is Rh⁻ and the fetus is Rh⁺. (Section 14.3)

The two Rh blood types are Rh⁺ and Rh⁻.

8 HIV infects cells of the immune system and can lead to AIDS. (Section 14.4)

The virus that causes AIDS attacks and kills cells that trigger the immune response.

9 Allergies are becoming a serious problem in our society. (Section 14.5)

Allergens are part of our environment and can trigger an inappropriate immune response if they enter our bodies.

10 Individuals who want to donate organs should make their wishes known to their families. (Section 14.6)

If individuals wish to donate organs, but they do not communicate that fact, they may not end up as donors.

Review Questions

1. Name three differences between antigens and antibodies.

2. When a person needs an organ transplant, a plea is often made on television or the internet asking many people to come and be tested. What test is done, and what happens when a match is found?

3. Who is the best match for an organ transplant?

4. How are blood transfusions and organ donation similar? How are they different? Give three reasons for each.

5. What stops HIV from entering cells?

6. Make a list of the steps in an immune response.

7. An HLA match isn't easy to find; what might happen if an organ didn't match and was transplanted anyway?

8. Antirejection drugs are very important in transplantation. How do they work?

9. Blood types were once used to determine paternity. Find out how.

10. Why are blood types not always successful in determining paternity?

11. What are some of the functions of proteins on the cell surface?

12. HIV is transmitted via bodily fluids. Why is this infection so difficult to stop?

13. What is the HLA complex?

Application Questions

1. We are exposed to many potential infectious agents in a single day. Make a list of some of these agents that your immune system works against.

2. Search the internet for a story of a donor who gave an organ to a stranger. Write a short report and share it with the class.

3. Do you think people would be willing to receive an animal donor organ in a transplant? Would you be willing if it were a matter of life or death? Why or why not?

4. Can a person who is HIV positive donate blood? Why or why not?

5. Research the most recent work being done on the *CC-CKR5* gene and write a short report.

6. A new procedure being advertised on television freezes the cord blood of newborns. Do some research on this process and write a paragraph discussing who could benefit from the use of cord blood and what it costs to store cord blood.

7. If a patient's body rejects a donated kidney after transplant, it becomes more difficult to find a match for this person. Why?

8. Who is the better match in each case for a person needing a transplant?
   a. an identical twin or a fraternal twin
   b. a fraternal twin or a sibling
   c. a mother or a father
   d. a complete stranger or a friend

   Give reasons for your answers.

9. Fast-tracking of drugs by the FDA began more than ten years ago, and it has now been applied to many more drugs than just those used to treat HIV. Compile a list of problems that have occurred because of fast-tracking.

10. Give your opinion on fast-tracking of drugs. Have there been any lawsuits based on this process?

Online Resources

Preparing for an exam? Log on at www.cengagebrain.com for study tools to help you assess your understanding. If assigned by your instructor, the Case B and Spotlight on Society activities for this chapter, "Prisoners Used as Guinea Pigs" and "Drug Development and Public Outcry," will also be available.

In the Media

USA Today, April 28, 2011

Boston Hospital Performs Second Full-Face Transplant

A team of more than 30 doctors, nurses, and other staff at Brigham and Women's Hospital worked for more than 14 hours last week to replace the full facial area of 30-year-old Mitch Hunter, of Speedway, Indiana.

The procedure replaced Hunter's nose, eyelids, lips, facial animation muscles, and the nerves that power them and provide sensation. Hunter suffered his injuries from a high-voltage electrical wire following a 2001 car accident. The donor family requested anonymity.

The lead surgeon, Dr. Bohdan Pomahac, said the procedure went smoothly and physicians expect Hunter to have a successful recovery and new life.

To access this article online, go to www.cengagebrain.com.

QUESTIONS

1. Why do you think the donor's family did not want their name revealed? List three reasons.

2. Two complete facial transplants done at Brigham and Women's Hospital were performed on patients who had much of their face destroyed by electric shock. Do some research on the partial face transplants and discuss how their injuries were caused. List them.

3. What will the recipient look like after the transplant? Himself? The donor? Does it matter?

Sydney Morning Herald, May 1, 2011

Fragile Life Saved in the Womb
Tim Barlass

Blood transfusions to fetuses in the mother's uterus are dangerous. In 2% of the cases there are deaths. In Sydney, Australia, Kellie Bush, who is Rh⁻, had the transfusion into her fetus with Rh⁺ blood. The physician used a fine needle to inject 90 ml of donated blood into a vein in the baby's liver, with an ultrasound screen in the darkened theater to guide him.

Finding a matching donor was also a challenge—the transfusion was planned for last week and blood was located in Victoria but

WHAT WOULD YOU DO IF...?

. . . someone in your family asked you to donate a kidney?

. . . someone you didn't know asked you to donate a kidney?

. . . you were asked to donate bone marrow? If you were paid, would your answer change? Why or why not?

. . . you were asked to move to a different seat in an airplane because you were eating a peanut butter and jelly sandwich?

. . . you were asked to participate in a trial for an AIDS vaccine?

then the transfusion was postponed. Only eight suitable donors were found.

It was the third transfusion during the pregnancy for Ms. Bush, whose husband, Paul, was by her side.

To access this article online, go to www.cengagebrain.com.

QUESTIONS

1. If there is such a danger to the fetus, why do you think the mother allowed it?

2. Whose blood type will this fetus have? The donor's? The mother's?

3. Using what you know about Rh factor, what would happen if they did not do the transplant and the baby was born?

Voice of America, August 31, 2007

Scientists Search for Allergy-Free Peanut
Paul Sisco

With peanut allergies increasing, scientists like Mohamed Ahmedna at North Carolina Agricultural and Technical State University are searching for a way to eliminate the allergens found in peanuts. He found a way to process peanuts after they are harvested that removes the allergy-related proteins. This means that farmers can continue to grow the strains they are familiar with, but the peanuts will be processed differently. These processed peanuts have not yet been tested on humans.

In another effort, Maria Gallo at the University of Florida is trying to grow peanuts that do not contain the proteins that cause allergic reactions.

To access this article online, go to www.cengagebrain.com.

QUESTIONS

1. If these scientists succeed in eliminating the allergen in peanuts, do you think it would be worth all the money spent in research? Why or why not?

2. One physician is giving patients small amounts of peanut dust to test subjects who have a known peanut allergy. Knowing what you do about the allergic reaction, explain how this may work.

15

Genetics and Populations

REVIEW

What can we learn from human pedigrees? (4.2)

How are autosomal dominant traits inherited? (4.2)

How does DNA repair itself? (BB2.8, 6.6)

CENTRAL POINTS

- Genetic conditions can become common in a specific population.

- Huntington disease affects large numbers of people in two villages in Venezuela.

- The frequency of traits can vary from one population to another.

- Calculations can determine the frequency of an allele in a population.

- Population genetics is used in cases involving DNA forensics.

CASE A: A Marker in the Blood

It began with a study of one family in Venezuela. This family and a much larger extended family live in two small villages around Lake Maracaibo. These villages have a higher frequency of Huntington disease (see Chapter 4 for a description of this disorder) than anywhere else in the world.

Nancy Wexler is a scientist who studies genetic disorders, including Huntington disease. When she learned of these villages in the 1970s, she began traveling there to learn why HD is so common.

She and her colleagues began by creating a pedigree of the families and collecting blood samples to test their DNA. The pedigree has grown to include almost 19,000 people, and more than 4,000 blood samples have been collected. These blood samples were used to identify a DNA marker in the people with HD. This marker led to the identification of the HD gene and, finally, to the development of a genetic test for HD.

For her work in identifying the gene for HD, Nancy Wexler won the Lasker Award, often called the U.S. equivalent of the Nobel Prize. However, she also had a personal reason for her interest in HD:

her mother, Lenore, and three of her uncles died from HD. In the later stages of the disease, Nancy's mother had lost her memory and had to be fed and cared for by Nancy's father, Milton. Later, Milton Wexler started the Hereditary Disease Foundation to study Huntington disease and related disorders, and Nancy made HD her life's work.

Nancy Wexler holding a photo of a Venezuelan man with HD.

Although a genetic test for HD was developed in 1993 and Nancy and her sister Alice are at risk for HD, Nancy has decided not to take the test. Alice was tested and found that she does not carry the HD gene.

Some questions come to mind when reading this case. Before we can address those questions, let's look at how the study of populations can open our eyes to certain genetic conditions.

Nancy Wexler and the Pedigree of a Large Venezuelan Family

15.1 Why Do Geneticists Study Populations?

In small communities such as those around Lake Maracaibo (Figure 15.1, page 302), people rarely travel very far and tend to marry others in the same village or nearby villages. Over many generations, inherited disorders can become more common because the populations are so isolated.

In these situations, the mutant allele causing the disorder usually was brought into the community many generations ago. In the Venezuelan villages with HD, the mutant allele arrived at the beginning of the 19th century. Almost all of the affected individuals in the pedigree can trace their ancestry to one woman, Maria Concepción Soto. She had 10 children and became the founder of the population of more than 18,000 people. Populations whose origins can be traced to a small number of people are an example of a phenomenon geneticists call the **founder effect**.

When constructing pedigrees of large families with a founder effect, geneticists discovered that these populations often have a high frequency of one or more genetic disorders. Wexler (Figure 15.2, page 302) and her team used pedigrees, along with blood and tissue samples, to identify, map, and isolate genes responsible for a number of genetic disorders. The gene for Huntington disease was one of those success stories.

founder effect populations whose origins can be traced to a small number of people

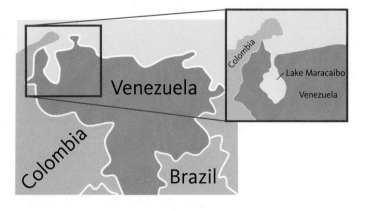

Figure 15.1 **Map of Lake Maracaibo, Venezuela** Nancy Wexler studied a large family with Huntington disease at Lake Maracaibo.

Figure 15.2 **Nancy Wexler Greeting a Boy on Her Return to Venezuela** During her trips to the Lake Maracaibo region, Nancy became very close with many of the families.

Acey Harper/Time Life Pictures/Getty Images

What causes Huntington disease?

Huntington disease is caused by the expansion of a repeated DNA triplet in the HD gene, located on chromosome 4. Normal copies of the gene carry between 10 and 27 copies of the CAG repeat (Table 15.1). People who carry fewer than 27 CAG repeats do not get HD. People who carry 27 to 35 copies of the repeat do not get HD themselves, but their children are at risk if the HD allele they inherit expands the number of repeats (Figure 15.3). Those who carry 36 to 39 copies of the repeat may or may not get HD, but those with 39 or more repeats almost always get HD. Anyone who inherits an allele carrying more than 60 repeats develops severe symptoms of HD as an adolescent.

The number of repeats in this gene is not always stable, and as the HD gene is passed to new generations, the number of repeats can increase, putting more people at risk.

Recall that in Chapter 4, Alan and his family worried about getting HD. If Alan or other members of his family carry the mutant HD allele, the disease will appear in midlife, usually after people have already had children.

The population in Venezuela sparked the curiosity of geneticists. Other populations around the world with high frequencies of other genetic disorders have been studied to discover what they have in common.

Table 15.1 **CAG Repeats and Huntington Disease**

Number of CAG Repeats	Allele Classification	Disease Status
< 27	Normal	Unaffected
27–35	Intermediate, expandable allele	Unaffected
36–39	Variable phenotype	Some affected, some unaffected
> 39	Full HD	Affected
> 60	Juvenile onset	Affected

Figure 15.3 **An Allele of the HD Gene Containing 29 CAG Repeats**

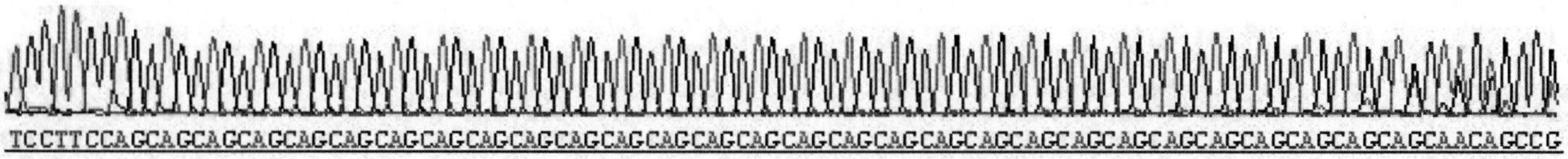

TCCTTCCAGCAGCAGCAGCAGCAGCAGCAGCAGCAGCAGCAGCAGCAGCAGCAGCAGCAGCAGCAGCAGCAGCAGCAGCAGCAGCAGCAGCAACAGCCG

15.1 Essentials

- **Certain genetic conditions can occur at a high frequency in a single community.**
- **Pedigrees can be helpful in tracking a condition through a large population.**
- **Huntington disease is caused by an increased number of copies of a repeated DNA sequence.**

15.2 What Other Genetic Disorders Are Present at High Frequencies in Specific Populations?

When looking at the map of Africa, Europe, and Asia (Figures 15.4 and 15.5), investigators saw that the geographic distribution of sickle-cell anemia almost completely overlapped with areas affected by malaria. They wondered if this overlap might be more than a coincidence and whether there might be a link between sickle-cell anemia and malaria.

Figure 15.4 Geographic Distribution of Sickle-Cell Anemia Notice how the frequency increases in certain areas of sub-Saharan Africa. Might this distribution change over time as people migrate?

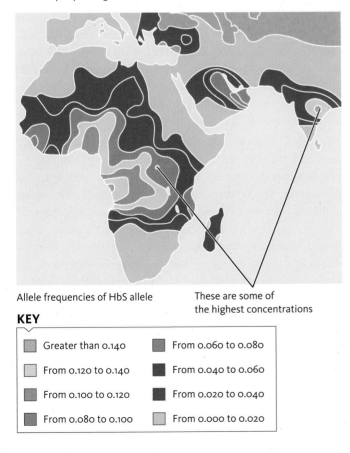

Allele frequencies of HbS allele

These are some of the highest concentrations

KEY

Greater than 0.140	From 0.060 to 0.080
From 0.120 to 0.140	From 0.040 to 0.060
From 0.100 to 0.120	From 0.020 to 0.040
From 0.080 to 0.100	From 0.000 to 0.020

Figure 15.5 Geographic Distribution of Malaria This map shows that regions with malaria overlap those with high frequencies of the sickle-cell allele.

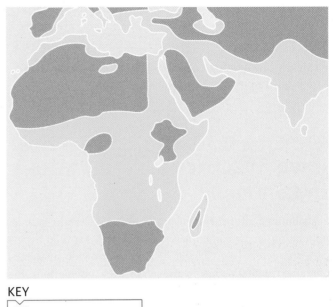

KEY

Regions with malaria

People with malaria experience recurring debilitating episodes of illness throughout life and often die at a young age. Although malaria may seem like an exotic disease to many of us, it affects more than 500 million people worldwide and kills more than 3 million people each year.

Malaria is caused by a parasite that infects red blood cells as part of its life cycle. Mosquitoes (Figure 15.6) that bite infected humans take in a small amount of their blood. These insects become infected with the malaria parasite and spread the disease when they bite uninfected people.

Researchers discovered that people who either are carriers of sickle-cell anemia (heterozygotes, *Ss*) or have the disorder (homozygotes, *ss*) are resistant to infection by the malaria parasite. In these individuals, the membrane of red blood cells is altered, making it difficult for the parasite to enter the cells. Although those with sickle-cell anemia are resistant to malaria, they still have sickle-cell anemia, and they often become very ill.

Heterozygous carriers of the mutant allele have an advantage over the other genotypes because they are resistant to malaria *and* do not have sickle-cell anemia. As we will see when we discuss natural selection, this

James Gathany/CDC Photo Library

Figure 15.6 The Anopheles mosquito carries malarial parasites and transmits them into a human's bloodstream when it bites.

heterozygote advantage led to the spread of the sickle-cell allele across populations in areas with malaria. In a later section of this chapter, we'll discuss the role of natural selection in the relationship between sickle-cell anemia and malaria.

15.2 Essentials

- **Sickle-cell anemia is an example of a disorder with a high frequency in certain populations.**

15.3 What Other Populations Have Specific Genetic Traits in Common?

Distributions of genetic disorders among certain human populations show some interesting patterns. Familial hypercholesterolemia (see Chapter 4) has a worldwide frequency of about 1 in 500 people. However, in a few populations, the frequency of this disease is much higher. For example, a small number of Dutch migrants (Afrikaners) moved to South Africa in the 1600s. These settlers carried with them a mutant allele for hypercholesterolemia. Because of the small size of the immigrant population and subsequent inbreeding, the frequency of hypercholesterolemia in this population is much higher than average; about 1 in 70 people are affected with this disorder.

As seen in Table 15.2, some populations have a much higher frequency of heterozygous carriers for certain recessive traits than others, called a **carrier frequency**. You can see there are some huge differences among various populations. In populations at high risk for certain disorders, the chances are greater that members of the population will be heterozygous carriers.

What can cause an increase in allele frequency in a specific population? Most scientists agree that there are several factors that affect allele frequency in a population. The first is mutation (see Chapter 6), which is the source of all new alleles. However, in the

heterozygote advantage situation where being heterozygous for a recessive trait confers an advantage for survival and reproduction

carrier frequency frequency of heterozygous carriers of a recessive gene in a specific population

genetic drift significant changes in allele frequency that occur by chance

natural selection differential reproduction among members of a population that is the result of a genetic advantage

Table 15.2 Frequencies of Carriers of Tay-Sachs and Cystic Fibrosis within Populations

Genetic Condition	Population Affected	Frequency of Carriers in a Population
Tay-Sachs	Ashkenazi Jewish in the United States	1 in 30 people are carriers of the mutant allele for Tay-Sachs
	All people in the United States	1 in 300 people are carriers
Cystic fibrosis	People in United States of Northern European descent	1 in 22 are carriers of the mutant allele for CF
	African Americans	1 in 62 are carriers
	Asian Americans	1 in 90 are carriers

long run, by themselves, mutations don't have much effect on allele frequency.

Second, *founder effects* can cause alleles to spread through small populations, as illustrated by the frequency of hypercholesterolemia in the Afrikaner population of South Africa. In addition, significant changes in allele frequency can also occur by chance, a process called **genetic drift**. The impact of drift is magnified in populations that are small, isolated, inbred, and stable over long periods of time. Yet, as we have seen, even large populations that do not live on remote islands and are not inbred have significant differences in allele frequency, demonstrating that founder effects and genetic drift are not the only factors that can change allele frequencies.

Third, **natural selection** is an evolutionary process in which the differential survival and reproduction of the best-adapted genotypes can change allele frequencies in a population. The relationship between sickle-cell anemia, an autosomal recessive disorder, and malaria is a classic example of how natural selection can change allele frequencies. Even though many untreated individuals with sickle-cell anemia die before reproducing, in some populations between 20 and 40% of the population is heterozygous for this disorder. If homozygotes carrying two copies of the mutant allele die before they reproduce, how can the frequency of the allele be so high? The answer is natural selection.

Those who carry one or two copies of the sickle-cell allele are resistant to malaria. This resistance gives heterozygotes a better chance of survival and the opportunity to have more children than those with a homozygous dominant genotype, who are susceptible to malaria. Those with a homozygous recessive genotype are resistant to malaria, but also have sickle-cell anemia and thus have a much lower survival rate. In this case, natural selection favors the survival and differential reproduction of heterozygotes (*Ss*) at the expense of the homozygous genotypes (*SS* and *ss*). Because of heterozygotes' resistance to malaria, and the ability to leave more offspring, the sickle-cell allele spread through populations in areas with malaria and is maintained at high levels by the serious effects of malaria infection and sickle-cell anemia.

In addition to natural selection, humans impose their own form of selection, called **artificial selection**, on many species of plants and animals by choosing individuals with desirable traits and using them as parents of the next generation. For example, the hundreds of different dog breeds are all derived from wolves by artificial selection. These breeds are then bred for certain traits (Figure 15.7).

15.3 Essentials

- **A number of genetic disorders are present at high frequencies in certain populations.**
- **There are several forces that change allele frequencies in population.**

15.4 Do Environmental Conditions Change the Frequency of Genetic Traits in Populations?

As discussed earlier in this chapter, the frequency of a trait may vary from population to population. Cystic fibrosis is an example of this. CF is an autosomal recessive genetic

disorder (see Chapter 4) that is common in some populations but relatively rare in others. In the United States, about 1 in 3300 children of European ancestry have CF, but for children of African descent or Asian descent, the frequencies are 1 in 15,000 and 1 in 32,000, respectively. Medical centers in many areas of the United States and other countries specialize in helping families with CF. Looking at Figure 15.8, you can see where these centers are located in the United States. Do you think that the location of these centers necessarily corresponds to places with the highest frequency of CF? Why or why not?

Figure 15.8 **Cystic Fibrosis Centers in the United States** Do you see any pattern of where they are located?

KEY

⭐ Coordinating center
● Research centers

artificial selection selection for desirable traits in plants and animals by human intervention

Figure 15.7 **Dogs Are Bred for Certain Qualities**
This miniature pinscher has given birth to six puppies. The puppies will closely resemble their mother when they are fully grown. This is an example of artificial selection.

Utekhina Anna/Shutterstock.com

Cystic fibrosis affects glands that produce mucus, digestive enzymes, and sweat, causing far-reaching symptoms. Most individuals with cystic fibrosis develop obstructive lung disease and infections, leading to premature death.

There is some evidence that people heterozygous for the CF gene (*Cc*) are more resistant to typhoid fever, an infectious disease caused by a bacterium (*Salmonella typhi*), which infects cells of the intestinal lining. When transgenic mice carrying copies of the human cystic fibrosis allele were injected with typhoid fever bacteria, scientists found fewer bacteria in their intestinal cells than were found in those of normal mice.

If heterozygous carriers of CF (*Cc*) are protected from one or more infectious diseases such as typhoid fever or related diseases, the mutant allele may be maintained in the population by natural selection in a way similar to that seen in the relationship between sickle-cell anemia and malaria in Africa and other regions.

15.4 Essentials

- **The frequency of traits may vary from one population to another.**

15.5 How Can We Measure the Frequency of Alleles in a Population?

If a genetic disorder such as sickle-cell anemia or cystic fibrosis is caused by a recessive allele, we cannot directly count those who carry that allele in a population. Why? We simply cannot tell the difference between heterozygous carriers of the mutant allele and those who are homozygous for the normal allele. Someone who is heterozygous for CF has the genotype *Cc* and has no symptoms. A person who does not carry the CF allele has the genotype *CC* and also has no symptoms. In other words, they have different genotypes but the same phenotype. The genotypes can be distinguished from each other only by a genetic test. Because over 1600 mutations have been identified in the CF gene, genetic testing for all these mutations is not practical. If population researchers cannot tell who is a heterozygote (*Cc*) and who is homozygous for the normal allele (*CC*), there is no direct way of counting how many heterozygotes (*Cc*) are in the population being studied.

But early in the 20th century, Godfrey Hardy and Wilhelm Weinberg (Figure 15.9) independently developed a simple formula that allows geneticists to measure the frequency of genotypes and alleles in a population without DNA testing of the population itself.

Hardy-Weinberg law mathematical formula for determining allele and genotype frequencies in a population

Now that we understand how certain genetic disorders interact with the environment and populations, let's look at some of the issues raised in the case of HD in Venezuela.

Nancy Wexler holding a photo of a Venezuelan man with HD.

1. A small part of the Venezuelan population has many individuals affected with HD. As we discussed earlier, the cause of HD is an increase in the number of copies of a CAG repeat sequence in the HD gene, carried on chromosome 4. As members of one of these populations reproduce, there is a significant chance they will pass the HD allele on to their children.

   Although most cases of HD appear in adults, those who inherit an HD allele with more than 60 repeats (see Table 15.2) may develop symptoms during their teen years. Could you or should you use prenatal diagnosis to determine your child's number of repeats?

2. What might be done to help the Venezuelan families plan for their future?

3. As discussed in Alan's case in Chapter 4, HD symptoms do not appear until later in life. But Nancy Wexler was born in 1945. Should Nancy Wexler take the test?

4. Draw the pedigree of Nancy and her sister. Include her father, her mother, her mother's three brothers who died of HD, her mother's mother and father, and her father's mother and father.

This formula, now called the **Hardy-Weinberg law**, is widely used by geneticists, clinicians, population biologists, evolutionary biologists, and others to study genes in populations.

To show how the Hardy-Weinberg law works, let's look at a gene with two alleles, the dominant allele (*A*) and the recessive allele (*a*). In their calculations, Hardy and Weinberg used *p* to represent the allele *A* (therefore, *p* = *A*) and *q* to represent the allele *a* (therefore, *q* = *a*).

Follow these statements below and it will become clear how the Hardy-Weinberg law works:

1. If a genotype has two copies of the dominant allele (*AA*), it is represented as p^2 ($p \times p = p^2$).

2. If a genotype has one copy of each allele (*Aa* or *aA*), it is represented as $2pq$ ($pq + qp$).

3. If a genotype has two copies of the recessive allele (*aa*), it is represented as q^2 ($q \times q = q^2$).

Figure 15.9 **Godfrey Hardy (left) and Wilhelm Weinberg (right)**
These two individuals discovered a way for geneticists to calculate allele and genotype frequencies in populations.

BY THE NUMBERS

1 in 10,000
Number of babies born each year with phenylketonuria (PKU) in the general U.S. population

1 in 4000
Number of babies born each year with PKU in the Turkish population

1 in 2000
People of European ancestry with cystic fibrosis in the United States

1 in 15,000
People of African ancestry with CF in the United States

1 in 32,000
People of Asian ancestry with CF in the United States

4. These are the only possible genotypes in a population (*AA*, *Aa*, and *aa*).

5. When added together, these three genotypes equal all the genotypes for this gene in the entire population.

6. Therefore, the equation $p^2 + 2pq + q^2 = 1$ represents 100% of the genotypes *AA*, *Aa*, *aa* in the population.

How can we use the Hardy-Weinberg equation? Measuring the frequency of alleles and genotypes in a population can be used to provide information about the gene pools of the population and the people in the population. For example, a couple can learn the risk factors for having a child affected with a genetic disorder. When a genetic disorder (such as sickle-cell anemia or cystic fibrosis) has a much higher frequency in one population than another, the Hardy-Weinberg law can be used to calculate how common the alleles are in these populations and how common a heterozygous carrier might be. From these calculations, someone from one of these populations can know his or her risk of having a child with a specific condition.

How are allele frequencies and heterozygote frequencies calculated? Suppose we want to discover the frequency of the allele for cystic fibrosis in a population, such as people with northern European ancestry in the United States, and also calculate how many heterozygotes (*Cc*) there are. Here is how we would do it:

1. Count the number of people in the population with cystic fibrosis. We can do this because we can identify everyone with the phenotype associated with this disorder. This will tell you how many people have the *cc* genotype. Let's say that in the population studied, we find 1 in 2500 people have cystic fibrosis. That is the frequency of CF in this population. Again, because CF is a recessive condition, we know the genotype of these individuals (*cc*).

2. Convert the frequency (1 in 2500) to a decimal. So, if 1 in 2500 people in a population has CF, as a decimal it is expressed as 0.0004 (1 divided by 2500). So, the genotype $cc = q^2 = 0.0004$.

3. To calculate q (the frequency of c), we take the square root of 0.0004. When you do that, you find that $q = 0.02$ or 2%. Therefore, 2% of all the CF alleles in this population are the mutant allele c; this means that 98% of the CF alleles in this population are the normal allele C.

Knowing the frequency of the alleles, you can calculate the percentage of heterozygous carriers (Cc) in the population by substituting the calculated allele values into the Hardy-Weinberg equation. In this equation the heterozygous carriers (Cc) are written as $2pq$. This is $2 \times p \times q$, or $2 \times 0.98 \times 0.02$, which equals 0.03 or 3%.

In other words, about 3% or 1 in 23 people of northern European ancestry in the U.S. population carry the gene for cystic fibrosis in the heterozygous condition.

How else can we use the Hardy-Weinberg equation?
Even though malaria is no longer found in the United States, some African Americans with West African ancestry still carry the sickle-cell allele as part of their genetic heritage. By counting the number of African American children with sickle-cell anemia born in the United States and using the Hardy-Weinberg law, we can calculate that about 8% or 1 in 12 African Americans with West African ancestry are carriers (Ss) of the mutant allele that causes sickle-cell anemia (s). This information may be of interest to couples such as the Johnsons (see Case A, Chapter 5) before they decide to get genetic testing.

In areas such as West Africa, where malaria is still a serious health problem, we can get useful information about the frequency of sickle-cell anemia using this same technique. Studies have shown that in some areas of West Africa, 20 to 40% of the population is heterozygous (Ss) for the sickle-cell allele, a striking difference from the 8% frequency for African Americans. This difference shows the impact of natural selection on allele frequencies.

15.5 Essentials
- One can measure the frequency of an allele in a population.
- Mathematically determining information about the genetic structure of a population can benefit all its members.
- Even if an advantage for having a specific allele existed in the past, changes in the environment may reduce the need for those benefits.

15.6 What Are the Legal and Ethical Issues Associated with Population Genetics?

The storing of genetic information in databases and the use of this information present a number of problems. When DNA

CASE B:
Girl Seeks Ancestors

Carina Jones was always different. Most of her fellow students could self-identify their race by looking in the mirror. She never could. On top of that, she was teased about her heritage.

Her father, Dan, an African American artist, met her mother while he was stationed with the army in Japan. He always said that the first time he saw his wife, Yoshi, he was in love. They married in Japan and came back to live in California, where Dan had grown up. Dan's mother, Carina's grandmother, always told her proudly that her great-grandmother had been a Sioux, and one of the first Native Americans to be college educated.

Right before Carina left for college, her mother's father died, and they all went to the funeral in Japan. Her grandmother was very nice to her, and even though she didn't speak English they seemed to bond. But Carina thought the Japanese looked at her strangely. Her father said she was imagining it.

On the return flight, Carina read a magazine article about Oprah Winfrey, who had just returned from a trip to Africa and confessed that she had had her DNA tested to investigate her African origins. The company she used offers tests to African Americans to see which of the 400 ethnic groups in Africa they are related to.

Carina thought she might major in biology, and making it personal seemed to make sense. She looked over at her father and wondered if she should ask him about the test.

1. **What should Carina do? Does she need her father's permission to take this test?**

2. **If Carina asks her father for permission, should he allow her to take the test? Why or why not?**

3. **Research some of the companies doing these tests and report on their prices and what tests they do.**

4. **Draw Carina's family's pedigree.**

5. **When Carina finds out her racial makeup, what might happen?**

is collected to construct DNA profiles from a large population, the results are usually stored in databases for later use. The use of this and other information from DNA databases raises several legal and ethical questions:

- Should anyone be forced to provide a DNA sample? What about those who are suspects, those who are arrested, or those who are convicted of a crime?

- Who has the authority to order that a DNA sample be taken? Does a police officer at the scene have that right? Would a judge, based on a subpoena, or a trial judge, after someone is convicted of a crime, have that power?

- Should the DNA profiles of those arrested but found innocent of a crime remain in the database, or should they be deleted?

- What crimes should be included in the database? All 50 states in the United States require those convicted of sex crimes to provide a DNA sample, and most states require convicted felons to provide a sample. Should this requirement be extended to lesser crimes such as fraud or income tax evasion?

- Private information unrelated to crimes, such as family relationships, paternity, and genetic disease susceptibility, can often be deduced from the samples. Police, forensic scientists, and researchers have access to people's DNA profile without their permission. Is this a violation of privacy rights?

- In the United Kingdom, the DNA database collects samples from all suspects, and ethnic minorities are overrepresented in the population of arrestees. Does this show that the criminal justice system is racially or ethnically biased?

The Hardy-Weinberg equation can be used to measure the frequency of DNA markers used in profiles after they are placed in databases. Using allele frequencies from various populations to sort DNA results can cause problems. Defense attorneys can argue that defendants do not receive fair testing if their samples are compared to a large population and not to their own ethnic group.

The opposite can also be argued. An attorney can ask that a larger population be used to match samples if this would give his or her client a fairer trial.

Table 15.3 addresses some of the ethical and legal questions related to DNA databases and population genetics.

discovery rule legal term that requires all evidence to be given to the defense before a trial

Table 15.3 **Ethical and Legal Questions in Populations**

Question	How Are These Questions Decided?	Related Case or Legal Issue
Comparing a defendant's DNA to samples in a database that includes members of a different ethnic group might not result in a match. Should a defendant have access to his or her DNA test results as well as the frequency of database alleles in his or her population subgroup?	If this information is available, the defense must be given all of the data. Courts demand that all evidence be disclosed by the prosecution and some by the defense.	This rule, called the **discovery rule**, has been in contention for a number of years. Some legislatures (Connecticut and Rhode Island) have passed laws that restrict what the prosecution has to give the defense.
Has the problem of which population databases to use in DNA matching been solved?	The FBI is compiling databases that take into consideration certain population subgroups and geographic backgrounds.	
Should states separate their DNA databases according to ethnicity or race?	Many think that this would be prejudicial to defendants because it narrows the number of suspects for comparison.	
Should DNA samples be taken from anyone suspected of a crime and placed in a state's database?	Some states have tried to institute this type of law.	So far New York and New Jersey have these laws on the books. Some legal experts believe this trend will continue.

Spotlight on Society

The Heritage of an African Tribe Is Revealed

An interesting combination of history and science came into play a few years ago when a sociologist and a geneticist got together to discover the truth behind the oral history of an African tribe.

In England, University of London scholar Dr. Tudor Parfitt was studying the background of the Lemba people, many of whom live in southern Africa. The Lemba's oral tradition tells of Jewish ancestors who left the Middle East thousands of years ago. Present-day members of the Lemba believe they are Jews descended from this population. Parfitt was interested in tracing their background to prove or disprove their claims (Figure 15.10).

He first met members of the Lemba when he gave a lecture at the University of South Africa about another African tribe that had immigrated to Israel. After his talk, two men wearing yarmulkes, or Jewish skullcaps, came up to him. They said that they were Jews and their ancestors had emigrated from the Middle East centuries before. Parfitt found this rather intriguing but difficult to believe. But he was curious about their claims and visited Lemba settlements.

He was amazed to find that their religious practices were very similar to those of Orthodox Judaism. They had kosher food restrictions, didn't allow marriage outside the faith, and had their sons circumcised.

Soon after his visit to South Africa, Parfitt was introduced to a group of scientists from the Genetic Anthropology Center at the University College of London. They were tracing the inheritance of the Y chromosome found in a group of males in Israel. These men were called the *Cohanim*, or Jewish priests. They claimed descent from a single Jewish ancestor, Aaron, the son of Moses, who lived 3000 years ago.

As discussed previously (see Chapter 1), the Y chromosome passes from father to son virtually unchanged. The University College group had identified a distinctive pattern of markers on the Y chromosome in some of these Israeli men. They called this pattern the **Cohen model haplotype**.

Parfitt suggested they work together to investigate the back ground of the Lemba and determine whether their oral hist about Jewish origins could be confirmed by analyzing the markers on their Y chromosomes. A group of geneticists tra eled with him to southern Africa and took DNA samples fro the Lemba leaders and their rabbis (Figure 15.11).

Analysis showed that 1 in 10 males of the Lemba carried th Cohen model haplotype. The DNA analysis in Figure 15.12 compares the Y chromosome haplotype of the Lemba and t *Cohanim*.

Figure 15.11 This Lemba leader is called a rabbi, as in oth Jewish congregations.

Figure 15.10 Members of the Lemba Preparing for Worship Many of the Lemba's religious rituals are similar to those of Jews in other countries.

Cohen model haplotype pattern of markers on the Y chro some found in some Israeli men

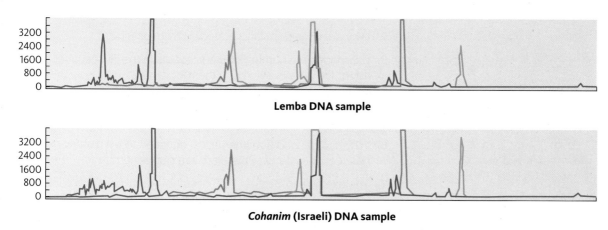

Figure 15.12 The pattern of Y chromosome DNA markers in a Lemba male (top) compared with the Cohen haplotype from an Israeli male (bottom). Do you think these match up?

Lemba DNA sample

Cohanim (Israeli) DNA sample

The analysis seems to prove that the Lemba do have Jewish ancestors who migrated from the Middle East. Could they use this information to become citizens of Israel? Research this question, and write a short report on what you find.

NOVA, the public television show, had a program on this topic. It was titled "The Lost Tribes of Israel" because legend has it that 2 of the original 12 tribes from the Bible have never been found. Could the Lemba be one of the tribes? Why or why not?

3. If you look at Figure 15.11, you might not immediately identify this man as being a Jew, even though he is wearing a traditional Jewish prayer shawl. What can you infer about the genetics of this group after it migrated to Africa?

1. Certain genetic diseases can be present at a high frequency in a population. (Section 15.1)

This is because one person many generations ago helped found the population and passed the allele for this disorder on to his or her offspring.

2. Pedigrees can be helpful in tracking a condition through a large population. (Section 15.1)

Because of language and cultural differences, pedigrees are often difficult to obtain.

3. Huntington disease is caused by an expanded number of copies of a repeated DNA sequence. (Section 15.1)

This CAG sequence can be repeated dozens of times in a mutant HD allele.

4. Sickle-cell anemia is an example of a disorder that has spread through a population. (Section 15.2)

There is a benefit to carrying the mutant allele, which helped a population survive.

5. A number of genetic disorders are present at high frequencies in certain populations. (Section 15.3)

One of these conditions involves an elevated amount of cholesterol in one's blood.

6. We can measure the number of heterozygous carriers of an allele in a population. (Section 15.3)

This information can help members of the population determine the chances of having a child with a genetic disorder.

7. The frequency of traits may vary from one population to another. (Section 15.4)

An example is cystic fibrosis.

8. One can measure the frequency of an allele in a population. (Section 15.5)

Allele frequency can be mathematically determined using the Hardy-Weinberg law.

9. Researching the genetic structure of a population can benefit its members. (Section 15.5)

This information can be used to determine the origin of the population.

10. Even if an allele confers an advantage on those who carry it, the environment may change and reduce the need for that advantage. (Section 15.5)

If a population moves to a different area, it may not need that advantage anymore.

Review Questions

1. What is the founder effect?

2. Why are large populations studied by geneticists?

3. What does the number of CAG repeats in the HD gene tell us about the people who carry them?

4. How are malaria and sickle-cell anemia related?

5. What is an allele frequency?

6. Why do we need a mathematical equation to tell us how many heterozygous carriers there are in a population?

7. The example in Section 15.5 does not reveal what the condition is. Do we need to know that to figure out the allele frequency?

8. Name two ways that using the Hardy-Weinberg equation can help people in a population.

9. What is the cause of malaria?

10. Explain genetic drift.

Application Questions

1. Research other communities with a large number of people with HD, and write a short report.

2. Groups of people with leukemia and other cancers have been found all over the world. These are called *cancer clusters*. Do you think this is the same situation as with the people in Venezuela? Why or why not?

3. Do some research to find out how many repeats are found in those who develop HD in their teens and twenties, and write a short report.

4. How are people with the following genotypes affected by malaria (S = normal, s = sickled): SS, Ss, and ss?

5. How might the information you have learned in this chapter help the following people in the cases in previous chapters?
   a. Martha Lawrence and her husband (Chapter 3)
   b. Chris Crowley and his son, Mike (Chapter 6)
   c. The attorney for William Bern (Chapter 8)
   d. Maria Sanchez's family (Chapter 13)

6. Physicians sometimes use a blood type called MN in their diagnoses. It is similar to the ABO blood type discussed in Chapter 13 but has only two codominant alleles (M and N). In testing all 1-year-old children in a small population, you find genotype frequencies of MM = 0.25, MN = 0.5, and NN = 0.25. Using the Hardy-Weinberg law, determine the allele frequencies for M and N.

7. In adults in the population described in question 6, the genotype frequencies are MM = 0.3, MN = 0.4, and NN = 0.3. Using the Hardy-Weinberg law, determine the allele frequencies for M and N.

Figure 15.13 Populations are groups of individuals belonging to the same species that live in the same geographic area.

Jonathan and Angela

8. Some scientists believe that lactose tolerance was spread by natural selection. Research this and write a paragraph explaining the origin and spread of this gene variation.

9. Kurt Vonnegut's book *Galápagos* tells the story of a group of people on a boat who visit the Galápagos Islands while a war destroys the rest of the world. These people must repopulate the human race. What happens? Which term in this chapter would best apply to the plot?

10. Explain how prenatal diagnosis and abortion can be considered examples of artificial selection.

11. Knowing the frequency of a specific allele in a population can help prosecutors and defense attorneys. Compile a list of ways this knowledge can be helpful, mentioning at least three.

12. Look at Figures 15.4 and 15.5 again. Using these maps, would it be possible to determine where one's family came from based on sickle-cell carrier status? Why or why not?

13. The photo in Figure 15.13 shows a population of zebras. It is obvious that this population has many similarities, but what phenotypic differences can you observe among individuals, and what factors contribute to these differences?

▶ Online Resources

Preparing for an exam? Log on at www.cengagebrain.com for study tools to help you assess your understanding. If assigned by your instructor, the Case A and Spotlight on Society activities for this chapter, "A Marker in the Blood" and "The Heritage of an African Tribe Is Revealed," will also be available.

In the Media

BBC News, April 24, 2008

Human Line Nearly Split in Two
Paul Rincon

The Genographic Project, run by the National Geographic Society, has been tracking human migration by DNA samples. Based on analysis of mitochondrial DNA of Africans living today, researchers found that two different populations existed in isolation between 50,000 and 100,000 years ago. The two populations seemed to have come together in the Stone Age, forced by population expansion.

To access this article online, go to www.cengagebrain.com.

QUESTION

If the population of early humans had split into two, what do you think might have happened?

The Wall Street Journal, April 15, 2011

The Mother of All Languages
Gautam Naik

The world's 6000 or so modern languages may have all descended from a single ancestral tongue spoken by early African humans between 50,000 and 70,000 years ago, a new study suggests. The finding in the journal *Science* could help explain how the first spoken language emerged, spread, and contributed to the evolutionary success of the human species.

Quentin Atkinson, the author of the article, based his research on phonemes, distinct units of sound such as vowels, consonants, and tones, and an idea borrowed from population genetics known as the founder effect. That principle holds that when a very small number of individuals break off from a larger population, there is a gradual loss of genetic variation and complexity in the breakaway group.

Dr. Atkinson figured that if a similar founder effect could be discerned in phonemes, it would support the idea that modern verbal communication originated on that continent and only then expanded elsewhere.

In an analysis of 504 world languages, Dr. Atkinson found that, on average, dialects with the most phonemes are spoken in Africa, while those with the fewest phonemes are spoken in South America and on tropical islands in the Pacific.

"Our own observations suggest that it is possible to detect an arrow of time" underlying proto-human languages spoken more than 8000 years ago, said Murray Gell-Mann of the Santa Fe Institute in New Mexico, who read the *Science* paper and supports it. The "arrow of time" is based on the notion that it is possible to use data from modern languages to trace their origins back 10,000 years or even further.

To access this article online, go to www.cengagebrain.com.

QUESTION

How does this relate to the population in Venezuela with Huntington disease?

16

Human Evolution

REVIEW

How can we measure the frequency of alleles in a population? (15.4)

How are populations studied with genomic techniques? (BB3.9)

CENTRAL POINTS

- Evolution is the study of changes in the gene pools of populations and the process of species formation.

- Migration of early humans has been documented.

- Our genome is very similar to those of both Neanderthals and chimpanzees.

- Genomics has shed much light on the evolution of our species.

CASE A: Darwin's Dilemma

Early in life, Charles Darwin was a religious man. After he developed the theory of evolution by natural selection, he worried about it. The world was different then, and he wrote, "I fear great evil from vast opposition in opinion on all subjects of classification." Because of the social and scientific environment in which he lived, he was afraid that if he didn't back his theory with a great deal of evidence, it would never be accepted.

For most of his adult life, Darwin suffered from a number of illnesses (see Spotlight on History) that some experts believe were stress related, perhaps because of his concern about how the scientific community would accept his theory. At one time he wrote, "I am not so well as I was a year or two ago. I am working very hard on my book, perhaps too hard."

After working on a sketch of his theory, and finally writing a 50,000-word essay over a 10-year period, Darwin was still suffering from anxiety. In 1858, a scarlet fever epidemic swept across England. Two of Darwin's children became ill; one survived, but the youngest child, Charles, died. At about the same time, Darwin received the manuscript of a paper from Indonesia entitled "On the Tendency of Varieties to Depart Indefinitely from the Original Type." After reading it, he realized that the author, Alfred Russel Wallace, had independently discovered the role of natural selection

in species formation. He wrote to Charles Lyell at Oxford University, saying, "I never saw a more striking coincidence . . . all my originality, whatever it may amount to, will be smashed."

Charles Darwin

© Bettmann/CORBIS

Lyell and others suggested that Darwin hurry to publish his work before Wallace could publish his. Darwin wanted to wait to publish until he had accumulated a mass of overwhelming evidence, but he was persuaded to have a joint paper with Wallace presented at a meeting of the Linnaean Society, an important scientific group, in 1858. After that meeting, Darwin pressed forward and published his book *On the Origin of Species* in 1859. In a later book, *The Descent of Man*, Darwin applied evolutionary theory to human evolution and the origin of our species from other primates.

Some questions come to mind when reading about Charles Darwin's dilemma and the decisions he faced. Before we can address these questions, let's look at some information about human evolution and what is known today.

Artist's Interpretation of a Neanderthal

16.1 What Is Evolution?

We can define **evolution** as a change in the genetic composition of a population over successive generations, as a result of natural selection acting on the genetic variation present among individuals, often resulting in the development of new species. From the definition, you can see that the study of evolution takes in much more than just the origin of our species. Here in this chapter, however, we will limit our discussion to the role of human genetics in exploring the evolution of human species, including our own.

Understanding human evolution begins with an understanding of geological time. Our species has only existed for less than 200,000 years, whereas the evolutionary history of life on Earth spans about 3.6 billion years. To put this in perspective, Figure 16.1, (page 318) shows a way of expressing the dimensions of this time frame.

16.1 Essentials

- **Understanding geologic time is important in understanding evolution.**

evolution change in the gene pool of a population

© Federico Gambarini/dpa/Corbis

Figure 16.1 **A Clock Representing the Evolutionary Time Scale** If we begin the formation of the earth at 12 a.m. and work around the clock, we can see that man is present in only a tiny section of the clock (shown in red).

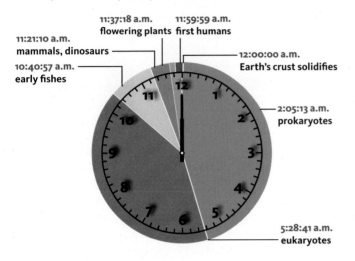

11:37:18 a.m.
flowering plants

11:59:59 a.m.
first humans

11:21:10 a.m.
mammals, dinosaurs

10:40:57 a.m.
early fishes

12:00:00 a.m.
Earth's crust solidifies

2:05:13 a.m.
prokaryotes

5:28:41 a.m.
eukaryotes

16.2 How Do We Study Human Evolution?

Modern ways of studying human evolution use the tools of genomics, anthropology, paleontology, taxonomy, archaeology, and even satellite mapping from space. These methods have helped us reconstruct the origins and ancestries of our species, *Homo sapiens*; establish our genetic relationships with other primates and other human species; and reconstruct the migrations that dispersed us across the globe.

What are fossils? **Fossils** are the preserved remains of once-living creatures (Figure 16.2). Different kinds of fossils are formed in different ways. No matter how they are formed, they need to be extracted very carefully. The age of the fossil can be determined by dating both the fossil and the layer of sediment or rock in which it is found. As fossils are categorized and analyzed using techniques that include isotope dating, x-ray examination, and now DNA analysis, they become part of the **fossil record**.

What have we learned from the fossil record? In many ways, our lineage began with the **primates**, a group of animals that first appeared in the fossil

record about 65 million years ago. Primates have eyes on the front of the face and color vision. They also have hands and feet with flattened nails instead of claws, and their hands are able to grasp objects. In addition, they have large brains relative to their body size and show complex social behavior, including parental care of offspring.

The primates consist of two main groups:

- the *prosimians*, which include lemurs, lorises, and tarsiers (Figure 16.3a). Most of these primates are small tree-dwellers that are active at night.

- the **anthropoids** (the word means "human-like"), which include New World monkeys of Central and South America, the Old World monkeys of Africa and Asia, and the **hominids**, including gorillas and chimpanzees (Figure 16.3b, c).

Africa is home to many living species of hominids, and the African fossil record traces our lineage from extinct hominids. However, there are many gaps in this part of the fossil record. With parts of the fossil record missing, is there another way to reconstruct the evolutionary relationships between other hominids and our species and provide a time scale for these events? As we will see later in this chapter, the genomes of living hominid species contain clues to those evolutionary relationships. These genomic differences can be used to reconstruct evolutionary relationships among species, producing a phylogenetic tree that shows the relationships among species with common ancestors. Figure 16.4 depicts a tree showing the evolutionary relationships among humans and other hominids based on genomic analysis. The tree shows that the hominid lineage began about 25 million years ago, and that chimpanzees and humans last shared a common ancestor about 7 million years ago, making chimpanzees our closest relative among the hominids. Until recently, the evolutionary events in the groups leading to our species after the split from the chimpanzee line were reconstructed mainly from the fossil record.

Figure 16.2 **Fossils Preserve the Remains of Organisms That Lived in the Past** This fish is an example.

Homo sapiens modern humans

fossil preserved record of extinct animals and plants

fossil record all fossils that have been described and catalogued

primates the group of mammals that includes humans

anthropoids the group of primates to which humans belong

hominids the group humans and great apes belong to

hominins humans and extinct related forms characterized by upright posture and walking on two legs

Figure 16.3 **Some Primates** (a) A tarsier is a prosimian (small enough to fit in your hand). Anthropoids include (b) the gorilla and (c) the chimpanzee. Humans are also anthropoids.

BlueOrange Studio/Shutterstock.com

a.

Mike Price/Shutterstock.com

b.

© imagebroker/Alamy

c.

16.2 Essentials

- **Humans belong to a group of mammals known as primates.**
- **Humans last shared a common ancestor with chimpanzees about 7 million years ago.**

16.3 When Did Early Humans Appear?

Over a period of several million years after the split from the line that led to chimpanzees, a large species group (14 or more species), collectively called **hominins**, appeared in Africa and led to our species (Figure 16.5).

As a group, hominins are defined by their upright posture and the fact that they walk on two legs. The oldest, from about 4.4 million years ago, is a species known as

Figure 16.4 **Hominids** This phylogenetic tree shows the evolutionary relationships among hominids. Note that chimpanzees and humans last shared a common ancestor about 7 million years ago, and that chimpanzees are our closest hominid relative.

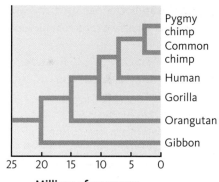

Pygmy chimp
Common chimp
Human
Gorilla
Orangutan
Gibbon

25   20   15   10   5   0
Millions of years ago

Figure 16.5 **Hominins** Estimates of the dates of origin and extinction of the three main groups of hominins (green, blue, and orange). The australopithecines split into two groups about 2.5 million to 2.7 million years ago. One of those groups, the genus *Homo*, contained the ancestors to our species, *Homo sapiens*.

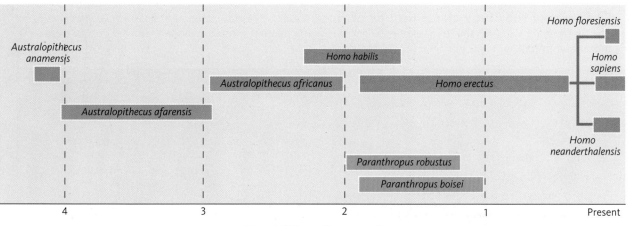

Australopithecus anamensis

Homo floresiensis

Homo habilis

Homo sapiens

Australopithecus africanus

Homo erectus

Australopithecus afarensis

Homo neanderthalensis

Paranthropus robustus

Paranthropus boisei

4   3   2   1   Present
Time (millions of years ago)

Figure 16.6 **Reconstruction of *Ardipithecus ramidis* ("Ardi")** Members of this species were about 1 meter tall.

Figure 16.7 **Hominid Fossils** (a) Fossil skeleton of *A. afarensis* (nicknamed Lucy) that lived in Africa 3.2 million years ago. (b, c) Footprints left by hominids about 3.7 million years ago.

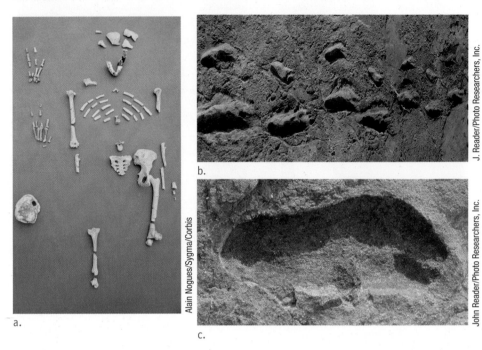

Ardipithecus ramidis, discovered as a group of 36 fossils in 2009. The most complete skeleton was nicknamed "Ardi," and a reconstruction is shown in Figure 16.6. Later hominin fossils were organized into three groups. Fossils of the first of these groups, ***Australopithecus***, first appeared just over 4 million years ago. These individuals were short (1–1.5 meters tall), with small brain cases, and walked upright (Figure 16.7). About 2.5 million to 3 million years ago, the australopithecines diverged in at least two directions. One line led to the *Paranthropus* species group, a genetic dead end that became extinct about 1 million years ago. The other line led to the genus *Homo*, our ancestral group, all members of which are called humans.

One early human species, known only from fragmentary fossils, is called *Homo habilis* ("handy man") because this species is thought to be the first to manufacture and use tools. In response to as-yet-unknown selection pressures and/or advantageous mutations, *Homo habilis* had a brain about one-third larger than australopithecines and used simple tools. A successor species, ***Homo erectus***, was taller than *H. habilis* and had a still larger brain (Figure 16.8). *H. erectus* made relatively sophisticated tool

Australopithecus an early relative of *Homo sapiens* that evolved in Africa

Homo erectus an early human species

Figure 16.8 **Skulls of the Three Major Groups of Hominins** (a) *Australopithecus africanus*, (b) *Paranthropus boisei*, and two species of *Homo*: (c) *H. habilis* and (d) *H. erectus*. Evidence suggests that our species, *H. sapiens*, evolved from *H. erectus* somewhere in Southwest Africa.

a. *Australopithecus africanus*   **b.** *Paranthropus boisei*   **c.** *Homo habilis*   **d.** *Homo erectus*

Table 16.1 Some Hominins

Genus	Species	Fossil Dates (millions of years ago)	Brain Size (cm³)
Ardipithecus	A. ramidis	4.4	325
Australopithecus	A. afarensis	4	450
Paranthropus	P. boisei	2.4	510
Homo	H. habilis	2.6	550
	H. erectus	2	1000

kits, used fire, and may have hunted animals. About 2 million years ago, this very successful species migrated out of Africa to the Middle East, Europe, and Asia, reaching as far as China and Indonesia. Table 16.1 summarizes some features of these early hominins.

16.3 Essentials

- **Ancestral hominins appeared about 4 million years ago.**
- ***Homo erectus* migrated out of Africa into Eurasia about 2 million years ago.**

16.4 When and Where Did Our Species, *Homo sapiens*, Originate?

Until recently, there were two opposing views on how and where our species originated: the multiregional hypothesis and the out-of-Africa hypothesis. Both agree that *H. erectus* originated in Africa and spread from there into Europe and Asia between 1 and 2 million years ago. Those supporting the multiregional hypothesis used fossil evidence to argue that, after leaving Africa, *H. erectus* formed a network of inter-breeding populations that gradually transformed into our species, *H. sapiens*.

The other school of thought, called out-of-Africa, used a combination of genetic and fossil evidence to argue that our species originated in Africa from *H. erectus* between 200,000 and 100,000 years ago. Genetic evidence shows that small groups of *H. sapiens* migrated from Africa 60,000 years ago and spread into Europe and Asia, eventually replacing populations of other human species, including *H. erectus* and Neanderthals (*H. neanderthalensis*), that lived in these regions. The two opposing ideas can be summarized as follows: one favors evolution and transition within a single species (the multiregional model), and the other favors the origin of our species in Africa, followed by migration and replacement of other species on other continents (the out-of-Africa model).

The evidence now overwhelmingly supports the out-of-Africa model and identifies Southwest Africa as the most likely place where modern humans originated and Eastern Africa as the point of migration from Africa. According to this model, modern human populations are all derived from a single event that took place in a restricted region of Africa. As a result, the human populations in Africa have the highest degree of genetic diversity, and all populations outside Africa show a high degree of genetic relatedness because they are derived from small groups that left Africa (Figure 16.9, page 322).

Now that we know something about human evolution, we can understand some of the issues Darwin might have been worried about in Case A. Look at the questions below.

CASE A: QUESTIONS

1. Knowing there would be a great deal of controversy when his book was published, did Darwin do all he could to prevent this?

2. The manuscript from Wallace caused Darwin a great deal of anguish. If he could have proved that he had the idea first, would that make a difference?

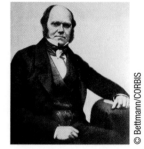

Charles Darwin

© Bettmann/CORBIS

3. If Wallace had published his paper first, what could Darwin have done? Should he have invited Wallace to contribute to the book, or even coauthor it? Keep in mind that royalty payments are made to authors of most books.

4. If Darwin and Wallace lived today, could Darwin have sued Wallace? What arguments might he have used to prove his case against Wallace?

Can we trace our ancient migrations across the globe? There is strong evidence that *H. sapiens* originated in Africa and spread from there to other parts of the world. There may have been one primary migration or several from a base in Eastern Africa. These emigrants carried a subset of the genetic variation present in the African population, consistent with the finding that present-day non-African populations have a small amount of genetic variation compared to African populations.

A logical question is, how can we map out migrations that occurred thousands of years ago and left no written records? The answer is written in the genomes of present-day populations. To work out these routes, geneticists use SNP haplotypes (see "Biology Basics: Genes, Populations, and the Environment") from the Y chromosome and mitochondria

Figure 16.9 **A phylogenetic tree** This tree shows that all non-African humans share a closer genetic relationship with each other than with African populations (Sub-Saharan and Paleoafrican). The arrow shows the branching of the tree that separates all non-African from African populations.

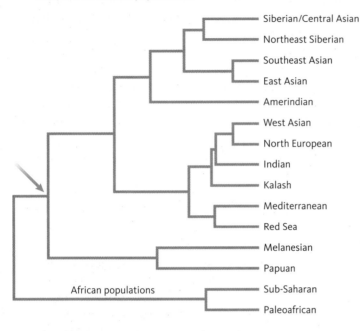

and trace them. Y chromosomes and their haplotypes are passed directly from father to son, whereas mitochondria and their haplotypes are passed from a mother to all her children. Using Y chromosome SNPs, men can trace their paternal heritage, and using mitochondrial SNPs, both men and women can trace their maternal heritage.

Over time, as mutations arise in Y chromosomal and mitochondrial DNA, they are passed on to future generations as part of SNP haplotypes and become identifiable markers. If people leave that region, they carry those haplotypes with them, making their path of migration traceable. Ancient migration routes are traced by cataloging the haplotypes in existing indigenous populations. DNA samples donated by about 10,000 members of indigenous and traditional people from around the world formed the database. Using this catalog, scientists can work backward to track the haplotypes through different populations. Each haplotype we carry represents an ancient point of origin, and an end point (where we are

now) along a path of migration. By surveying many people in present-day populations, the track of each haplotype can be reconstructed and gives us a history of the spread of our species across the globe (Figure 16.10).

16.4 Essentials
- **There is strong evidence that our species, *Homo sapiens*, originated in Africa.**
- **Migrations of ancient populations can be traced using SNP haplotypes.**

16.5 What Can Genomics Tell Us about Our Evolutionary History?

The fossil record provides an outline of the events in the history of our species and enables us to catalog the major phenotypic changes that have taken place during human evolution. However, fossils cannot supply us with any information about the genetic changes that took place after the separation of the human and chimpanzee lines. Now, however, genomic techniques including DNA sequencing, bioinformatics, the development of SNP markers, and microarray technology have revolutionized many areas of genetics, including evolutionary genetics. Once the human genome sequence was completed, the genomes of closely related primates, including the gorilla and the chimpanzee, were sequenced. Recently, scientists completed the genome sequence of another human species *(H. neanderthalensis)* with whom we coexisted for

Figure 16.10 **The Origins and Paths of Migrations of *H. sapiens* Reconstructed from Fossil and Genetic Evidence**

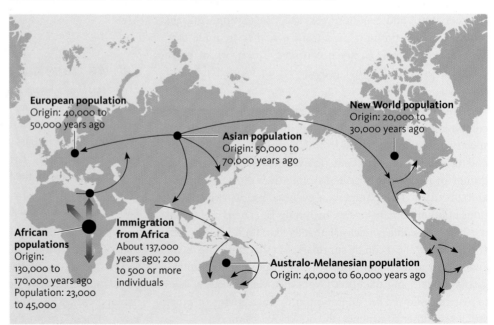

over 30,000 years. These advances now allow us to examine what genetic changes took place after the split from the chimpanzee line and to determine what genetic changes separate us from Neanderthals.

In the following sections, we will discuss what we have learned by comparing the genomes of the chimpanzee, our closest primate relative; the genome of our closest human relative, the Neanderthals; and our genome.

Is our genome similar to that of the chimpanzee?

The chimpanzee and human genomes have been separated for about 7 million years. Now that the genome sequences of these two species are available, analysis shows many similarities, but also many subtle differences:

- In spite of their long separation from a common ancestor, the sequences of the human and chimpanzee genomes are 98.8% identical.

- A number of important genome differences, including insertions, deletions, and duplications, exist between the species. Many of these are present in one species but not the other, potentially changing the effect of genes and helping explain phenotypic differences between the species.

- Phenotypic differences between humans and chimps cannot be explained only by differences in coding sequences and dosage; they probably also involve changes in gene expression and regulation. Significant differences between the two genomes in promoter sequences and transcription factors have been identified, meaning that there may be important differences in mechanisms of gene regulation.

Comparative genome analysis has identified about 14,000 protein-coding genes that have changed since humans and chimpanzees shared a common ancestor. Most of these changes are single nucleotide substitutions, indicating that although our genome sequences are very similar, small differences in amino acid composition of proteins can make big differences in phenotypes.

Are we descended from the Neanderthals?

Three known species followed *Homo erectus*. They are *Homo neanderthalensis*, *Homo sapiens*, and *Homo floresiensis*. Fossil evidence indicates that Neanderthals lived in the Middle East, Western Asia, and Europe 300,000 to about 30,000 years ago. For at least 30,000 years, Neanderthals lived alongside *H. sapiens*, raising several questions: (1) Were Neanderthals our ancestors? (2) What differences are there between our species and Neanderthals? (3) Was there interbreeding between the two species? In other words, do we carry some Neanderthal genes?

To answer the first question, researchers sequenced more than 1 million nucleotides of Neanderthal DNA. Comparison of the small number of differences detected showed that we are not descended from the Neanderthals and that the two species last shared a common ancestor about 400,000 years ago (Figure 16.11).

Are we closely related to the Neanderthals?

Knowing the two species separated about 400,000 years ago, how much has our genome changed in that time? Comparison of our genome to the Neanderthal genome shows that

Figure 16.11 Genomic and fossil evidence was used to estimate the times of divergence of human and Neanderthal genomes relative to landmark events in both human and Neanderthal evolution.

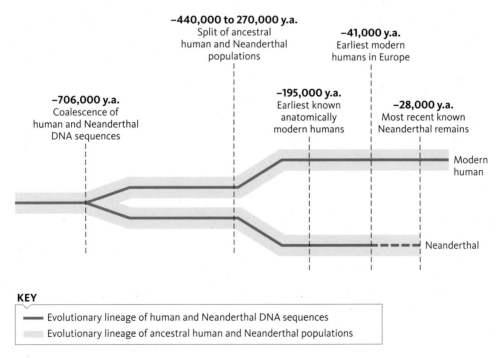

−440,000 to 270,000 y.a.
Split of ancestral human and Neanderthal populations

−41,000 y.a.
Earliest modern humans in Europe

−706,000 y.a.
Coalescence of human and Neanderthal DNA sequences

−195,000 y.a.
Earliest known anatomically modern humans

−28,000 y.a.
Most recent known Neanderthal remains

Modern human

Neanderthal

KEY
— Evolutionary lineage of human and Neanderthal DNA sequences
▨ Evolutionary lineage of ancestral human and Neanderthal populations

they are the same size, and that the DNA sequences are 99.7% identical. This means that there are very few differences between our genome and that of the Neanderthals. Although these differences are small, could any of these changes show up as differences in protein-coding genes that may help explain the phenotypic divergence between our species and Neanderthals?

The answer came from a cross-genome analysis of nucleotide substitutions that changed the amino acid sequence in protein-coding genes. If our genome and the Neanderthal genome carry the same nucleotide substitution, then this change must have occurred before the Neanderthal and modern human lines split from each other. If the substitution is present in our genome, but not the Neanderthal genome, then the substitution occurred after the split between the two species. The results show that almost all of the 11,000 substitutions identified among the chimp's, Neanderthal's, and our genome took place before the split between Neanderthals and modern humans. Comparison with the chimpanzee genome shows that there was significant evolutionary change after the split from the chimpanzee line, but little change after the split between our species and the Neanderthals.

The researchers found only 88 substitutions in 83 genes that separate our genome from that of Neanderthals. In other words, we are nearly genetically identical to Neanderthals. However, although the differences are few, some of these changes are in important genes that control cognitive development and formation of the skeleton. These gene differences may help explain the phenotypic differences between the two species.

Do we carry any Neanderthal genes?
Members of our species migrated out of Africa between 60,000 and 100,000 years ago and encountered Neanderthal populations in the Middle East. The two species coexisted there and in other areas for thousands of years, before Neanderthals became extinct.

As part of the Neanderthal genome project, researchers compared sequences in the Neanderthal genome with those of five living humans from different parts of the world: a San tribesman from South Africa, a member of the Yoruba tribe from West Africa, a Western European, a Han Chinese, and a Pacific islander. The researchers found that the genomes of those living outside of Africa contained from 1 to 4% Neanderthal sequences, but that no traces of the Neanderthal genome were detected in people from Africa. Researchers concluded that interbreeding between the two species took place after prehistoric humans migrated out of Africa, probably in the Middle East (Figure 16.12), and before they dispersed to Europe and Asia.

Have we identified all our human relatives?
When our species (H. sapiens) migrated out of Africa, parts of Europe and Asia were already occupied by humans of other species. Until recently, it was thought that when this migration took place, Neanderthals (H. neanderthalensis) were the only other existing human species. Now, however, it is clear that a number of species coexisted in Eurasia before and after our species appeared in these regions.

The story began in 2004 with the discovery (Figure 16.13) of fossil human skeletons on the island of Flores in Indonesia. Members of this new species, called **Homo floresiensis**, were only about 1 meter tall (called "hobbits" by the popular press) and were adapted to life on an island (Figure 16.14). The fossil record extends from 38,000 years ago to about 13,000 years ago, when this species apparently became extinct. Other evidence indicates that Flores may have been populated by this species as far back as 95,000 years ago. This means that for

Figure 16.12 Possible Migration Paths out of Africa for Neanderthals (blue) and H. sapiens (red) Neanderthals left Africa much earlier than our species and were already in the Middle East before members of our species arrived. Interbreeding probably took place in this region, before the dispersal of our species across Europe and Asia. The genetic legacy of these interactions is the presence of Neanderthal genes in the genomes of non-African populations.

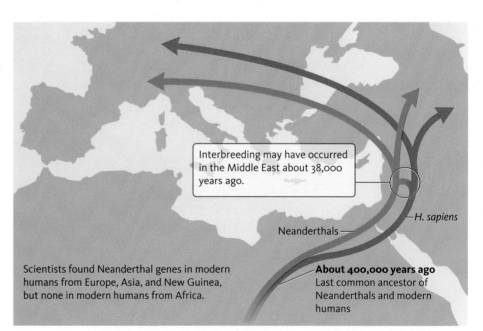

Interbreeding may have occurred in the Middle East about 38,000 years ago.

Scientists found Neanderthal genes in modern humans from Europe, Asia, and New Guinea, but none in modern humans from Africa.

H. sapiens

Neanderthals

About 400,000 years ago Last common ancestor of Neanderthals and modern humans

Homo floresiensis a human species found on the island of Flores

Figure 16.13 The cave on the island of Flores where the fossils of *H. floresiensis* were discovered.

Suzanne Long/Alamy

7 million years ago
When our species and chimpanzees last shared a common ancestor

150,000 years ago
Homo sapiens appeared

200,000 years ago
Neanderthals lived in Europe, Asia, and the Middle East

18,000 years ago
Homo floresiensis lived on the island of Flores

Photo courtesy Ryan Somma

about 80,000 years, at least three human species co-existed on this planet. Whether *H. floresiensis* originated from *H. erectus* or from early forms of *H. sapiens* is still being investigated.

More tantalizing evidence that we had other relatives was uncovered in a cave near Denisova, Siberia, in 2008. Researchers found a finger bone that dated to around 40,000 years ago. Mitochondrial DNA from this finger showed that these humans diverged about 1 million years ago from the common ancestor of modern humans and Neanderthals (recall that Neanderthals and modern humans diverged from a common ancestor about 400,000 years ago).

Later, the complete genome sequence from this fossilized Denisovan was obtained. Comparisons with the genomes of other humans confirm that Denisovans and Neanderthals share a common ancestry following separation from the ancestor of modern humans (Figure 16.15, page 326).

Although present-day populations of Europe and Asia contain 1 to 4% Neanderthal DNA, there is no direct contribution from the Denisovans to these genomes. However, genomic analysis indicates that Denisovans interbred with the ancestors of present-day Melanesians, whose genomes are 4 to 6% Denisovan.

Whether the Denisovans should be regarded as a separate species or a branch of the Neanderthals remains to be determined. What is clear is that we now know that the story of human evolution is more complex than originally thought. As more human fossils are recovered and have their genomes sequenced, the story may get far more interesting.

Is our species still evolving?
Many questions about human evolution are still being considered. Due to the development of culture, medicine, and technology, we no longer face early death (due to illness) and natural predators. Therefore, we may ask ourselves, are we still evolving?

Using SNP haplotypes, scientists have searched for regions of our genome that have changed under the influence of selection. The results show that regions carrying about 1800 genes have been changed by natural selection over the last 50,000 years, indicating that, as a species, we are still evolving.

Figure 16.14 The skulls of *H. sapiens* (left) and *H. floresiensis* (right) show the relative size differences between these species.

Homo sapiens   *Homo floresiensis*

© Peter Brown

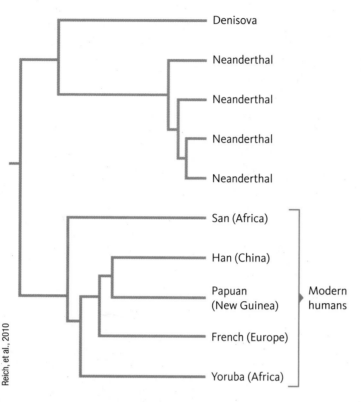

Figure 16.15 A phylogenetic tree shows that Denisovans were more closely related to Neanderthals than to members of our species. This confirms that Denisovans split from the line leading to Neanderthals before the split between Neanderthals and *H. sapiens* occurred.

Reich, et al., 2010

What kinds of genes might be involved in this evolutionary change? Several types of genes have changed, including those involved in defense against infectious diseases, probably as a result of the domestication of farm animals that carry diseases that can be transmitted to humans. Other changes include genes that control protein and carbohydrate metabolism, presumably related to changes in diet as populations made the transition from hunter-gatherers to agricultural societies. Genes that control the structure and function of the nervous system have also undergone changes. To many researchers, it is clear that interactions between culture and our genes have helped shape these and other changes in our genome.

A clear-cut example of this interaction is the development of lactose tolerance in humans. All infant mammals are nourished by milk, and the principal sugar in milk is lactose (human milk is 7% lactose). Lactose is digested by the enzyme lactase, and in all mammals the lactase gene is turned off soon after weaning, making the adults lactose intolerant. In humans, lactose-intolerant individuals develop cramps, gas, and diarrhea after consuming milk or other dairy products.

Unlike all other mammals, some human populations keep the lactase gene active in adults, and these people can use milk and other dairy products as food. This trait developed as part of the transition to agriculture and the use of dairy animals. People who could use milk when other foods were scarce were better able to survive and reproduce. They and their descendants have a high level of lactase production as adults and can digest milk and other dairy products. In other words, a cultural practice (dairy herding) was an agent of natural selection for lactase persistence (Figure 16.16). Archeological and genetic evidence dates the beginnings of this genetic change to about 7500 years ago.

Figure 16.16 **Pattern of Lactose Intolerance in Populations around the World**

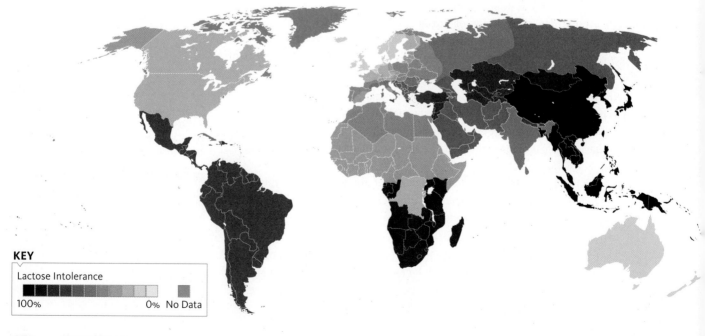

KEY

Lactose Intolerance

100%   0% No Data

16.5 Essentials

- **The human and chimpanzee genomes have very similar sequences but small differences in protein-coding genes, and regulatory genes may be responsible for the large differences in phenotype.**
- **Neanderthal genes make up 1 to 4% of the genomes of non-African human populations.**
- **Our species is still evolving, and about 7 to 10% of our genome is being influenced by natural selection.**

16.6 Ethical and Legal Aspects of Human Evolution

One of the interesting aspects of the study of human evolution has been its place in classrooms across the United States. For over 80 years, this has been a difficult and serious problem. As we have learned throughout earlier chapters, laws and court decisions often solve serious problems.

With respect to the teaching of evolution, a number of states in the United States have addressed the issue of evolution in classrooms. Oddly enough the same questions that plagued Darwin still worry us today.

Many Christian denominations teach the biblical creation story as history, placing it at odds with many lines of scientific evidence that favor evolution, including the fossil record, developmental biology, and genomics. Attempts to prevent the teaching of evolution or the requirement that alternatives be taught have been a contentious issue in American education for over 85 years.

The most famous case involving the teaching of evolution was *The State of Tennessee v. John Thomas Scopes* in 1925. John Scopes (Figure 16.17), a high school teacher, was tried for violating the Butler Act, which prohibited the teaching of evolution in the state of Tennessee. This trial and the personalities involved in it spawned many books and at least one movie, *Inherit the Wind*.

The case itself was a setup to challenge the law that, even then, many believed went against the scientific evidence in favor of evolution. Scopes was asked and volunteered to teach a lesson on evolution, was arrested, and became the defendant in one of the most important trials of our time. At the end of the trial, Scopes was found guilty, but the verdict was overturned on a technicality, and he was never brought back to trial. However, there was

Figure 16.17 John Scopes was tried for teaching evolution in a Tennessee classroom.

Watson Davis/Smithsonian Institution Archives

CASE B:
What Should a Teacher Do?

A teacher pauses to consider something.

© Torbjorn Lagerwall/Alamy

Ms. Allison was just back from lunch. She was sitting at her desk waiting for the bell to ring for her 6th period class. Just then, an announcement came over the intercom: "An important communication has just been sent to us from the Board of Education that affects all science teachers. All teachers of biology, chemistry, earth sciences, and physics should send a student to the office to get it immediately."

Because she had a few more minutes before class, Ms. Allison decided she would walk to the office and get the notice herself. On the way, she tried to figure out what it might be about. There had been some problems with the costs for science supplies and by the time she got to the office she decided they had cut her budget. But that wasn't it at all . . .

The notice read:

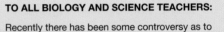

TO ALL BIOLOGY AND SCIENCE TEACHERS:

Recently there has been some controversy as to whether students enrolled in biology and science classes should be required to learn about alternatives to evolution as part of the curriculum. The Board of Education has decided to mandate that a short section on "intelligent design" be taught along with the evolution unit.

All principals will be required to send a memo to us stating that this rule is being carried out in all such classes.

Next Friday at 3:00 the Board of Education will hold an open forum for teachers to discuss this issue. At the forum, teachers will receive a booklet explaining how to cover this material.

Superintendent of Schools

As she walked back to her classroom reading the memo, Ms. Allison was furious. She decided right then and there she would go to that forum and give her opinion.

1. **What four arguments against the rule could Ms. Allison use at the forum?**

2. **List four arguments the Board of Education or other teachers and parents could use in favor of the mandate.**

3. **How could a group (teachers, parents, etc.) go about getting this mandate suspended?**

intense national publicity surrounding the trial, as reporters from all over the world flocked to the small town of Dayton, Tennessee. The big-name lawyers representing each side were William Jennings Bryan, a three-time Democratic presidential candidate who argued for the prosecution, and Clarence Darrow, the famed defense attorney, who spoke for Scopes (Figure 16.18).

In 2005, a case in Pennsylvania revolved around a similar question. *Kitzmiller v. Dover Area School District* was one of the recent important cases since *Scopes* to address the issue of teaching evolution. The question raised by *Kitzmiller* was whether it is constitutional for public-school classes to present the argument of "intelligent design" in addition to evolution by natural selection. Intelligent design is based on the idea that certain features of the natural world are so complex and intricately put together that they must have been deliberately fashioned by an intelligent being. The plaintiffs, represented by the American Civil Liberties Union (ACLU), were parents of children in the Dover, Pennsylvania, school system.

The argument of the plaintiffs claimed this was a violation of the First Amendment's establishment clause, which allows states to make their own laws. After a lengthy trial, the judge issued a decision that the Dover school system's

Figure 16.18 **The Attorneys at the Scopes Trial**
Clarence Darrow (left) and William Jennings Bryan (right)

Hulton Archive/Getty Images

mandate, which required that a statement about intelligent design be read in class, was unconstitutional. He went on to bar the teaching of intelligent design in the Dover school district's public school science classrooms. Table 16.2 lists some of the questions that apply to teaching human evolution.

Table 16.2 **Legal and Ethical Questions Raised by Teaching Evolution**

Question	How Are These Questions Decided?	Related Case or Legal Issue
Who decides what is taught in local classrooms?	These issues are usually decided by local school boards or state boards of education. There has been work to centralize curriculum in some areas.	In 1976, *Daniel v. Waters* struck down the Tennessee law making it illegal to teach evolution. Then in 1987, *Edwards v. Aguillard*, 482 U.S. 578, was heard by the Supreme Court of the United States. The Court ruled that a Louisiana law requiring that "creation science" be taught in public schools along with evolution was unconstitutional.
Can we trust that our children are learning the "correct" things in school?	Individual parents can either opt out of the public school system or demand changes locally. Often cases are brought up and never actually heard by a court of law.	Most antievolution cases have not been successful and usually have no effect.

Spotlight on History: Darwin's Illness

After reading Case A and Section 16.6 in this chapter, we can see that questions about evolution have been asked for centuries. History plays an important part in our understanding of our own evolution, as well as how Charles Darwin lived.

In Case A we learned that Darwin suffered throughout his life from some illnesses that often overwhelmed him. Over the years, Darwin scholars have attributed his illness to many things, including insect bites or seasickness, but others propose that his illness might have been hereditary or psychosomatic. Figure 16.19 shows a timeline of Darwin's life related to his illness and his publications. It will help in answering some of the questions at the right.

Review Chapter 4 (how to read a pedigree) and study the pedigree of Darwin's family in Figure 16.20 and Table 16.3.

QUESTIONS

1. How many generations are shown on the pedigree?
2. Identify one affected relative and his or her symptoms.
3. What do the double lines mean between the symbols representing Charles and Emma?
4. Why is the Wedgewood family on the pedigree?
5. Find other members of the Darwin family who married members of the Wedgewood family.
6. Who was Francis Galton? What is his relationship to Darwin and what did he do that was important?

Figure 16.19 Timetable of Events in Darwin's Life and the Dates of His Publications

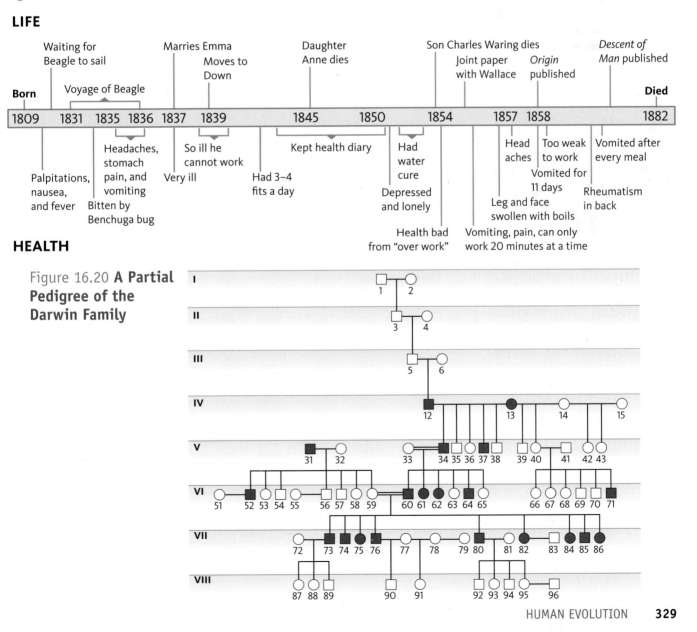

LIFE

Born	1809
Waiting for Beagle to sail / Voyage of Beagle	1831
	1835 1836
Marries Emma	1837
Moves to Down	1839
Daughter Anne dies	1845
	1850
Son Charles Waring dies	1854
Joint paper with Wallace	1857
Origin published	1858
Descent of Man published	
Died	1882

HEALTH

Palpitations, nausea, and fever — Bitten by Benchuga bug — Headaches, stomach pain, and vomiting — Very ill — So ill he cannot work — Had 3–4 fits a day — Kept health diary — Had water cure — Depressed and lonely — Health bad from "over work" — Vomiting, pain, can only work 20 minutes at a time — Head aches — Leg and face swollen with boils — Too weak to work — Vomited for 11 days — Vomited after every meal — Rheumatism in back

Figure 16.20 A Partial Pedigree of the Darwin Family

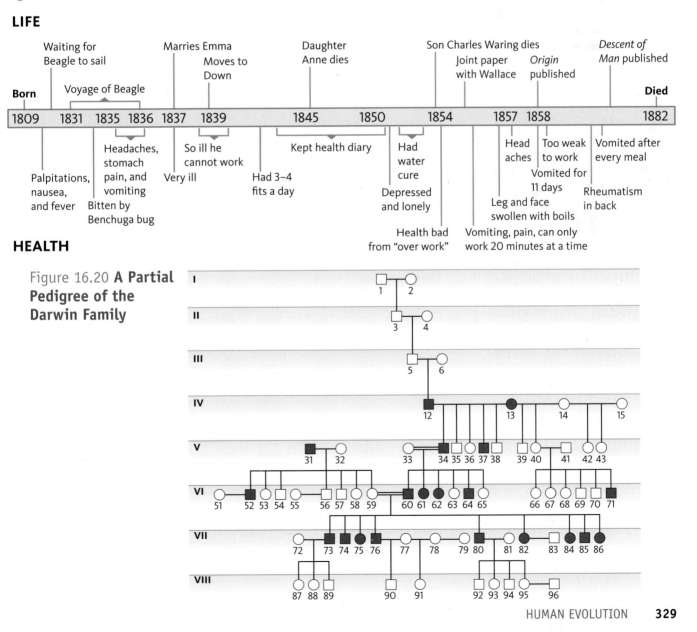

Table 16.3 Charles Darwin's Family Pedigree

The descriptions in this table use terminology in common use in the nineteenth century. Marriage was considered normal an[d] "odd" was a term of some medical significance.

#	Name	Known Condition	#	Name	Known Condition
1	William		64	Erasmus Darwin	Asexual, depressed, melancholy, never married, stuttered, committed suicide
2	Anne		65	Emily Darwin	
3	William		66	Elizabeth Galton	
4	Anne		67	Lucy Galton	
5	Robert		68	Milly Galton	
6	Elizabeth Hill		69	Darwin Galton	
12	Erasmus Darwin	CD brother; asexual, depressed, eccentric, hypersensitive, never married, never worked.	70	Erasmus Galton	
			71	Francis Galton	Melancholy and depressed
13	Mary Darwin	Sickly	72	Ida Darwin	
14	Elizabeth Pole	Odd, reserved, depressed	73	Horace Darwin	Frail, retardation, always ill
15	Ms. Parker		74	Leonard Darwin	Obsessively conscien[t]ious, hypochondriac
31	Josiah Wedgewood		75	Elizabeth Darwin	Clumsy, hypersensiti[ve,] never married, odd, stout
32	Elizabeth Wedgewood				
33	Susana Darwin Wedgewood		76	Francis Darwin	Melancholy, depresse[d]
34	Robert Darwin	Obese, high falsetto voice, eccentric, rude	77	Amy Darwin	
			78	Ellen Darwin	
35	William Darwin		79	Flo Darwin	
36	Elizabeth Darwin		80	George Darwin	Ill after parties, very nervous, eccentric, phobias, abdominal problems
37	Erasmus Darwin	Eccentric, stuttered, obese			
38	Charles Darwin				
39	Edward Darwin		81	Wife of G. Darwin	
40	Francis Darwin		82	Etty Darwin	Eccentric, hypersens[i]tive, hypochondriac, odd, phobia of germs, had breakfast in bed, her entire life
41	Sam				
42	Susan Darwin				
43	Mary Darwin				
51	Carolyn Wedgewood				
52	Josiah Wedgewood III	Depressed, odd, reserved	83	Richard Litchfield	
			84	Annie Darwin	Abdominal problems, died at age 9
53	Charlotte Wedgewood				
54	Hensleigh Wedgewood		85	Charles Waring Darwin	Died early, retardatio[n], never walked or talk[ed]
55	Jesse Wedgewood				
56	Henry Wedgewood		86	Amy Darwin	
57	Francis Wedgewood		87	Nora Darwin	
58	Fanny Wedgewood		88	Ruth Darwin	
59	Emma Wedgewood Darwin		89	Erasmus Darwin	
60	**Charles Darwin**	**Phobias, sickly, eccentric, hypochondriac, melancholy, abdominal problems, depressed**	90	Bernard Darwin	
			91	Francis Darwin	
			92	William Darwin	
			93	Margaret Darwin	
61	Mary Ann Darwin	Rude, difficult child, unpleasant	94	Charles Darwin	
			95	Gwen Darwin Raverat	
62	Carolyn Darwin	Unpleasant, rude	96	J. Raverat	
63	Susan Darwin				

1 Evolution studies changes in living things over time. (Section 16.1)
The time frame can involve millions of years.

2 The study of evolution is interdisciplinary. (Section 16.2)
Anthropology and paleontology are used along with genomics.

3 The fossil record is an important part of the evidence for evolution. (Section 16.2)
Fossils left a record of organisms alive in a specific time period.

4 Beginning about 4.4 million years ago, hominins appeared in Africa. (Section 16.3)
Members of our genus, *Homo*, appeared about 2.5 million years ago.

5 Members of the genus Australopithecus were among the earliest hominins. (Section 16.3)
Fossils of Australopithecus are found in Africa.

6 *Homo sapiens* may have developed from *H. erectus* somewhere in Southwest Africa. (Section 16.4)
Later, members of our species migrated from Africa.

7 Neanderthals were already in other parts of the globe when our species migrated out of Africa. (Section 16.5)
Homo sapiens interacted with Neanderthals.

8 Study of the human genome shows we are still evolving. (Section 16.5)
The genome of *Homo sapiens* is adapting to its present environment.

9 Recently, other relatives of our species have been uncovered. (Section 16.5)
The evolution of human species is far more complex than previously thought.

10 There has been controversy about the theory of evolution for hundreds of years. (Section 16.6)
In the United States, the teaching of evolution in schools is still being debated.

Review Questions

1. What is evolution?

2. How do fossils contribute to our information on human evolution?

3. Molecular and genomic techniques also add information to our ancestry. Name three.

4. List all the species groups of fossil hominins up through *Homo sapiens*.

5. What is one difference between prosimians and anthropoids?

6. From which species group did *Homo sapiens* evolve?

7. As human ancestors evolved, did their brains get bigger or smaller? How can scientists tell?

8. What is the largest similarity between humans and chimps?

9. How many years ago did chimps and humans diverge from a common ancestor?

10. How are we related to Neanderthals?

Application Questions

1. Researchers have found that Neanderthals were more closely related to humans than was originally thought. Should they still be classified as a separate species?

2. Would our evolutionary history change if researchers found that Neanderthals were not related to humans at all?

3. What might happen if scientists stopped sequencing DNA from newly discovered human fossils? How would we be able to distinguish ancestors from relatives?

4. Is it possible that the development of culture has itself been a selective force in our evolution?

5. Draw a phylogenetic tree showing the relationship of the species in Table 16.1 to our species.

6. The drawing at the beginning of the chapter is an artist's conception of what a Neanderthal looked like. Do you agree with this drawing? Why or why not? See if you can find other drawings of this human.

7. *National Geographic* magazine had an issue in October 2008 on the Neanderthal genome. Find it online and read the article.

8. Reread the Spotlight on History. If a genetic counselor were consulted today, knowing the family tree of Charles and Emma Darwin, how would he or she counsel them about having children?

9. Why wasn't there a clear diagnosis of Darwin's medical condition?

10. What ideas in the early 1800s do you think dictated that members of wealthy and prominent families intermarried?

11. What other trial was Clarence Darrow famous for?

12. Watch *Inherit the Wind* and do some research to see how accurate it is by creating a chart comparing the actual Scopes trial to its depiction in the movie.

WHAT WOULD YOU DO IF...?

. . . you were asked to speak about evolution at a church?

. . . you were talking to a friend who said, "We evolved from monkeys"?

. . . you were asked to vote on halting the study of evolution in high schools?

Online Resources

Preparing for an exam? Log on at www.cengagebrain.com for study tools to help you assess your understanding. If assigned by your instructor, the Case A and Spotlight activities for this chapter, "Darwin's Dilemma" and "Spotlight on History," will also be available.

In the Media

The Cornell Sun, April 15, 2011

Ancestry Project Reveals Results
Jamie Myerson

At a panel discussion on Thursday, the Cornell University Genetic Ancestry Program released its results, which tell the genetic lineages of 200 randomly selected undergraduates. This project was the first part of Cornell University's partnership with the National Geographic Society's Genographic Project. It revealed that the genetic lineage of 200 randomly selected undergraduates is more diverse than the residents of Queens, New York—arguably one of the most diverse places on earth.

National Geographic's researchers had previously tested the DNA of 200 random Queens residents in a single day on a single city block and found that the DNA represented all of humanity's ancient migratory paths.

Results of the Cornell study reveal regional lineages of 57% European, 14% East Asian, 11% South Asian, 8% Middle Eastern, 5% Native American, and 5% African ancestry within the sampling.

To access this article online, go to www.cengagebrain.com.

QUESTION

Could this discovery be used to determine who will be susceptible to migraines? How?

ASU News, May 9, 2011

Anthropologists Promote Understanding of Human Origins

After a 2009 Gallup Poll found that only 39% of Americans say they "believe in the theory of evolution," famed paleoanthropologists Professor Donald Johanson and Richard Leakey came together at the American Museum of Natural History in New York to discuss human evolution. They were concerned but not surprised that the poll also found a strong relationship between education and belief in Charles Darwin's theory.

The world-famous scientists discussed the overwhelming evidence in the hominid fossil record, and why understanding our evolutionary history is of such critical relevance today.

Johanson and Leakey have each spent decades hunting for hominid fossils in east Africa, working to unearth the physical record of human evolution and piece together the complex biological narrative that explains—in very real terms—who we are as a species. Their work has led to some of the most important paleoanthropological discoveries of the last half-century, including Johanson's discovery of the "Lucy" specimen in Ethiopia's Awash Valley in 1974 and Leakey's discovery of the 1.6-million-year-old *Homo erectus* youth skeleton known as "Turkana Boy" in 1984.

The day before the talk, Johanson and Louise Leakey, daughter of Richard and Meave Leakey and a paleontologist with experience in Kenya and east Africa, led two special educational sessions at the museum with a group of 125 high school students and over 250 teachers from the New York City area to promote the importance of science education and teaching human evolution.

To access this article online, go to www.cengagebrain.com.

QUESTIONS

1. How does this tie into Case B in this chapter?

2. Do you think using famous anthropologists to bring evolution to the public will change people's minds? Why or why not?

3. How would you do it?

Different Worlds: The Past, Present, and Future of Human Genetics

CENTRAL POINTS

- We can learn a lot from the eugenics movement of the 1920s.

- Genetics is moving ahead using many new technologies.

- We face a future full of important questions to be decided by society and individuals.

Human Genetics Has Changed Dramatically in the Last 50 Years

17.1 What Can We Learn from the Past?

Thinking back on the cases discussed in this book, we can make some interesting observations. The Carters, the Crowleys, the Franklins, and others were worried about the genes that had been passed on in their families. In the recent past, some scientists and governments thought they could control which genes were passed on to the next generation and, over time, create a "better" human. The following is an incredible, but true, account of events from the early 20th century.

Fitter families win prizes. In the early part of the 20th century, scientists at the U.S. Eugenics Office, a privately funded organization, began running contests around the country to pick the families with the "best genetics." These families were given medals and awards and encouraged to have more children (Figure 17.1).

This program, part of the *eugenics movement*, existed in the United States for more than 40 years. A number of scientists, led by Charles Davenport, began by applying Darwin and Wallace's theory of natural selection to humans. This was based on

Figure 17.1 **The Eugenics Movement** The eugenics building at a fair in the 1920s. Fitter family contests were run annually to encourage families such as these to have more children.

Figure 17.2 Winners of a fitter family contest.

American Philosophical Society Library

the idea that those best adapted to their environment would survive and therefore reproduce. They would have more offspring than those who are less well adapted, and these good genes would be passed down for generations.

In England, Darwin's cousin, Francis Galton, theorized that selection could be used to improve our species by encouraging the fittest individuals to reproduce and discouraging the less fit from reproducing. This idea took strong hold in the United States, and eugenics became a major force in shaping laws about reproduction and immigration.

Eugenicists decided which traits were desirable and encouraged people with those traits to have many children (Figure 17.2). This was called **positive eugenics**. At the same time, they compiled a list of traits that were not desirable. To eliminate those traits from the population, it was necessary to prevent those with these traits from reproducing. To expedite this, eugenicists helped pass laws that forced the sterilization of people with undesirable traits to prevent them from reproducing. In addition, laws were passed limiting immigration from certain countries whose people were

positive eugenics movement to increase desirable traits in a population by encouraging people with those traits to have many children

negative eugenics using eugenics to remove certain genes or traits from a population by preventing those with unwanted genes or traits from reproducing

considered to have undesirable traits. This was called **negative eugenics**.

As part of the eugenics movement, many states passed laws to sterilize criminals, "imbeciles," and women who were "promiscuous." Remember that little was known about genes or which traits were inherited at that time. The U.S. Supreme Court upheld these laws in the case of *Buck v. Bell*, which challenged the sterilization of a woman named Carrie Buck (Figure 17.3). According to officials in Virginia, Carrie was said to be suffering from "feeblemindedness and promiscuity." She already had an illegitimate child, who was presumed to be "feeble-minded," as was Carrie's mother. The Supreme Court, led by Justice Oliver Wendell Holmes, stated that her sterilization was proper because "It is better for all the world, if instead of waiting to execute degenerate offspring for crime or to let them starve for their imbecility, society can prevent those who are manifestly unfit from continuing their kind. . . . Three generations of imbeciles are enough."

In reality, Carrie Buck was neither "feeble-minded" nor promiscuous. Although she only completed the sixth grade, she was an average student, and her pregnancy resulted from a rape by a relative of her foster parents. In addition, later

Figure 17.3 Carrie Buck (left) and her mother after Carrie's sterilization. At that time, they were living in an institution for the "feeble minded."

American Philosophical Society Library

investigation showed that her child, Vivian, was not "feeble-minded" either.

Using U.S. eugenics laws as a model, in the 1930s, Nazi Germany passed laws that forced the sterilization of people who were regarded as undesirables, including people with epilepsy, physical deformities, and alcoholism. Later, the Nazi regime expanded their efforts to include the sterilization of whole groups of people, including Jews and Gypsies, using the science of genetics as the basis for these atrocities. Laws were amended to allow mercy killing of newborns who were incurably ill with genetic disorders. Gradually this program of mercy killing was expanded to include adults in mental institutions and from there to include whole groups of people in concentration camps, most of whom were Jews, Gypsies, homosexuals, and political opponents of the Nazi regime.

The argument the Nazis used to rationalize this killing was genetics. To rid a population of bad genes, you must stop people with these traits from reproducing. Killing them was the fastest way to do this. After these killings were revealed, the eugenics movement in the United States rapidly declined. As a result, fear of how genetics can be misused still exists today in the United States and other countries.

17.2 What Are Some of the Newest Technologies of the Present?

Today our society has taken a real interest in science. This may be because discoveries are available to scientists and the public. Every newspaper and television network has researchers who read the latest issues of scientific journals and then write articles for the media. Because of this, the general public has access to the newest findings.

In addition, new medical treatments, prescription drugs, medical tests, and genetic tests are advertised directly to the public in the media and on the internet. This direct-to-consumer marketing is fairly new and assumes a certain knowledge of one's medical condition. Some of the new technologies available to the public are discussed in the following pages.

How drastic should treatment be? In Chapter 8, we explored the types of genetic testing that can be done today and how the use of these tests may expand in the future. One test mentioned earlier is for a breast cancer gene. If a woman tests positive for the mutant alleles of the *BRCA1* or *BRCA2* gene (see Chapter 12) and breast cancer runs in her family, what are her options? What should her physician recommend (Figure 17.4)?

After a positive test for a mutant allele, a woman may continue monitoring herself with self exams and mammograms,

WHAT MIGHT HAPPEN IF...?

. . . the federal government proposed a law that would require sterilization of people who carry a "lethal" gene, such as Tay-Sachs?

. . . preimplantation genetic diagnosis were available to all parents using *in vitro* fertilization?

. . . we could manipulate the genes of fetuses so they would all be smart, tall, and good-looking?

Figure 17.4 Often physicians are unsure about how to proceed after a patient tests positive for mutant alleles of the *BRCA1* or *BRCA2* gene. They usually give patients the information about their test results and the options available to them, or refer the patient to a genetic counselor, who is trained to interpret the results of genetic tests and who works to explore options with the patient and the family.

Zdorov Kirill Vladimirovich/Shutterstock.com

but the results of this genetic test can have a lingering psychological effect, as well as a physical one. She may continue to be worried about a future diagnosis and will be exposed to x-rays in periodic testing for breast cancer. These factors may only increase her fears.

In addition, if lumps are detected in her breasts or lymph nodes, she will have to undergo biopsies, and there may be several of those. Some lumps may be benign, but others may be cancerous—her greatest fear.

As discussed in Chapter 12, the anti-cancer drug tamoxifen is used to treat breast tumors that have estrogen receptors on the surface of their cells. The drug can be given to a woman before she actually has any tumors. Because the U.S. Food and Drug Administration (FDA) has approved this drug for prevention of breast cancer, women who test positive for *BRCA1* or *BRCA2* mutations can begin taking it. However, the drug has side effects, and some women may not want to take such a drug every day.

As an option, some physicians have proposed removing all breast tissue before cancerous cells can begin growing. A procedure called a **subcutaneous mastectomy** may be an answer for women carrying the mutant *BRCA1* or *BRCA2* gene. But is this treatment too drastic? Some surgeons and some women are not happy with the removal of healthy tissue because it *may* become cancerous. They may ask, isn't this true of all cells?

Should an insurance company pay for this type of surgery? If the company is looking at the long-term costs, it may see that this surgery would save money in the long run. But in the short run, it is an expense that may not be necessary. If Harriet Abeline (see Chapter 12, Case A) could have identified herself as *BRCA* positive, she might have been able to take some precautions such as a mastectomy.

Figure 17.5 **An Over-the-Counter Genetic Test**

The results of the genetic test are sent to your physician or to your home when you order it from Myriad Genetics. An ad for the test often runs in major magazines.

Courtesy Myriad Genetics, Inc.

Get the genetic test. In Chapter 8, we learned about genetic testing and how it is changing the way physicians analyze patients' conditions. Recently, companies and private labs that offer genetic tests have been advertising directly to consumers. Television and internet ads have been popping up to inform the public that these tests exist and to encourage patients to ask their physicians about having these tests.

The first step in this type of advertising is to give a recognizable name to the test. In the case of the breast cancer gene, the test is called **BRACAnalysis**®. The ads for this (Figure 17.5) and other tests place the decision about whether to get tested squarely in the hands of the individual. BRACAnalysis® and other genetic tests are now available over the counter in drug stores. Anyone can purchase a test kit, scrape the inside of his or her cheek (Figure 17.6), and send the swabs to a lab for the results. Whether positive or negative, the results are sent back to the patient, often with little or no explanation.

Some physicians feel that if patients use mail-order genetic tests, they do not get the information and advice they need to interpret the results and develop a list of options. If the result is positive, is a piece of paper with an explanation of the result the same as a physician's explanation or a session with a genetic counselor?

Twenty years ago, prescription drugs were marketed only to physicians, and drug companies sent their sales representatives into physicians' offices to encourage them to prescribe their medicines. Health care practitioners have been surprised at how many patients ask for the drugs they see on television. This

WHAT MIGHT HAPPEN IF...?

. . . a woman who tests positive for a *BRCA1* mutation applied to her insurance company for a precancerous mastectomy?

. . . a woman with a family history of breast cancer (but no test results) applied to her insurance company for a precancerous mastectomy?

. . . a woman with no history of breast cancer and no genetic test applied to her insurance company for a precancerous mastectomy?

subcutaneous mastectomy removal of the breast tissue before it is cancerous

BRACAnalysis® brand name for a test for mutant alleles of *BRCA1* and *BRAC2*

method of marketing has definitely been successful, as you can see by noting how many of these commercials are on television.

These television ads have targeted the actual purchasers of the prescription drugs: patients. The ads made drugs such as Prilosec and Ambien household names. Patients who didn't really understand what these drugs treated went to their physicians asking for them by name. After this form of advertising became widespread, more legislation was enacted to ensure that ads clearly identify side effects and include other information.

Many questions have been raised about advertisements for genetic tests because they target an entire population when only a few people may carry the gene the test is detecting. The ads do not disclose that both false positive and false negative results are possible.

In addition, some surprising legal decisions have come from the patenting of the *BRCA1* gene and its test. In 2010, the decision in the case *Association of Molecular Pathology v. USPTO (United States Patent and Trademark Office)* turned gene patenting on its head. In this case, heard in the Federal Court for the Southern District of New York, the judge ruled that a gene is not changed when it is removed from the DNA, and that the patent on the *BRCA1* test is not valid. This case will probably go to the U.S. Supreme Court within a few years. If the decision of the lower court is upheld, then many biotechnology companies will have to rethink the patenting issue.

WHAT MIGHT HAPPEN IF...?

. . . every woman demanded a genetic test for breast cancer?

. . . hundreds of tests for different genes were developed?

. . . physicians had trouble convincing their patients they don't need these tests?

. . . the U.S. Supreme Court upholds *Association of Molecular Pathology*?

a. What might happen to gene patenting in the future?

b. What might happen to patients who are tested using the breast cancer gene kit?

c. What might happen to Myriad Genetics, which developed the test?

Figure 17.6 An oral swab is used to take a DNA sample for breast cancer testing.

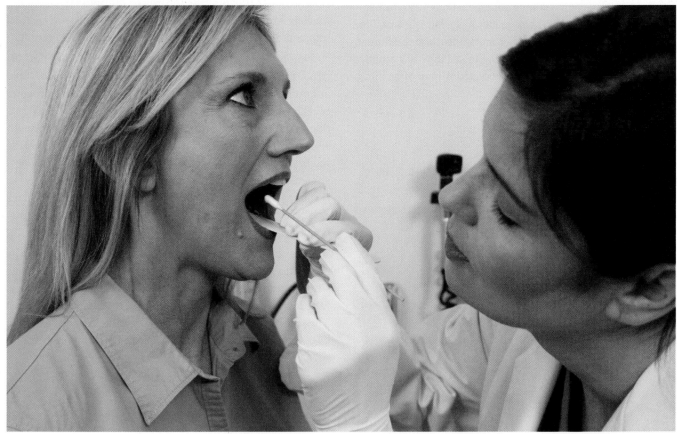

© Custom Medical Stock Photo/Alamy

The dog genome. Geneticists have recently identified a mutant allele of the **myostatin gene** in whippets, a type of racing dog (see Figure 17.7). The normal myostatin gene product regulates muscle growth, and the mutant allele encodes a protein that is only partially functional. As a result, the muscle mass in affected dogs doubles in size. A dog with one copy of the mutated myostatin gene (heterozygous) can run much faster than a dog with two normal myostatin alleles. A dog that carries two mutant copies (homozygous) of the gene is known as an overmuscled **"bully" whippet**.

If two fast whippets (each carrying one copy of the mutant gene) are bred (*Mm × Mm*), some of the offspring may be fast (*Mm*), but they can also produce "bully" puppies (*mm*). This outcome is identical to the inheritance of recessive traits in humans.

Mutations in the myostatin gene have been found in more than a dozen breeds of cattle, where it is called *double muscling*. This condition has attracted attention as commercially useful. Cattle with this mutation produce more muscle (meat) to sell at the market, and offspring of these cattle would be a benefit to agribusiness (Figure 17.8).

In 2004, a German couple adopted a baby who immediately showed physical abilities far beyond what is normal for his age. It turned out he carried a mutant form of the human myostatin gene. It will be interesting to see if he has a future in competitive athletics, where his genetics may offer him a definite advantage.

Figure 17.7 **Myostatin Gene Mutations** The "bully" whippet is the dog on the right. As you can see, it has double musculature.

Stuart Isett Photography

Figure 17.8 This steer has a naturally occurring mutant allele of the myostatin gene that doubles its muscle mass.

© 1997 National Academy of Sciences, USA

Research on the dog genome also involves studying more than one gene at a time. A number of years ago a couple who wanted their dog cloned offered $2 million to anyone who would work on the project. Scientists at Texas A&M University had already set up a lab to collect cell samples from cats and dogs around the country. In 2002, they succeeded in cloning a cat by transplanting DNA from Rainbow, a female three-colored (tortoiseshell or calico) cat, into an egg whose nucleus had been surgically removed. This embryo was then implanted into Allie, a cat that served as a surrogate mother.

Cloning dogs, however, has turned out to be more difficult. Finally, in 2005, a Korean scientist, Hwang Wook-Suk, announced the cloning of a dog he called Snuppy. Although other cloning work by Dr. Hwang has been discredited, his dog has been confirmed as a clone. In 2007, RNL Bio, a South Korean company associated with Dr. Hwang, began advertising that it will clone dogs for a $150,000 fee, to be paid when a cloned dog is delivered to the customer. In 2008, a California company also associated with Dr. Hwang announced plans to offer a series of online auctions for dog cloning, with the opening bid set at $100,000.

On December 7, 2005, an international team led by researchers at the Broad Institute of MIT and Harvard University announced the publication of the dog genome sequence. Dog lovers have been breeding dogs for hundreds of years. Through this process of artificial selection, many traits have been bred in and out of dogs. This includes "personality traits"

myostatin gene gene that regulates the growth of muscle fibers

"bully" whippet dog with a mutated myostatin gene

such as gentleness or aggressiveness, as well as physical traits such as size and color. Now with the dog genome sequence known, clones of specific genes available, and the availability of cloning, dog breeding may increasingly rely on genomic information.

Knockout mice win big prize. The Nobel Prize for Physiology or Medicine in 2007 was awarded to three scientists whose work led to the creation of the **knockout mouse** (Figure 17.9). The ability to isolate stem cells from mouse embryos made it possible to target certain genes in mice and then turn them off in stem cells. This is called *knockout technology*. Many labs have succeeded in using this technology, and about 10,000 genes, almost half of the genes in the mouse genome, have been knocked out in various mouse strains. One example was discussed in Chapter 13, where the behavior of mice was changed when one gene was knocked out. The knockout mice became very aggressive and couldn't be kept in a cage with normal mice.

With a gene switched off by knockout technology, scientists (Figure 17.10) can deduce the function of the normal gene by observing phenotypic changes in the mice. When behavior or other traits change in the knockout mouse, we know the effect of the gene that was turned off. Studying how the phenotypes of knockout mice develop gives researchers an insight into how abnormal conditions progress. This technology has proved invaluable in developing and testing new drug therapies for human diseases. Gene targeting has already

produced more than 500 different mouse models of human disorders, including cardiovascular and neurodegenerative diseases, diabetes, and cancer.

In March 2011, a knockout mouse was created to study the genetics of deafness. A gene called *CACNA1D*, which encodes a protein that regulates the flow of calcium in and out of cells, was knocked out in these mice. Although the knockout mice did not have a functional copy of the *CACNA1D* gene, they still lived a normal life span but had slower than average and arrhythmic heart rates. The knockout mice were also completely deaf. Using these mice and other animal models, much can be done to study this type of genetic deafness.

Figure 17.9 **A Knockout Mouse** The mouse on the left is the knockout mouse. In this case, knocking out a specific gene causes the mouse to gain weight. The mouse on the right carries two normal alleles of this gene.

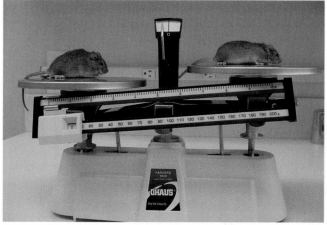

Lexicon Pharmaceuticals, Inc.

knockout mouse mouse with a specific gene turned off for scientific study

AP Photo/Douglas C. Pizac, file

Figure 17.10 Mario Capecchi from the University of Utah shared the 2007 Nobel Prize for Physiology or Medicine with Oliver Smithies and Martin Evans for developing knockout mice.

17.3 What Is in Store for the Future?

Some of the people in our case studies were concerned about the impact of genetic technology on future generations. Closely examining human genetics reveals an ever-expanding outlook for this branch of genetics. To emphasize the importance of being educated and able to make informed decisions about both the present and future genetic technology, this book poses important genetic questions about these issues to you, your family, and society.

Even more remarkable is how the results from the Human Genome Project and future scientific discoveries may be applied to our near and distant future. When genetically engineered food becomes even more widespread on grocery store shelves, will we become used to it? When everyone's unborn fetus can have its genome sequenced and analyzed for all genetic traits and diseases it carries, will insurance pay for it? When a woman who has had a hysterectomy wants to have a baby, will she be able to? The answer to all of these questions is probably yes. All of these things and much more will influence and change our genetic future.

Here are just a few of the possible future scenarios in human genetics, some of which are already beginning to come true.

Non-invasive prenatal testing for Down syndrome.
As discussed in Chapter 8, extracting fetal cells from the mother's blood can be used as a noninvasive method of prenatal testing. A recent large-scale study used fetal DNA recovered from the mother's blood to test for Down syndrome. The group, which includes Rossa Wai Kwun Chiu (Figure 17.11), showed that the test is very accurate. The cost is much less than more invasive testing, such as amniocentesis, and has no risk to the fetus.

But even with these positive results, some questions remain. For example, Chui's study showed a 2.1% false positive rate. Some physicians have stated that this error rate is too high to make the test available to the general public.

The discovery that cell-free fetal DNA fragments can be detected in maternal blood during pregnancy is now almost 15 years old. Although this type of testing is being used in physicians' offices for some traits such as fetal sex and Rh blood type, the development of a noninvasive prenatal test for Down syndrome could ultimately transform prenatal screening and care for all pregnant women.

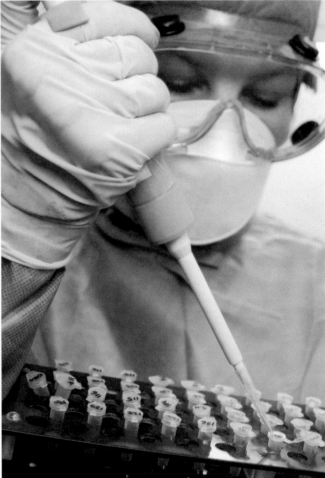

Figure 17.11 **Rossa Wai Kwun Chiu** Chiu, from the University of China, Hong Kong, has developed a noninvasive test for Down syndrome using fetal DNA fragments from maternal blood.

Bork/Shutterstock.com

WHAT MIGHT HAPPEN IF...?

... a woman comes to her physician for this new test because she already has one child with Down syndrome? Could her blood still carry fetal DNA fragments from her Down syndrome child and affect the results of a test for her second child?

... a woman age 25 comes to her physician for this test because she read about it in an article and is worried about her pregnancy?

... a woman age 45 comes to her physician for this test because she read about it in an article and is worried about her pregnancy?

Artificial uterus developed. In 1999, Yoshinori Kuwabara and his colleagues at Juntendo University in Tokyo began an experiment in which they constructed an **artificial uterus** using a clear acrylic tank. The tank was filled with eight quarts of amniotic fluid kept at body temperature. As shown in Figure 17.12, the umbilical cord of a goat fetus (with its artery and vein) was threaded into two heart-lung machines to supply oxygen and food for the fetus and to clean the blood of waste products. Although the experiment lasted only 10 days, it is clear that with only minor modifications this technology could be adapted to growing humans in artificial wombs. Some people think that this type of assisted reproduction is unethical. If this technology becomes available, these machines might be in laboratories and parents could come to visit their babies and watch them grow.

The use of this technology will pose some legal questions. In cases dealing with parental rights, courts usually consider motherhood to be determined either biologically (giving birth) or legally (by adoption). If a child is adopted, legal papers determine who the parents are. Genetics usually determines a child's biological parents. DNA testing for paternity is often used to help the court make this determination. In some states, the woman who gives birth to a child is automatically considered the mother, even though she may be a surrogate hired by a couple to bear the child. If the artificial uterus becomes available, motherhood will be difficult to determine without DNA samples from the egg and sperm donors and the fetus.

Figure 17.12 **Artificial Uterus** The acrylic uterus, developed in Japan, is shown here with a goat fetus close to the time of birth.

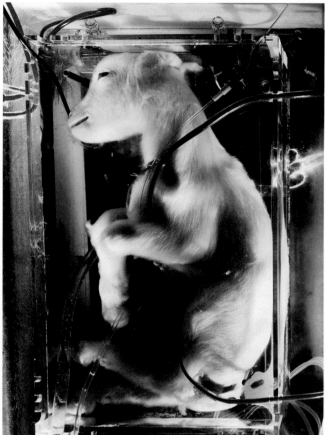

Tom Schierlitz/Getty Images

artificial uterus machine that will mimic the uterine environment

WHAT MIGHT HAPPEN IF...?

. . . the artificial uterus were available for human babies?

. . . large numbers of women began choosing to use an artificial uterus to carry their babies? Would this affect the definition of who is a parent? How?

. . . a private company developed this technology and offered it at a price lower than that for usual prenatal care and hospital expenses associated with the birth? Would you consider having a child using this uterus?

ASSIGNMENTS

1. Develop a marketing plan for the artificial uterus.

2. List four types of people who might use this technology.

A mother at any age. In Chapter 2, Brian and Laura wondered how to address their infertility problem. But in other cases, what might happen if a woman was fertile but decided only later in life to have a baby? Might this result in society deciding that women of a certain age should have to follow nature's plan and not have children?

As part of work on infertility, scientists have been studying human eggs to find out why older eggs are difficult to fertilize. If an older woman (Figure 17.13) wants to conceive a child, she will have some problems. If she is undergoing or has completed menopause, her eggs are not released on a regular basis and she will have trouble conceiving. However, if she is still producing eggs, they can be removed in a surgical procedure (see Chapter 2) and analyzed for their ability to be fertilized. Those eggs can be used for *in vitro* fertilization.

The media gives a lot of coverage to celebrities who have a baby. If the mother is over 40 there is a great deal more attention. Some celebrities who have brought public awareness to older women being pregnant include Cheryl Tiegs (at age 52), Madonna (at age 41), J. K. Rowling (at age 39), and Helen Hunt (at age 40).

Many older women have become mothers using donor eggs if they are not producing eggs or their own eggs are not viable. Many of these women want to have their own genetic children. How would this be possible?

Recently it has been found that the problem with older eggs is in the cytoplasm, which may cause failure of the embryo to divide by mitosis. Using a technique called **nuclear transfer** (Figure 17.14), Drs. James Grifo and John Zhang of New York University Medical Center removed the nucleus from an older woman's egg and transferred it to a younger woman's egg that had its nucleus removed. They taught this technique to a group of Chinese physicians who used it to initiate a pregnancy in China in 1995. Even though the twins conceived by this method later died, there were no chromosomal problems, indicating that with further refinement, this method may become a way for older women to have genetically related children.

In the United States, a similar version of this technique has been successful, leading to a live birth. At St. Barnabas

Figure 17.13 Aleta St. James of New York gave birth to twins just before her 57th birthday.

Gregory Bull/AP/Wide World Photos

Medical Center in Livingston, New Jersey, Drs. Richard Scott and Jacques Cohen removed cytoplasm from a young woman's eggs and injected it into the eggs of older women. Two live births have resulted from this process, and others will occur in the future.

Using this same technique, some labs in Britain transferred a human nucleus into the egg of another mammal to study early stages of human embryo formation. The development of this technique raises the possibility that a resulting child might carry the cytoplasm of another species.

Figure 17.14 **Nuclear Transfer** Here a nucleus (arrow at the tip of the pipette) is being transferred to an egg that has had its own nucleus removed.

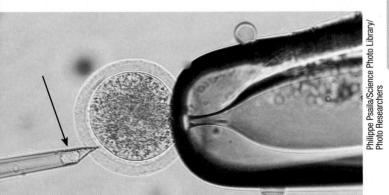

Philippe Psaila/Science Photo Library/
Photo Researchers

WHAT MIGHT HAPPEN IF...?

. . . state legislatures enacted laws limiting insurance payments for infertility treatment for women under the age of 40?

. . . legislatures decided that a law was needed to control the age at which a woman could become a mother?

. . . women could freeze their own eggs at an early age to use later?

. . . a woman with no eggs, no access to sperm, and no uterus wanted a child?

QUESTION Some people call these embryos "three-parent embryos." Why?

nuclear transfer moving a nucleus from one cell to another

DNA of your baby. The discussion of genetic screening in Chapter 8 described how Al and Victoria's newborn was tested for a number of genetic conditions. Some, such as PKU, the genetic disorder their baby had, are treatable; others are not. Many parents wonder whether this type of screening may increase. As companies create more and more tests for genetic disorders, such tests are likely to be used in state testing programs or by parents who want them done on their newborn, even if no treatment for these disorders are available.

As we become better and better at analyzing the human genome, some people have suggested that we take and analyze DNA samples from every newborn. This would not be difficult because blood is routinely taken by a heel stick (Figure 17.15) from almost every newborn for tests that are done today. This blood could be used to sequence part or all of the newborn's genome to create a database of genetic markers similar to the ones used by law enforcement. Both positive and negative opinions exist about this type of sampling. Table 17.1 shows some of the issues related to this type of database.

Figure 17.15 **Collecting a Blood Sample from a Baby**

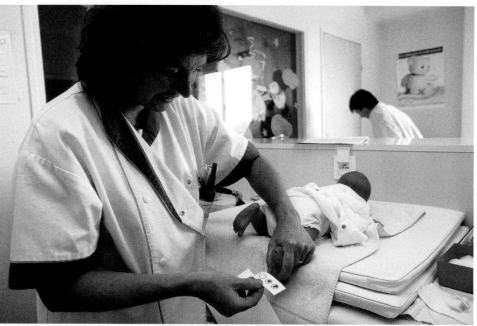

Phanie Agency/Photo Researchers, Inc.

WHAT MIGHT HAPPEN IF...?

. . . this type of DNA database existed?

. . . laws were passed that mandated this type of sampling?

. . . you were asked to vote on such a law?

Table 17.1 **Legal and Ethical Questions about Newborn DNA Databases**

Question	Pro	Con
Who will have access to this database?	Scientists, parents, physicians, and medical professionals would be able to do large population studies.	Lawyers, police, insurance companies, and potential employers would have access as well.
Will parents have the right to opt out of having their child's DNA in the database?	This would allow parents to make sure their child's DNA was not available to everyone.	If people were allowed to opt out, the database would be incomplete and possibly useless.
Can this database give us useful information about the genetics of our children?	When all or most human genes are identified and sequenced, we will know everything about someone's genome.	If we knew about the presence of only a few conditions, it would have little use.
Is this the future?	?	?

Figure 17.16 **Dolly the Sheep and Her Baby** Dolly was euthanized after she contracted a fatal viral infection.

AP Photo/John Chadwick

The ultimate question.
Many people feel that the ultimate question that scientists and ethicists may have to deal with in the future is whether humans should be cloned.

Dolly the sheep (Figure 17.16) was cloned in the late 1990s. Since then other animals have been cloned, including cats, dogs, monkeys, and cows. The method used is called *nuclear transfer*, also referred to as reproductive cloning. We discussed nuclear transfer earlier in this chapter when exploring how older women might have their own genetic children. Reproductive cloning is similar to this process.

To clone an organism using **reproductive cloning** (Figure 17.17) a body cell is removed from the "parent" organism and the diploid nucleus is transferred to an egg cell that has had its own nucleus removed. The egg cell is then stimulated to activate cell division. The resulting embryo is transferred to the uterus of a surrogate mother. Five mice are shown in Figure 17.18. Two are clones (#1 and #2), and one is the genetic mother because she contributed the body cell's nucleus (#3). The egg donor (#4) and the surrogate mother (#5) are also shown.

In the quest to clone Dolly the sheep, the failure rate was very high; many embryos were transferred into surrogates and only one sheep resulted. Although mice have been cloned with a much higher success rate using a method related to the one used to produce Dolly, cloning of animals by nuclear transfer has not been done on a large scale.

Most of the furor surrounding reproductive cloning involves scientists who have talked about cloning human embryos. On a smaller scale, human embryos have been created in the laboratory, usually for isolating stem cells. But full reproductive cloning of humans probably has not been done. Some groups have claimed to have created a cloned baby, but this has not been shown to be true.

To see what might happen at some point in the future, let's look at the case profiled in Spotlight on Law.

reproductive cloning cloning an organism by nuclear transfer

Figure 17.17 **Nuclear Transfer** This method was used to create Dolly the sheep. It could be used to clone any mammal.

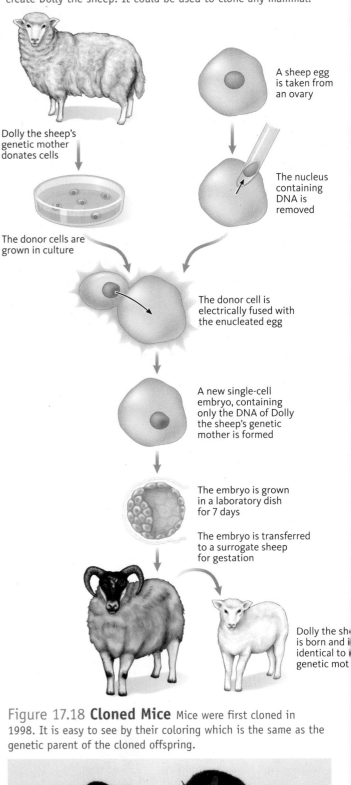

Dolly the sheep's genetic mother donates cells

A sheep egg is taken from an ovary

The nucleus containing DNA is removed

The donor cells are grown in culture

The donor cell is electrically fused with the enucleated egg

A new single-cell embryo, containing only the DNA of Dolly the sheep's genetic mother is formed

The embryo is grown in a laboratory dish for 7 days

The embryo is transferred to a surrogate sheep for gestation

Dolly the sh is born and i identical to genetic mot

Figure 17.18 **Cloned Mice** Mice were first cloned in 1998. It is easy to see by their coloring which is the same as the genetic parent of the cloned offspring.

#4 #3 #5 #1 #2

REUTERS/STR New

Spotlight on Science, Ethics and Law

Please Clone My Daughter: Decision-Making in the Future

Robert Denker made his money in real estate. He began with one apartment building, and now he owned enough land and buildings that he was the richest man in Charlotte, North Carolina. But being rich didn't keep him from tragedy. One day a drunk driver hit his 5-year-old daughter, Lucy, as she was crossing the street. By the time the ambulance arrived, she had suffered serious brain damage. In the emergency room, she was put on a respirator, and her heart was beating normally, but physicians told Mr. Denker that she probably would never regain consciousness. He was devastated.

In Charlotte, a group of scientists had recently cloned a cow from one that was a high milk producer. They leased Mr. Denker's land for their herd, and he had been following their work. As a businessman, he saw the potential for the process. Now he had another idea.

Dr. Meding-Smith was the head scientist on the project. Mr. Denker called her a week after the accident. He told her he was willing to spend every cent he had to bring his daughter back. He wanted Dr. Meding-Smith to clone his daughter from one of her body's cells.

1. **What should Dr. Meding-Smith do?**

2. **Give three reasons why she should not clone Lucy.**

3. **Give three reasons why she should clone Lucy.**

4. **If Mr. Denker offered Dr. Meding-Smith not only money now but also future funding for her research, would this make a difference if you were Dr. Meding-Smith?**

5. **Dr. Meding-Smith would be the first to clone a human being; she could have the most modern laboratory and everything she ever wanted. Should this play a part in her decision?**

6. **Should scientists do everything they *can* do? Give four reasons for your answer.**

▰ Analysis and Research Questions

1. If analysis of the dog genome was used to define all breeds, what effect might this have on the classification of dog breeds and on dog shows?

2. Drugs are being developed to treat patients based on their genetic makeup. List three and tell their uses.

3. In the past, scientists tried to improve our species by using eugenics. Is it wrong to encourage genetically healthy families to have more children? Why or why not? Who determines what "genetically healthy" means?

4. Think back through the chapters of this book. Are any of the new techniques used today forms of eugenics? Which ones and why?

5. Research the knockout mouse. List four different strains of mice created with this technology and discuss how they are used.

6. Because we can manipulate genes using recombinant DNA and other techniques, we may be able to use this technology to change eggs and sperm and pick the genes our children will have. How do you see this being done scientifically? Will there be people who want this? Who, and why?

▰ Large Research Assignments

1. Create a reading notebook of articles on any or all of the topics in this book from newspapers, magazines, and the internet.

2. Set up a blog of opinions on the topics that have been covered throughout the semester, and share it with either the public or members of the class.

3. Pick a topic covered this semester and survey 100 people to get their opinions. Analyze the results and prepare a talk with Power-Point slides.

4. Take a survey on any of the controversial topics we covered this semester and report the results.

5. Do research on the newest work being done in gene therapy, the human genome, gene testing and marketing, or another topic from the scientific journals available online and through the book's Web site or in your library. Write a short report on your findings.

6. Create a video that addresses one of our questions or cases. Place it on YouTube and collect comments.

7. Research the marketing strategy used to advertise Gardasil, a vaccine for human papillomavirus.

8. Research how the United Kingdom has handled laws controlling genetics and assisted reproduction. Write a summary of its laws and discuss whether they could be enacted in the United States.

Learn by Writing: The Continuation of the Blog

Throughout the book, we have suggested that you begin a blog with members of your class or others. Now is the time to look back at what you and others have written and see how these writings can relate to the past, present, and future of genetics.

Don't just think about the genetics of humans or the genetics of your family; try to remember what we covered throughout the book and relate it to the future. The cases we've asked you to look at are a good starting point. Make a list of them, pick the most intriguing, and write about them in your blog. Ask others to comment.

How do you think the future of genetics will turn out? Will laws and consensus about the uses of genetic technology keep up with rapid developments in this field?

Do you think you will have some personal connection to genetics or genetic testing?

Even if you didn't start a blog, now is the time to talk to people you know (such as your friends, your parents, or your grandparents) about what you learned from this book. You'll be surprised at how interested they will be. Do this in person or via e-mail, phone, or blogging.

Glossary

5-HTT a gene that encodes a protein that regulates nerve impulses

ABO blood types the major blood types in humans

acute myeloblastic leukemia a condition of the blood where white blood cells proliferate, also called blood cancer

adenine a nitrogen-containing purine base found in nucleic acids

adenosine deaminase (ADA) the gene that is mutated in one form of severe combined immunodeficiency disorder

adult stem cells cells found in the adult human body that give rise to new but specific cell types

albinism a genetic condition with a lack of the pigment melanin

alleles different forms of the same gene

allergens antigens in the environment that cause allergic reactions

allergies an abnormal immune response to allergens in the environment

alpha globin one of the proteins in hemoglobin

alternative splicing incorporating different combinations of exons into mRNA during splicing of pre-mRNA, creating a number of different proteins from the same genetic information

Alu sequences a highly repetitive sequence that makes up 10% of the human genome

amicus curiae a legal term translated as "friend of the court," applied to briefs submitted to a judge providing new information before a decision is rendered

amino acid one of the 20 subunits found in proteins

amniocentesis a prenatal test for genetic conditions done by analyzing cells in the amniotic fluid

amniotic fluid a fluid that surrounds the fetus and offers protection

anaphylactic shock a potentially fatal allergic reaction involving major organ shutdown

androgen receptor (AR) gene a gene that encodes the protein that acts as a receptor for male hormones

androgens male sex hormones

anencephaly neural tube defect in which major parts of the brain and skull do not form

aneuploidy an abnormal number of chromosomes that is not an exact multiple of the haploid number

animal models animals used in laboratory work to mimic human diseases

anthropoids the group of primates to which humans belong

antibodies proteins synthesized by B cells that inactivate antigens

anticodon a group of three nucleotides in a tRNA molecule that recognize a specific codon in mRNA

antigens molecules that enter the body and trigger an immune response

apoptosis genetically programmed cell death

arrhythmias irregular heartbeats

artificial selection selection for desirable traits in plants and animals by human intervention

artificial uterus machine that mimics the uterine environment

Ashkenazi Jews an ethnic and religious group of Eastern European ancestry studied for the breast cancer gene

aspermia a condition in which no sperm are produced

assembler software used in genomic research to assemble sequenced segments of DNA into the sequence of the whole genome

assumed consent legal term meaning that consent for treatment is presumed

Australopithecus an early relative of *Homo sapiens* that evolved in Africa

autoimmune disease condition in which a person's immune system attacks his or her own tissues

autoimmune disorders conditions in which the body's immune system attacks tissues and organs in the body

autonomy a legal and ethical term meaning the right to make decisions about yourself

autosomal dominant a trait that appears when only one copy of a gene is mutated

autosomal recessive a trait that appears when both copies of a gene are mutated

autosomes all chromosomes except the X and Y

B cells antibody-secreting cells of the immune system

background radiation radiation in the environment that we are exposed to in our everyday lives

Barr body the inactivated X chromosome in human females (and other mammals)

base analogs chemicals that are structurally similar to nucleotide bases (A, G, T, C)

behavioral genetics a branch of genetics that studies the genetic basis of behavior

bell curve a frequency distribution that resembles the outline of a bell

benign tumors that are not cancerous

beta globin one of the proteins in hemoglobin

biohistorian a person who specializes in the history of biology

bioinformatics a field of genomics that uses computers and software to collect, analyze, and store genomic sequence information

biotechnology process that uses recombinant DNA to make commercial products

blastocyst a stage of development in the early embryo

blastomere one of the early embryonic cells

BRACAnalysis® brand name for a test for mutant alleles of *BRCA1* and *BRAC2*

BRCA1 the first gene discovered that can, if mutated, increase a woman's risk of breast cancer

BRCA2 the second gene discovered that can, if mutated, increase a woman's risk of breast cancer

bulbourethral glands male organs that create a fluid that lubricates intercourse

"bully" whippet dog with a mutated myostatin gene

Burkitt's lymphoma a cancer of the lymphatic system thought to be associated with a chromosomal translocation

cancer cluster population or group where cancer is widespread

cancer condition where cells in the body multiply uncontrollably and spread to new sites

carcinogens environmental agents associated with cancer

carrier frequency frequency of heterozygous carriers of a recessive gene in a specific population

carrier testing a genetic test for carrier status of a recessive disorder

causation a legal term meaning that one thing causes another

CC-CKR5 gene that encodes a cell surface protein that signals the presence of an antigen

cell cycle series of events within a cell that results in cell division; also called the cell division cycle

centimorgans (cM) unit that measures how far apart genes are on chromosomes

central nervous system the portion of the nervous system consisting of the brain and spinal cord

cervix opening of the uterus

CFTR cystic fibrosis transmembrane conductance regulator, the protein encoded by the cystic fibrosis gene

chimeric immune system an immune system created in the recipient when bone marrow is transplanted between species

chorion the outermost membrane surrounding an embryo, part of which gives rise to the placenta

chorionic villi finger-like projections of the fetal chorion that work their way into the wall of the uterus

chorionic villus sampling (CVS) a procedure that removes a small sample of villi for genetic testing

chromosomes DNA-containing threadlike structures in the nucleus that carry genetic information

chronic myelogenous leukemia (CML) a form of leukemia

class-action lawsuits lawsuits that involve many plaintiffs

clinical trial use of human volunteers to test the effectiveness and potential side effects of drugs and treatments

codominant a condition where two alleles are both fully expressed in the heterozygote

codon a group of three nucleotides in mRNA that encode information for a specific amino acid

Cohen model haplotype pattern of markers on the Y chromosome found in some Israeli men

Combined DNA Index System (CODIS) panel a national database of DNA profiles begun by the FBI

complete androgen insensitivity (CAI) a genetic condition that causes an XY karyotype child to appear female

complex traits inherited traits and conditions caused by any combination of genes and environmental factors

concordance in twins, a condition in which both twins have or do not have a given trait

consanguinity a close family relationship between two individuals

conversion legal term meaning removal of someone's property

cystic fibrosis a recessively inherited genetic disorder with abnormal mucus production

cytokinesis division of the cytoplasm in cell division

cytoplasm viscous material that is located between the inner surface of the plasma membrane and the outer nuclear envelope that contains organelles, each with specialized functions

cytosine a nitrogen-containing pyrimidine base found in nucleic acids

deoxyribonucleic acid (DNA) molecule consisting of antiparallel strands of polynucleotides that is the primary carrier of genetic information

deoxyribose a pentose sugar found in DNA

diethylstilbestrol (DES) a synthetic estrogen used as a drug and as a food additive

diploid number (2*n*) condition in which chromosomes are present as pairs; in humans, the diploid number is 46

discovery rule legal term that requires all evidence to be given to the defense before a trial

DNA fingerprinting a method of DNA forensics used to identify individuals

DNA polymerase enzyme that catalyzes the synthesis of DNA using a template DNA strand and nucleotides

DNA profile the pattern of DNA fragments that can identify individuals

DNA sequencer a machine that analyzes the nucleotide sequence of DNA fragments

dominant a trait expressed in the homozygous or heterozygous condition

Drosophila a fruit fly used as a model genetic organism

dystrophin the protein that is abnormal in muscular dystrophy

egg the female reproductive cell

embryonic stem cells cells found in the inner cell mass of a developing embryo

endometrium lining of the uterus

endoplasmic reticulum (ER) series of cytoplasmic membranes arranged as sheets and channels that function in the synthesis and transport of gene products

endorphin neurotransmitter associated with happy feelings

endoxifen antiestrogen compound used in cancer chemotherapy

enfuvirtide a drug in clinical trials that prevents entry of HIV into a cell

epididymis an area of the testes that stores sperm

epigenetic trait phenotype that is produced by epigenetic changes to DNA

epigenetics the study of heritable changes in the genome other than changes in DNA sequence

epigenome the epigenetic state of a cell

erectile dysfunction (ED) an inability of the penis to become erect

essential amino acids those amino acids that cannot be synthesized and must be in the diet

estrogens female hormones

evolution change in the gene pool of a population

exome the fraction of the genome that contains only exons

expressed genes the set of genes that are active in transcription

factor VIII a blood-clotting protein that is abnormal in hemophilia

fallopian tubes the structure that carries the embryo to the uterus

fertilization when the sperm and egg fuse to form a zygote

folic acid a B vitamin that is a key factor in cell growth and development

follicles a structure surrounding the developing egg

fossil preserved record of extinct animals and plants

fossil record all fossils that have been described and catalogued

founder effect populations whose origins can be traced to a small number of ancestors

fragile X syndrome genetic disorder caused by a mutation in the *FMR-1* gene, which maps to the tip of the long arm of the X chromosome

frameshift mutation a mutation caused by the insertion or deletion of nucleotides

free radicals reactive molecules that have unpaired electrons

fructose a sugar that supplies energy to the sperm

G1 the first stage of interphase

G1/S checkpoint a regulatory point in the cell cycle

G2 the last stage of interphase

G2/M checkpoint a regulatory point in the cell cycle

gene regulation the mechanisms that control the activity of genes

genes carriers of genetic information in the form of DNA

gene therapy the transfer of a normal allele to treat a genetic disease

genetic counselor a person who works with patients to explain genetic testing and its results

genetic drift significant changes in allele frequency that occur by chance

genetic imprinting selective expression of either the maternal or paternal copy of a gene

genetic liability in complex traits, the number of genes in an individual's genome that affect phenotype along with environmental influences

genetic map diagram showing the order of and distance between genes on a chromosome

genetic screening large-scale genetic testing

genetic testing testing for genetic conditions

genome set of DNA sequences carried by an individual

genome-wide association studies (GWAS) large-scale studies of populations to find common haplotypes

genomic library collection of DNA sequences that contains an individual's genome

genomics study of the organization, function, and evolution of genomes

genotype the genetic constitution of an individual

germ cells cells that divide to form sperm and eggs

Golgi apparatus (GA) membranous organelle composed of a series of flattened sacs; it sorts, modifies, and packages proteins produced in the ER

gonad an embryonic organ that gives rise to the testes or ovaries

guanine a nitrogen-containing purine base found in nucleic acids

hairy cell leukemia a cancer of white blood cells; the cancerous cells have hair-like projections from their surface

haplogroup groups of haplotypes from a specific region of the world

haploid number (*n*) condition in which each chromosome is present once, unpaired; in humans, the haploid number is 23

haplotype combination of alleles carried close together on a chromosome

Hardy-Weinberg law mathematical formula for determining allele and genotype frequencies in a population

hemizygous genes on the X chromosomes in males that have no alleles on the Y chromosome

hemoglobin the oxygen-carrying protein in red blood cells

hemolytic disease of newborns (HDN) blood condition that occurs when antibodies from an Rh⁻ mother destroy blood cells in an Rh⁺ fetus

hemophilia a genetic disorder of blood clotting

herbicides poisons that kill plants

heterozygote advantage situation where being heterozygous for a recessive trait confers an advantage for survival and reproduction

heterozygous having two different alleles of the same gene

histone protein that DNA coils around to form nucleosomes

HLA complex a group of genes on chromosome 6 that determine tissue type for transplants

hominids the group humans and great apes belong to

hominins humans and extinct related forms characterized by upright posture and walking on two legs

Homo erectus an early human species

Homo floresiensis an extinct human species discovered on the island of Flores

Homo sapiens modern humans

homologous chromosomes members of a chromosome pair

homozygous having two identical alleles of the same gene

human chorionic gonadotropin (hCG) a hormone made by the chorion that holds the lining of the uterus in place for implantation

Human Genome Project the project that is deciphering the organization, sequence, and function of all human genes

human papillomavirus (HPV) virus that is one cause of cervical cancer

huntingtin the protein that is defective in Huntington disease

Huntington disease (HD) an autosomal dominant neurodegenerative disease

hyperacute rejection a rapid and massive reaction to a transplanted organ from another species

I isoagglutinin, a gene whose three alleles (A, B, and O) determine blood type

I^A an allele that determines the A blood type

I^B an allele that determines the B blood type

immunogenetics the study of the genes that play a part in the immune system

immunosuppressive drugs drugs given to transplant recipients to suppress their immune system to prevent rejection of a transplanted organ

implantation the process by which the fertilized egg attaches to the uterine wall

in vitro **fertilization (IVF)** a method of assisted reproduction done in a laboratory

innate behavior behavior that is present at birth

inner cell mass group of cells in the early embryo

Institutional Review Boards (IRB) groups of physicians, patients, and clergy that oversee clinical trials

intelligence quotient (IQ) a score derived from a standardized test of intelligence

interphase the stage of the cell cycle between divisions

I^O an allele that, when homozygous, determines the O blood type

junk science legal term meaning science that has yet to be generally accepted by members of the scientific community

karyotype a way of displaying human chromosomes

knockout mouse mouse with a specific gene turned off for scientific study

laparoscopy a procedure using fiber optics to see into the body through a small incision

law of independent assortment random distribution of homologues into cells during meiosis

law of segregation separation of homologues into different cells in meiosis

learned behavior behavior that is acquired by experience or through teaching after birth

liability legal term meaning legally responsible

linkage map map derived from crossing-over studies showing the order of and distance between genes on the same chromosome

linkage two or more genes on the same chromosome that tend to be inherited together

luteinizing hormone (LH) a hormone produced at the time of ovulation

lymphokine a chemical in the body produced by cells of the immune system

lysosomes membrane-enclosed organelles that contain digestive enzymes

macromolecules large biological molecules composed of subunits

major histocompatibility complex (MHC) a gene set that encodes proteins that act as molecular identification tags on cells

malignant cancerous

markers alleles, SNPs, or other molecular characteristics that can be used in linkage and other studies

market share liability a legal term referring to the division of large settlements in class-action lawsuits among the manufacturers of the product

meiosis form of cell division in which haploid cells are produced

memory cells cells produced by activated B cells that protect an individual from future infections by the same antigen

meninges the membranes that surround the brain and spinal cord

menopause a condition where a woman no longer ovulates

mesothelioma form of lung cancer caused by exposure to asbestos

messenger RNA (mRNA) a single-stranded RNA carrying an amino acid–coding nucleotide sequence of a gene

metastasis migration of cancer cells from the primary tumor to other sites in the body

methylation addition of a methyl group to a DNA base or a protein

micro RNA (miRNA) a short RNA molecule that regulates gene expression by binding to mRNA

millirems unit of radiation effect that is 1/1000 of a rem

minisatellites DNA sequences composed of 10 to 1000 base pairs that are scattered throughout the genome

mitochondria (singular, mitochondrion) membrane-enclosed organelles that are the site of energy production

mitochondrial DNA (mtDNA) DNA found in the mitochondria of a cell

mitosis form of cell division that produces two daughter cells that are genetically and chromosomally identical to the parent cell

monoamine oxidase type A an enzyme that breaks down neurotransmitters such as serotonin

monosomy having one less than the diploid number of chromosomes

monozygotic (MZ) twins genetically identical twins derived from the fertilization of a single egg by a single sperm

multifactorial traits traits that are caused by the interaction of two or more genes and environmental factors

multiple myeloma cancer of the bone marrow cells thought to be associated with a chromosomal translocation

multipotent cells that can grow into only a limited number of cell types

muscular dystrophy a degenerative disease of muscles carried on the X chromosome

mutagens physical or chemical agents that cause mutations

mutation a heritable change in the nucleotide sequence of DNA

myometrium the muscle of the uterus

myostatin gene gene that regulates the growth of muscle fibers

nail-patella syndrome a dominantly inherited disorder with malformations of the nails and the kneecap

natural selection differential reproduction among members of a population that is the result of a genetic advantage

negative eugenics using eugenics to remove certain genes or traits from a population by preventing those with unwanted genes or traits from reproducing

neural tube defects a group of disorders that result from defects in the formation or development of the neural tube

neurofibromatosis an autosomal dominant condition with benign tumors

neurotransmission the process of transmitting a nerve impulse across a synapse between two cells in the nervous system, mediated by neurotransmitters

neurotransmitters chemicals secreted across synapses in cells of the brain

nondisjunction a failure of chromosomes to separate properly during cell division

nonexpressed sequences the set of genes that are transcriptionally inactive

nuclear transfer moving a nucleus from one cell to another

nucleoli (singular, nucleolus) region in the nucleus that synthesizes ribosomes

nucleosomes structures formed when DNA is coiled around histones

nucleotide basic building block of DNA and RNA; each nucleotide consists of a base, a phosphate, and a sugar

nucleus membrane-enclosed organelle in cells that contains the chromosomes

oncogene a gene, when mutated, that can cause cancer within a cell

OncoMouse the first genetically modified mammal patented

oocytes developing eggs

oogenesis development of eggs

organelles membrane-enclosed structures with specialized functions found in the cytoplasm of cells

ovaries female organs that make eggs

PAH a gene associated with PKU

pedigree a diagram showing inheritance of familial traits

peptide bond the chemical bond that links amino acids together

peripheral nervous system portion of the nervous system outside the brain and spinal cord

phenotype the observable properties of an organism that are produced by the genotype

phenylalanine an essential amino acid

phenylthiocarbamide (PTC) chemical that only certain members of a population can taste

Philadelphia chromosome a translocated chromosome created from parts of chromosome 9 and 22 that is associated with CML

phosphate group a compound containing phosphorus chemically bonded to four oxygen molecules

pigment a colored substance

placenta a disk-shaped organ that nourishes the developing embryo; formed in the uterus by the interaction of maternal cells and fetal cells

plasma membrane outer border of cells that serves as an interface between the cell and its environment

plasmid circular DNA found in bacteria

pluripotent stem cells from the embryo that can grow into any type of cell

polar body small cells produced in meiosis I and meiosis II that do not become gametes and are ultimately discarded

polygenic trait a trait controlled by two or more genes

polymerase chain reaction (PCR) a molecular technique that allows production of many copies of a specific DNA sequence

polypeptides chains of amino acids

polyploidy having extra chromosome sets

population frequency a measurement that determines how often specific combinations of alleles are present in a population

positional cloning recombinant DNA-based method of identifying and isolating genes

positive eugenics movement to increase desirable traits in a population by encouraging people with those traits to have many children

preimplantation genetic diagnosis (PGD) a procedure that removes one cell from an early embryo for genetic testing

prenatal diagnosis diagnosis of genetic conditions in a developing fetus

presymptomatic testing genetic testing before symptoms manifest themselves

primates the group of mammals that includes humans

primers short nucleotide sequences used in PCR

proband the individual who is the focus of a pedigree

promoter regulatory region located at the beginning of a gene

prostaglandins chemicals secreted into the semen that causes contractions of the vagina during intercourse

prostate gland a male organ that makes a fluid for sperm viability

protein a cellular macromolecule composed of amino acid subunits

proteomics the study of the functions and interactions of the set of proteins in a cell

proto-oncogene a normal gene that works to initiate or maintain cell division; mutant alleles, called oncogenes, are associated with cancer formation

psychiatrists physicians who specialize in treating psychiatric disorders

psychologists non-physicians who specialize in treating psychological and psychiatric disorders

purine a class of double-ringed organic bases found in nucleic acids

pyrimidine a class of single-ringed organic bases found in nucleic acids

RAS a proto-oncogene involved in signal transduction

recessive a trait expressed only in the homozygous condition

recombinant DNA molecule a DNA molecule formed by joining DNA segments from different sources

recombinant DNA technology the process by which recombinant DNA is created

rejection an immune reaction to a transplanted organ that does not match the recipient

rems unit of radiation effect equal to a standard dose of x-rays

repetitive DNA regions of the genome that contain DNA sequences present in many copies

reproductive cloning cloning an organism by nuclear transfer

restriction enzyme a protein that, when mixed with DNA, cuts the DNA in a specific place

restriction fragment length polymorphism (RFLP) analysis one of the methods of DNA testing

Rh blood group a human blood type

Rh negative (Rh⁻) blood type lacking the Rh antigen

Rh positive (Rh⁺) blood type with the Rh antigen on blood cells

ribonucleic acid (RNA) a nucleic acid molecule that contains the pyrimidine uracil and the sugar ribose; the several forms of RNA function in gene expression

ribose a pentose sugar found in RNA

ribosomes cytoplasmic organelles that are the site of protein synthesis

RNA interference (RNAi) partial or complete gene silencing by small RNA molecules

RNA polymerase an enzyme that synthesizes single-stranded RNA molecules using a DNA template

S phase part of the cell cycle in which DNA is replicated

schizophrenia set of mental disorders characterized by hallucinations and loss of touch with reality

secondary amenorrhea a condition of not menstruating

secondary sex characteristics changes in the body that occur in early teens that relate to reproduction

semen the liquid that sperm travel in

seminal vesicles a set of male reproductive glands

seminiferous tubules the region of the testes where sperm production occurs

senescence a resting stage in which a cell does not divide

sequence the order of nucleotides in a DNA molecule

serotonin a neurotransmitter

settlement an agreement by parties to a lawsuit that resolves all issues

severe combined immunodeficiency disorder (SCID) a complex genetic disorder in which there is no functional immune system

sex chromosomes chromosomes involved in sex determination; in humans, the X and Y chromosomes are sex chromosomes

sex ratio the proportion of males to females in a population

sex selection a method of predetermining the sex of a child by using prefertilization or preimplantation methods

short tandem repeats (STRs) DNA sequences 2–9 base pairs long

sickle-cell anemia a genetic disorder with altered red blood cell shape

signal transduction transfer of signals from outside the cell to the nucleus that result in new patterns of gene expression

single nucleotide polymorphisms (SNPs) DNA sequence variations that are only one base pair long

small interfering RNA (siRNA) a short RNA molecule that regulates gene expression by destroying specific mRNAs

somatic cells cells that are part of the body of an organism but do not form gametes

sperm sorting a method of separating sperm that carry Y chromosomes from sperm that carry X chromosomes; separated sperm are used in fertilization to determine the sex of the offspring

sperm the male sex cell

spermatids cells in early stages of sperm formation

spermatocytes cells that divide by meiosis to form spermatids

spermatogenesis formation of sperm

spina bifida (SB) a neural tube defect caused by the failure of the neural tube to close

start codon the AUG codon that signals the location in mRNA where translation begins

stem cells cells in the embryo and adult that can divide to form many different cell types

stimuli environmental agents that generate a response from an organism

stop codon codons (UAA, UAG, UGA) in mRNA that signal the end of translation

Streptococcus pneumoniae bacterial species that causes pneumonia

subcutaneous mastectomy a procedure for removal of the breast tissue

surrogacy a method of assisted reproduction where one woman carries a fetus for another

synapse the cleft between nerve cells

T cells one of two types of immune system cells (B cells are the other)

T4 helper cells specialized T cells that serve as the on switch for the immune system

tamoxifen a drug given to breast cancer patients that controls the amount of estrogen in the bloodstream

Tay-Sachs disease a recessive genetic trait that is almost always fatal

telomeres DNA sequences on the ends of chromosomes that may be related to aging

termination sequence the nucleotide sequence at the end of a gene that signals the end of transcription

testes the male organ where sperm are formed

testosterone a steroid hormone produced by the testes; the male sex hormone

thymine a nitrogen-containing pyrimidine base found in nucleic acids

tort type of lawsuit based on an act that injures someone

transcription the transfer of genetic information from DNA to mRNA

transfer RNA (tRNA) a small RNA molecule that brings amino acids to the ribosome during protein synthesis

transformation a heritable change caused by DNA that originates outside the cell

transgenic crops crops that have been altered by introduction of DNA from another organism

transgenic organism animals, plants, or microbes that have been genetically altered using recombinant DNA techniques

translation the transfer of genetic information in mRNA into the amino acid sequence of proteins

transposons DNA sequences that can move from location to location in the genome

trisomy a condition where an organism is diploid except for one chromosome that is present in three copies

tumor suppressor genes genes that normally function to suppress cell division but when mutated cause cells to become cancerous and divide uncontrollably

ultrasonography, or ultrasound a prenatal test that uses sound waves to visualize an unborn fetus

uniparental disomy a condition in which both copies of a chromosome are inherited from one parent

universal donors individuals with type O blood; they can donate to all other blood types

universal recipients individuals with type AB blood; they can accept donations from all blood types

uracil a nitrogen-containing pyrimidine base found only in RNA

urethra a tube present in males and females that carries urine out of the body

uterus an organ in the female where an embryo grows

vaccination stimulation of the immune system and memory cell production by administration of an attenuated antigen

vaccine preparation containing a weakened or noninfective antigen used to confer immunity against that antigen

vagina the female organ that connects the uterus to the outside of the body

vas deferens a sperm-conducting duct in the testes

vasectomy a procedure that cauterizes the vas deferens

vector molecules used for carrying a DNA segment into a host cell, where it is copied

villi (sing. villus) finger-like projections composed of cells

wrongful-birth suit a lawsuit that is brought by parents against a physician or lab who did not correctly identify a birth defect

wrongful-life suit a lawsuit that is brought by the child against a physician or lab who did not correctly identify a birth defect

X chromosome the sex chromosome present in organisms (including humans) where female gametes carry one type of sex chromosome

X-linked genes carried on the X chromosome

X-linked recessive a recessive trait carried on the X chromosome

x-ray diffraction method of studying the structure of a crystal using x-rays

xenografts organ transplanted between species (same as *xenotransplants*)

xenotransplantation the process of transplanting organs between species

xeroderma pigmentosum a genetic disorder that results from defects in DNA repair

Y chromosome testing a process that maps sequence variations on the Y chromosome

Y chromosome the sex chromosome present in organisms (including humans) where male gametes carry one of two types of sex chromosomes

Y-linked genes carried on the Y chromosome

zygote fertilized egg that develops into a new individual

Index

Note: Page numbers followed by *f* and *t* indicate figures and tables, respectively.

genetic, 203–205, 205*t*
of human epigenome, 139
Marfan syndrome, 82*t*, 114
Marker(s), genetic, 204, 248
Market share liability, 142
Mastectomy, subcutaneous, 338
Maternal selection, 59
McCarty, Maclyn, 95
MCS. *See* Multiple chemical
sensitivities (MCS)
MD. *See* Muscular dystrophy (MD)
Medical genetics, 267
Medical record(s), access to, 87, 119, 120*t*
Medicine(s), response to, population
variations in, 220
Meiosis, 3, 15*f*, 33, 35*f*, 36, 37*f*, 52–53,
53*f*–55*f*, 54*t*
nondisjunction in, 53, 55*f*, 58
Meiosis I, 54*f*, 54*t*, 75, 75*f*, 76*f*
Meiosis II, 54*t*, 55*f*, 75, 75*f*, 76*f*
Melanoma, 254*f*
mutations in, 215
risk factors for, 253–254
MELAS syndrome, 86*t*
Memory cells, 282–283
Mendel, Gregor, 73–74
pea-plant experiments, 73–74, 74*f*, 75*f*
Meninges, 228, 228*f*
Menopause, 37
premature, 38
Menstruation, 36
lack of, 38
Mesothelioma, 139, 140*f*
Messenger RNA (mRNA), 107–108, 107*f*,
108, 109, 112*f*
Metaphase, 246, 246*f*
Metaphase I, 54*f*, 54*t*
Metaphase II, 54*t*, 55*f*
Metastasis, 241, 242*f*
Methionine, mRNA codon for, 110*f*–111*f*,
110*t*, 111
Methylation, 136, 137, 137*f*, 139
MHC. *See* Major histocompatibility
complex (MHC)
Mice
as animal models in study of human
diseases, 154–155, 154*f*
cloning of, 346, 346*f*
genetically engineered, 154–155, 154*f*
knockout, 270, 271*f*, 341, 341*f*
obese, 231, 232*f*
in study of human behavior,
267–269
Microarray(s), DNA, 187

Microcephalic osteodysplastic
primordial dwarfism type 1, 144
Micro RNA (miRNA), 113
Microtubule(s), 4*f*
Millirems (mrem), 128
Minisatellite(s), DNA, 184
miRNA. *See* Micro RNA (miRNA)
Miscarriage
age and, 37
amniocentesis and, 56
chorionic villus sampling and, 57
monosomy and, 59
trisomy and, 59
of 45,X embryos, 61
of 45,Y embryos, 61
Mitochondria (sing., mitochondrion),
4*f*, 9, 9*f*, 86
functions of, 10*t*
malfunction, disorders associated
with, 10*t*
Mitochondrial disorders, 10*t*, 86, 86*f*, 86*t*
Mitochondrial DNA (mtDNA), 9,
191, 191*f*
and mapping of ancient human
migrations, 321–322
Mitosis, 3, 241, 244, 244*f*, 245
stages of, 245–246, 245*f*, 246*f*
Mobley v. Georgia, 274–275, 275*f*
Monkey(s)
New World, 318
Old World, 318
Monoamine oxidase type A (MAOA),
deficiency, 274–275, 275*f*
and aggressive behavior, 270, 270*f*
Monosomy, 58
autosomal, 59
X chromosome, 61, 62*f*
Y chromosome, 61
Monozygotic (MZ) twins, 231, 267
intelligence studies in, 232
Moore, John, 212
*Moore v. Regents of the University of
California*, 212, 257
Mosquitoes, and malaria, 303, 303*f*
Motive, 272
Mouthwash, genetically engineered, 162
M phase. *See* Mitosis
mRNA. *See* Messenger RNA (mRNA)
MS. *See* Multiple sclerosis (MS)
mtDNA. *See* Mitochondrial DNA
(mtDNA)
Mullis, Kary, 185–186, 186*f*
Multifactorial trait(s), 218, 225, 226, 227*f*
fingerprints as, 231, 231*f*

genetics of, methods for studying,
227–228
intelligence as, 231–232
obesity as, 231
recurrence risk, 227–228
Multiple births, assisted reproductive
technology and, 44, 45
Multiple chemical sensitivities
(MCS), 136
Multiple myeloma, chromosomal
translocations and, 252
Multiple sclerosis (MS), 221, 283, 283*t*
infections and, 222*t*
male-female ratio of, 222*f*
Multipotent cell(s), 156
Murray, Joseph, 284, 284*f*
Muscle(s), malfunctions of, 10*t*
Muscular dystrophy (MD), 84–85, 175*t*
preimplantation genetic diagnosis
of, 170
Mutagen(s), 128
Mutation(s), 3, 114
and cancer, 127, 242–244, 247–248
and breast cancer, 248–249
causes of, 102, 103*f*, 128–130, 128*f*
chemicals causing, 128, 129–130
definition of, 127
detection of, 130–132
and DNA replication, 102, 128*f*
effects of, 127
frameshift, 130
frequency of, 132–133
in mitochondrial genes, 86
repair of, 102, 103*f*, 134–136
and sickle-cell anemia, 116, 116*f*, 117,
117*f*, 118
types of, 133, 134*t*
Myometrium, 36, 36*f*
Myostatin gene, 340
Myotonic dystrophy, 133
gene, mapping of, 205*t*

N

Nail-patella syndrome, 202, 203–204,
203*f*
Narborough (Leicestershire, England)
murders, 195
Nasopharyngeal cancer, EBV and, 220*t*
National Human Genome Research
Institute, 215–216
Natural selection, 304–305
and allele frequencies, 308
and continuing evolution of humans,
325–326

Placenta, 18, 19*f*

Plant(s), transgenic, 152*f*. *See also* Crops

Plasma membrane, 4*f*, 5, 5*f*, 5*t*
  functions of, 5, 6

Plasma membrane proteins, 5*f*, 284, 284*f*

Plasmid(s), 149, 149*f*

Pluripotent cell(s), 156

PNS. *See* Peripheral nervous system
  (PNS)

Polar body, 53

Polycystic kidney disease, 166*t*
  treatment of, 175*t*

Polygenic trait(s), 225–226

Polymerase chain reaction (PCR)
  definition of, 184
  discovery of, 185–186
  handheld machine for, 189*f*
  method for, 185–187, 186*f*, 187*f*,
    187*t*, 189*f*

Polynucleotides, 97

Polypeptide(s), 106, 106*f*, 111

Polyploidy, 58

Population(s)
  definition of, 313*f*
  study of, genomics used in,
    222–223

Population frequency, 184, 185*t*

Population genetics, 314
  legal and ethical issues associated
    with, 308–309, 309*t*

Population screening, for carriers of
  genetic disorders, 172–174

Porphyria, 82*t*

Positional cloning, 204, 205, 205*t*

Positive eugenics, 336

Prader-Willi syndrome (PWS), 61, 62*f*,
  138, 139*f*

Pregnancy
  maternal diet during, and weight of
    child, 144
  wrongful, 66

Pregnancy test(s), home, 18

Preimplantation genetic diagnosis
  (PGD), 15–17, 16*f*, 44, 44*f*,
  170, 170*f*

Pre-mRNA, 108, 109*f*

Prenatal diagnosis, 166

Presymptomatic testing, 166, 172

Primates, 318, 319*f*

Primer(s), 186, 186*f*

Prion(s), 114, 115*f*

Privacy of medical information, 87, 88*t*,
  119, 120*t*, 175–176, 176*t*

Proband, 78, 78*f*

Progressive external ophthalmoplegia
  (PEO), 86*t*

Promoter(s), 108, 109*f*, 137, 206
  and gene regulation, 113, 114*f*

Prophase, 246, 246*f*

Prophase I, 54*f*, 54*t*

Prophase II, 54*t*, 55*f*

Prosimians, 318, 319*f*

Prostaglandin(s) (PG), 35

Prostate gland, 34*f*, 34*t*, 35

Protein(s), 4, 5*t*, 105
  alteration, and sickle-cell anemia,
    115–117, 116*f*, 117*f*
  biosynthesis of, 105, 107–108, 207. *See
    also* Translation
  cellular transport of, 8*f*, 9, 9*f*
  changes in, and phenotypic
    changes, 114
  folding, 9
    and phenotype, 114–115, 115*f*
  folding of, 106
  functions of, 106
  human
    genes for, 207
    number of, 207
  mitochondrial, 9
  plasma membrane, 5*f*, 284, 284*f*
  posttranslational modification, 8*f*, 9
  shape of, changes in, and disease,
    114, 115*f*
  shortened, production of, 115
  structure of, 106, 106*f*
  subunits of, 106, 106*f*

Proteomics, 201

Proto-oncogenes, 243–244
  RAS, 247–248

Pseudomonas syringae, transgenic,
  strawberry plants treated with, 155

Psoriasis, 283*t*

Psychiatrist(s), 267

Psychologist(s), 267

PTC. *See* Phenylthiocarbamide (PTC)

Punnett square, 74, 75*f*

Purine(s), 96, 97, 97*f*

PWS. *See* Prader-Willi syndrome (PWS)

Pyrimidine(s), 96, 97, 97*f*

Q

Quan, Shirley, 212

Quinn, Pat, 194

R

Radiation exposure, 128, 129*f*, 129*t*
  and infertility, 40

RAS, 247–248

Receptor(s), 106
  plasma membrane, 6

Recessive trait(s), 74

Recombinant DNA molecule(s),
  149, 149*f*

Recombinant DNA technology, 147,
  149–150, 149*f*
  and gene therapy, 209–211
  and prenatal genetic testing, 170
  safety of, 153–154, 153*f*, 159

Reflex(es), 264

Rejection, of organ transplant, 284, 285

Re Kubin, 157

rem (unit), 128

Repetitive DNA, 206

Reproductive cloning, 346, 346*f*

Reproductive organs, development
  of, 21

RER. *See* Rough endoplasmic
  reticulum (RER)

Restriction enzyme(s), 149, 149*f*, 150*t*

Restriction fragment length
  polymorphisms (RFLPs)
  analysis of, 185, 185*f*, 186*f*
  definition of, 184

Retinoblastoma, gene, mapping of, 205*t*

Retrovir. *See* AZT

RFLP. *See* Restriction fragment length
  polymorphisms (RFLPs)

R group, 106

Rh blood group, 288, 289*f*, 297

Rheumatoid arthritis, 221, 283, 283*t*
  infections and, 222*t*
  male-female ratio of, 222*f*

Rh factor, 288, 289*f*

Rh negative (Rh⁻), 288, 289*f*, 297

RhoGAM, 288

Rh positive (Rh⁺), 288, 289*f*, 297

Ribonucleic acid, 97. *See also* RNA

Ribose, 96, 100

Ribosome(s), 4*f*, 6*f*, 7–9, 8*f*, 107,
  110*f*–111*f*, 111, 112*f*

Rice, golden, production of, by
  recombinant DNA
  techniques, 152*f*

RNA, 5, 5*t*, 96, 97*f*
  and DNA, comparison of, 101*t*
  messenger (mRNA), 107–108, 107*f*,
    108, 109, 112*f*
  micro (miRNA), 113
  molecular organization of,
    100–101, 101*f*
  nucleotides in, 96–97, 97*f*, 108

as multifactorial trait, 229–230
prenatal surgical treatment of, 237–238
prevention of, 229
Spinal cord, 264f
embryology of, 228, 228f
Splenectomy, 212
SRY gene, 21, 22f
Start codon(s), 109, 110t, 111
Stem cell(s), 3
adult, 156
in biotechnology, 155–156
in cancer treatment, 250
embryonic, 155–156, 156f
human, 18
Sterilization, forced, 336–337
Stern, Wilhelm, 232
Steroid(s), 4
Stimuli, 263
Stop codon(s), 109, 110t, 111f, 112, 115, 206
Stratified medicine, 220
Strawberry plants, treated with
transgenic bacteria, 155
Streptococcus pneumoniae, 95, 95f
Subcutaneous mastectomy, 338
Sudden infant death syndrome (SIDS),
prevention of, 238
Surrogacy, 42, 42f, 46t, 49
Synapse(s), 265, 265f
Systemic lupus erythematosus (SLE),
male-female ratio of, 222f

T

T4 helper cells, 281–282, 282f
HIV infection (AIDS) and, 289
Tamoxifen, 220–221, 221f, 250, 338
*Tarasoff v. Regents of the University of
California*, 88–89
Tarsier, 318, 319f
Taste preferences, 218–219
Tay-Sachs disease (TSD), 79t, 114, 166t,
172, 173f, 175t
carriers, frequency within
populations, 304t
genetic testing for, 172–174, 173f
preimplantation genetic diagnosis
of, 170
T cells, 281–282, 282f
helper, T4, 281–282, 282f
HIV infection (AIDS) and, 289
TCGA. *See* The Cancer Genome
Atlas (TCGA)
Telomere(s), 51, 52f
Telophase, 246, 246f

Telophase I, 54f, 54t
Telophase II, 54t, 55f
Tennessee v. John Thomas Scopes,
327–328, 327f, 328f
Termination sequence, 108, 109f
Testes, 33, 34, 34f, 34t
development of, 21, 22f
Testosterone, 4, 5t, 21, 33
androgen insensitivity and, 23
and fetal sexual differentiation, 21, 22f
in infertility treatment, 40
Thalassemia, 166t
The Cancer Genome Atlas (TCGA), 244
Thompson-Canino, Jennifer, 193, 193f
Thoracic nerves, 264f
Threshold model, 227, 227f
Thymine, 96, 97, 97f
Thyroid disease, male-female ratio
of, 222f
Tobacco companies, lawsuits against,
254–255
Tort cases, 254, 255
Trait(s), 73–74
autosomal, 78
autosomal dominant, 78, 79, 81–82,
81f, 82t, 130, 130f, 166t
autosomal recessive, 78–81, 79f, 79t,
131, 166t
preimplantation genetic diagnosis
of, 170
combinations in offspring, 74, 75f
complex, 225–226
dominant, 74
epigenetic, 136
multifactorial, 218, 225, 226, 227f
genetics of, methods for studying,
227–228
polygenic, 225–226
recessive, 74
X-linked, 78
X-linked recessive, 78, 79, 83–86, 85f,
85t, 131, 166t
Trait value, 226
Transcription, 107, 107f, 108,
109f, 112f
gene regulation in, 113, 113f
Transfer RNA (tRNA), 109–111,
110f, 112f
Transformation, 95–96, 95f
Transgenic crops, 150–153, 150f, 151f,
152f, 155
safety of, 153–154, 153f

Transgenic organism(s), 148–150, 149f,
285, 285f
patenting of, 156–157
safety of, 153–154
in study of human diseases,
154–155, 154f
Translation, 107, 107f, 108–112,
110f–111f, 112f
gene regulation in, 113, 113f
Translocation(s), chromosomal
and cancer, 251–252
Robertsonian, and Down syndrome,
61, 61f
Transposon(s), 206
Trinucleotide repeats, 133, 133f
Trisomy, 58, 175t
autosomal, 59, 60f
risk factor for, 59, 59f
Trisomy 13, 59, 60f, 168
Trisomy 18, 59, 60f, 168
Trisomy 21, 59–61, 59f, 60f, 167
maternal age and, 59, 59f
prenatal diagnosis of, 167f, 168
TSD. *See* Tay-Sachs disease (TSD)
Tumor suppressor genes, 243, 253f
Turkana Boy, 333
Turner syndrome, 25, 61, 62f, 63f
Twin(s)
concordance for selected traits in, 231,
267, 267t, 269, 269t
fraternal, 267
identical, 267. *See also* Monozygotic
(MZ) twins
fingerprint differences in, 231, 231f
obesity in, 231
organ transplantation between,
284, 284f
phenotype differences in, 136
"two Jims," 263, 270
Twin study(ies)
of autoimmune diseases, 221
of behavior and genetics, 267
of IQ, 232
of multifactorial traits, 229
of obesity, 231
Typhoid fever, cystic fibrosis carriers
and, 306

U

UDP. *See* Uniparental disomy (UDP)
Ulcerative colitis, male-female ratio
of, 222f
Ultrasonography. *See* Ultrasound